SCIENCE FOR THE SUSTAINABLE CITY

STEWARD T. A. PICKETT,
MARY L. CADENASSO, J. MORGAN GROVE,
ELENA G. IRWIN, EMMA J. ROSI, AND
CHRISTOPHER M. SWAN, EDITORS

Science for the Sustainable City

EMPIRICAL INSIGHTS FROM THE BALTIMORE

SCHOOL OF URBAN ECOLOGY

Yale UNIVERSITY PRESS   NEW HAVEN AND LONDON

This volume is a contribution of the Long Term Ecological Research (LTER) program funded by the National Science Foundation. The largest and longest-lived ecological network in the United States, the twenty-eight LTER sites encompass diverse ecosystems from Alaska and Antarctica to islands in the Caribbean and the Pacific. The Baltimore Ecosystem Study has been an LTER site since 1997.

NATIONAL SCIENCE FOUNDATION
LTER NETWORK
LONG TERM ECOLOGICAL RESEARCH

Published with assistance from Furthermore: a program of the J. M. Kaplan Fund

Furthermore:
a program of the J.M. Kaplan Fund

and from the foundation established in memory of Amasa Stone Mather of the Class of 1907, Yale College.

Yale University Press books may be purchased in quantity for educational, business, or promotional use. For information, please e-mail sales.press@yale.edu (U.S. office) or sales@yaleup.co.uk (U.K. office).

Set in Scala and Scala Sans types by Integrated Publishing Solutions.
Printed in the United States of America.

ISBN 978-0-300-23832-7 (hardcover : alk. paper)
ISBN 978-0-300-24628-5 (paperback: alk. paper)
Library of Congress Control Number: 2019933689

A catalogue record for this book is available from the British Library.

This paper meets the requirements of ANSI/NISO Z39.48-1992 (Permanence of Paper).

10 9 8 7 6 5 4 3 2 1

On behalf of the BES Community, we dedicate this book to the memory of Beth Strommen. Beth worked twenty-six years for the city and helped to start Baltimore City's Office of Sustainability. She was the office's first director. You would find Beth behind almost all of the city's environmental initiatives. And there are almost too many to count. Professional achievements aside, Beth was a friend to us all. She was smart, creative, and funny. A shoulder to cry on and a friend to laugh with. She never stopped loving the city and fighting for it, its people, and its environment. She is sorely missed and forever inspiring.

CONTENTS

It is a great pleasure for me to write a foreword to this volume on the Baltimore School of Urban Ecology. One of the most satisfying aspects of my ten years working in the Division of Environmental Biology at the National Science Foundation (NSF) was serving as the program director for Long-Term Ecological Research (LTER) from 1995 until 2000. During that time the number of sites in the LTER Network increased from eighteen to twenty-four. Sometime in early 1996 or so I was told to write a program announcement for two new LTER sites focused on urban ecosystems. This came as a complete surprise to me, and I was curious to know how this idea came about. As it turns out, the answer was in the LTER Ten-Year Review (the Risser Report), by far the most influential evaluation of the LTER Network in its now roughly forty-year history. That report recommended that the LTER Network expand to fifty sites, twenty-five of which should be run by NSF and twenty-five by other agencies. Suggested enhancements to the network included urban ecosystems. So, as instructed, I drafted a program announcement seeking proposals for two new LTER sites in urban ecosystems. In some form of twisted irony, I wrote the first draft of that announcement in the library building at the Sevilleta Field Station while my colleagues and I were working on the Konza Prairie LTER synthesis volume.

Later that year I was involved in a meeting at the National Center for

Ecological Analysis and Synthesis in Santa Barbara along with Steward Pickett. I knew Steward was keenly interested in urban ecology, and as we strolled down State Street one evening I mentioned to him our plans to establish urban LTER sites. I believe Steward's response was simply, "Well, it's about time." When we eventually issued the program announcement a very prominent population ecologist from an already established LTER site contacted me to let me know that this was the dumbest thing NSF had ever done. It is fair to say that even though NSF was interested in long-term research in urban ecosystems this was not a widely held view among ecologists at the time. Ecologists just did not work in cities.

A total of twenty-three proposals was submitted to NSF in response to the program announcement. It was evident that the Baltimore Ecosystem Study (BES) team had been thinking deeply about how to conceptualize urban ecology research. Thus it was no surprise that the BES proposal rose to the top of the pile. Two other northeastern cities were also considered to be competitive, along with the proposal from Arizona State (CAP LTER) for metropolitan Phoenix. Based on their conceptual frameworks it was clear that the BES and CAP teams really understood that NSF wanted to support projects that treated urban areas as integrated social-ecological systems, whereas most proposals focused on biological research inside urban areas. The reviewers also recognized that, as conceived, CAP and BES had the greatest potential for synergies as well as providing an interesting contrast between Sunbelt and rust belt cities. After the results of the competition were announced, a very prominent population ecologist from an already established LTER site contacted me to apologize and to let me know that these two projects were going to be great additions to the LTER Network. Just by NSF choosing BES and CAP some naysayers started to believe in urban ecology!

The theoretical, conceptual, and empirical study of urban ecosystems has advanced remarkably over the past twenty plus years, and the Baltimore School has been at the forefront of that development globally. The evolution of thought from ecology *in* the city to ecology *of* the city was transformative at the start. The new focus on ecology *for* the city brilliantly integrates components of the built environment, human institutions, and social systems with ecological principles to create more sustainable urban environments in the future. This book represents an exciting synthesis of empirical research in Baltimore. For me, it is a pleasure to witness how that little aside in the Risser Report has led to a massive paradigm shift in ecology, with the Baltimore School paving the way.

Scott L. Collins
University of New Mexico

The Baltimore Ecosystem Study, commonly referred to as BES, is widely recognized as a paragon of modern urban social-ecological science. After more than two decades of research, education, and community engagement in a complex, spatially extensive city-suburban-exurban system, there are insights to share, generalizations to examine, and gaps to highlight. We have therefore created this book to synthesize the key empirical findings, to meld the perspectives of different research traditions, and to celebrate the accomplishments of interacting with diverse communities and institutions in improving the understanding of Baltimore's ecology in the most inclusive sense. The insights from Baltimore are widely applicable and can both contribute to understanding the ecology of other cities and provide a node for comparison for the global process of urbanization.

Our story is distinctive for several reasons. One is the excitement about theory. Does the study of urban systems require new or altered theory? How has familiar, disciplinary theory been revised and advanced by studying the built environment? The concept of the built environment encourages us to unpack the term "urban." We have therefore considered not just a simple, empirical compilation but also one that links to conceptual advances. We link the empirical research to key concepts, assumptions, generalizations, and hypotheses suggested by the emerging theory of urban ecology.

Although ecological theory is central to our agenda for urban systems, in the contemporary world of seven billion persons, more than half of whom live in urban systems, studying cities is simply *necessary*. Social-ecological understanding has an immense practical benefit. BES represents an urban research program that aims to meet both theoretical and practical needs. The project is unusual and adaptive, and it has evolved. We present our recipe as both an example and a foil for others interested in urban social-ecological research and application.

If theory is to be a guide, it is important to say what that means. Theory is a conceptual device consisting of ideas, empirical content, and propositions that connect it with the material world or with as-yet-unexamined situations, whether they currently exist or not. Theory must state the spatial, temporal, and material domain to which it applies. It explicitly lays out its assumptions and highlights the boundaries beyond which it is not intended to apply. It encompasses accepted facts, generalizations, and laws. Theory also contains models, which are physical, quantitative, qualitative, or verbal representations of the components, interactions, outcomes, and limits of the dynamics in a system. Models can generate testable statements or hypotheses. To be most useful, theories must have explicit ways to translate their abstractions to real or simulated situations. Finally, theories are tied together by frameworks that are, in essence, models of the whole conceptual-empirical structure. This long list of requirements for urban theory has not been met by any single study to date. However, it provides a goal toward which the field must aspire.

A seminal contribution of BES has been the shift from urban ecology as studying ecology *in* the city to studying ecology *of* the city. This contrast points to the difference between answering primarily biological questions in the green spaces within urban systems versus addressing questions about the whole range of urban habitats and places, requiring attention to the hybrid social-ecological nature of urban areas. Although we take this insight as a significant expansion in the past development of urban ecology, how the field might evolve in the future is also important. What are the implications of the growth from ecology *in* to the ecology *of* urban systems? We must answer this broader question by examining a trajectory of knowledge: We thought this, we learned these new things, and this is where research and application need to go. These questions are especially important as urban residents and policy makers increasingly focus on sustainability and resilience in the systems they inhabit and shape.

Our synthesis points to and gives context and meaning to the rigorous and important data that BES has contributed. The book demonstrates insights that emerge from both the traditional approach to the study of ecology *in* cities, suburbs, and exurbs, as well as the more integrated understanding that emerges from the study of ecology *of* urban systems. These include not only refinements and advances in urban ecological concepts and theory but also new data on biophysical processes that exist in urban areas and respond to continuing urban changes, in addition to exposing social interactions and drivers that feed back over time with the biophysical dynamics. The book summarizes newly discovered mechanisms of interaction and change but also presents surprising ecological richness in the processes that persist or have emerged in urban systems. The book shows that there is truly an urban ecology as a body of knowledge responsive to the novelties of urban systems as well as to the firm foundations from biological ecology. Furthermore, this body of data, concept, and integrated understanding has demonstrated utility for management, policy, and urban design.

The book summarizes the theories examined, the approaches employed, the knowledge generated, and the empirical gaps identified by the BES. Readers will understand what theory and concepts have been used to advance urban ecology in Baltimore. In particular, the book details how BES allows existing ecological theory to be tested, and whether the theory changed due to application in a novel system; use in an interdisciplinary context; extension to different levels of complexity; or because of the availability of new data sources and scales. The book describes what was learned in the various topical areas and integrative pursuits of BES. The book also explores how the insights from Baltimore compare with information from other, contrasting cities or other ecosystem types. The book shows how the knowledge has been applied to such activities as education, public understanding, policy, and management. Finally, the chapters describe what research and application might be needed next.

A novel contribution of the book is to employ existing and emerging principles of urban ecology as a tool for synthesizing the empirical findings and conceptual structures. These principles summarize the major themes and needs that the research, education, and community engagement activities in BES have identified.

The summary of empirical insights from BES embodied in the book complements our earlier work, *The Baltimore School of Urban Ecology: Space, Scale, and Time for the Study of Cities,* by J. Morgan Grove, Mary L. Cadenasso, Steward T. A. Pickett, Gary E. Machlis, and William R. Burch Jr. Also published by Yale University Press, that volume clarifies the differences in key schools of urban ecology that have emerged over time and shows how BES provides an advance in terms of approach, conceptualization, and application. The current book fills in the empirical details that bring this new school of urban ecology to life and provide the firm foundation for its use in the changing urban realm.

ACKNOWLEDGMENTS

This book is the result of support and collaborations in the Baltimore region for more than twenty years. Given this long period of time, it is understandable that we have many organizations and persons to recognize. Extending recognition to cover such a long time period is a perilous endeavor, however. We may have missed some people, and if we have done so we apologize in advance.

The Cary Institute of Ecosystem Studies has been the institutional home of BES. Leadership at Cary Institute by Gene Likens, Bill Schlesinger, and Josh Ginsberg has been crucial. Importantly, Brenda Beach, Marie Smith, Holly Beyar, Maribeth Rubenstein, and Jonathan Walsh have been essential to the performance and success of BES. Cary Institute is where Mark McDonnell established the Urban-Rural Gradient Ecology program, which was the first scientific inspiration for BES. We are grateful to Mark for his leadership and his continued dedication to advancing urban ecology worldwide.

The USDA Forest Service has been an important partner to BES. Wayne Zipperer, Bob Neville, Mark Twery, Max McFadden, Bob Bridges, Robert Lewis, Tom Schmidt, Michael Rains, Lynne Westphal, Keith Nislow, Beth Larry, and Tony Ferguson have been crucial to the long-term support of BES. In urban ecology, one has to "think outside the box *and* the pipe," and we have been supported in doing so by the Forest Service.

The local home of BES has been the University of Maryland, Baltimore County. President Freeman Hrabowski has been a friend, enduring supporter, and inspiring voice for what BES could be. His example in the pursuit of greater inclusiveness in the research community and his personal encouragement have meant a great deal to us. To quote one of his undergraduate students, "Black Scientists Matter."

The early roots of BES grew at Yale University's School of Forestry and Environmental Studies. Bill Burch and his students at Yale had the vision through the Urban Resources Initiative to think that work in Baltimore could set the foundation for a long-term urban watershed program. Our thanks to Bill are profound and enduring.

Insights from the Baltimore Ecosystem Study, Long-Term Ecological Research program, supported by National Science Foundation grants DEB-1027188 and DEB-1637661, and from the Urban Sustainability Research Coordination Network, supported by grant DEB 1140077-CFDA 47.074, have contributed to the ideas and findings presented in this book. Early leadership for the urban sites in the long-term ecological research program came from Scott Collins, Tom Baerwald, and the late Henry Gholz. We truly appreciate their leadership. We deeply miss Henry.

Staff members from government agencies and nonprofit organizations were more than our "decision maker" partners. They taught us about the region, provided direction many times, and offered insights and interpretations in our collective research results. In Baltimore City, numerous public agencies played essential roles. In the Department of Planning, we are grateful to Beth Strommen, Anne Draddy, Alice Kennedy, Abby Cocke, Amy Gilder-Busatti, Jenny Guillaume, and Lisa McNeily. We miss Beth's vision, honesty, tolerance, and ability to laugh and spread joy. Kristin Baja participated while a member of the department and now as a climate resilience officer in the Urban Sustainability Directors' Network. In the Department of Public Works, we have depended on Kim Grove, Mark Cameron, and Bill Stack to understand the complexity of urban watershed dynamics more fully and in ways that we could never have imagined. In the Department of Recreation and Parks, we thank Charlie Murphy, Erik Dihle, Shaun Preston, and Bill Vondrasek.

Baltimore City has numerous nonprofit organizations that are crucial connectors. The Parks & People Foundation is an original member of BES, with participation by Jackie Carrera, Guy Hager, Mary Washington, Valerie

Rupp, Laura Connolly, and the current president, Lisa Schroeder. Baltimore Green Space's Miriam Avins and Katie Lautar pointed out and joined us in new research adventures in urban forest patches. Blue Water Baltimore's Halle van der Gaag, Darin Crew, David Flores, Ashley Traut, and Jenn Aisosa have been important watershed advocates.

Baltimore County's Don Outen and Steve Stewart from the Department of Environmental Protection and Resource Management and later the Department of Environmental Protection and Sustainability advanced our knowledge not only of ecological systems but also of the interactions with policy and planning.

The Maryland Department of Natural Resources provided insights about regional issues and informed us about connections to statewide policies. Rob Northrop, Michael Galvin, Jeff Horan, and Anne Hairston-Strang have been leaders in these connections.

The partnerships with administrators and teachers in the public school districts of Baltimore City and of Baltimore County, in addition to productive partnerships with several private schools, have been crucial to our success. It has been a pleasure to interact with these leaders and educators and with the many students in their charge. These partnerships have helped us better understand both teaching and learning as a part of a human ecosystem.

This long list of institutions and people who have made BES an effective, stimulating, and productive community of intellect and practice would be incomplete without mention of the engaged and hardworking undergraduates, graduate students, and postdoctoral associates who have worked with us. Unfortunately, there are too many to list individually here, but we are grateful to each and every one of them nonetheless. They have pursued many pathways as they have impacted the field of urban ecology. They have gone on to advanced training, private jobs in urban professions, positions in socially engaged not-for-profit organizations, employment and practice in government at all levels, and faculty positions in liberal arts colleges and research universities. Many are already leaders in the next generation of urban science and application. Others are among the informed citizens and public voices in the civic dialogue that drives democracy. You are a big part of the reason we do this work—and certainly a big part of its success. Thank you all.

Finally, our twenty-year anniversary is celebrated with the Central

Arizona–Phoenix LTER project, which has been our urban LTER twin. We have prospered and grown together. Nancy Grimm, Chuck Redman, and Dan Childers have shared the pains and joys of running an urban LTER project and contributed to our development. Thanks sis and bros!

In sum, we count ourselves fortunate to work with so many wonderful people and in such a great, quirky place as Baltimore. While we have worked together as professionals, we feel that we have been blessed to become a community. We look forward to the next twenty years and all the new challenges, opportunities, and new knowledge yet to be imagined.

Finally, we gratefully acknowledge a grant from Furthermore: A Program of the J. M. Kaplan Fund to support the editing, design, printing, and binding of the book.

SCIENCE FOR THE SUSTAINABLE CITY

PART ONE URBAN ECOLOGY AND BALTIMORE: AN INTIMATE RELATIONSHIP

CHAPTER ONE

Goals of the Baltimore School of Urban Ecological Science

Steward T. A. Pickett, Mary L. Cadenasso, Lawrence E. Band,
Alan R. Berkowitz, Grace S. Brush, Jacqueline Carrera,
J. Morgan Grove, Peter M. Groffman, Charles H. Nilon,
David J. Nowak, and Richard V. Pouyat

IN BRIEF

- The Baltimore Ecosystem Study (BES), a Long-Term Ecological Research (LTER) project, integrates natural and social sciences to understand metropolitan Baltimore as an ecological system.
- The fundamental research questions address the spatial structure and change in the system, the fluxes of matter and energy, and the role of ecological knowledge in education and in civic decision making.
- BES applied important approaches from mainstream ecology, including the ecosystem and the watershed, and combined these with key social science frameworks to improve the understanding of humans and their institutions as components of urban ecosystems.
- The program is evolving to address urban sustainability, resilience, and adaptation of Baltimore as a complex system.

INTRODUCTION

The Baltimore Ecosystem Study (BES) was designed at a time when urban research and education were far from the mainstream of ecological science in the United States. Although humans were among the concerns of ecological science in its earliest days, and leading ecologists and social scientists of the mid-twentieth century encouraged ecologists to include the city, few heeded this call. To foster the integration of urban ecosystems into the domain of ecology, the founders of BES deliberately structured our urban LTER program in a way that employed familiar ecological concepts and approaches (chapter 15), while at the same time going beyond the usual scope of the discipline.

The founders of BES also set out to study the entire city-suburban-exurban (CSE; box 1.1) system from an ecological perspective, rather than to focus only on the obviously "green" parts of the CSE, such as parks, vacant lands, and open waters. This decision suggested three questions to guide the establishment of BES (box 1.2). These questions are, in essence, specific forms of the questions that would be asked at any ecological frontier: What is the system structure? What processes occur in the system? How does information in the system feed back to influence subsequent system structure and functioning? The third question was novel for biological ecologists, as it required us to approach the CSE as a self-aware and culturally adaptive entity.

The three questions guiding the establishment of BES were an extension of the pioneering work of Mark McDonnell, who began the Urban-Rural Gradient Ecology (URGE) project in the New York metropolitan area. McDonnell realized that the study of a system as complex as the metropolitan area covering parts of New York, New Jersey, and Connecticut and encompassing thirty-one counties would need a tight focus. McDonnell assembled a team, which included Margaret Carreiro and Wayne Zipperer, along with several of the authors of this chapter, Mary Cadenasso, Rich Pouyat, and Steward Pickett, in establishing a study transect extending from the Bronx in New York to exurban or rural western Connecticut. The transect contrasted several mixed oak forests experiencing differing degrees of urbanization but sharing the same substrate, tree species, age, and canopy structure, and an absence of signs of natural disturbance. This transect yielded a gradient of contrasts in soils contamination, microbial ecology, nutrient processing, and botanical structure and composition. It was the solid and sometimes surprising body of knowledge that McDonnell and his colleagues discovered in the URGE project that allowed BES to initially focus its questions as broadly as it did.

GUIDING QUESTIONS: STRUCTURE, PROCESS, UNDERSTANDING
The three questions introduced above guided the establishment of BES and set its long-term research trajectory. The questions can be thought of as

BOX 1.1. KEY TERMS

Abiotic: The nonbiological components of a system, such as air, water, and mineral matter, or the physical regulators and signals.

BES: The Baltimore Ecosystem Study, both broadly as a program of research, education, and community engagement and more narrowly as an NSF-sponsored Long-Term Ecological Research site from 1997 to 2021.

Biodiversity: Short for biological diversity, which can include differences in genetics, species composition, or even the heterogeneity of ecological landscapes.

Biogeochemistry: The study of the patterns, causes, and consequences of the circulation of chemical elements in ecosystems and the biosphere. The word combines biology, geology, and chemistry to indicate the complexity of those patterns and processes.

Biotic: The biological components of ecological systems, including organic matter that is no longer alive. Hence "biota" refers to all the species and organisms and their direct products in ecosystems.

CSE: City-Suburban-Exurban systems. A term that indicates that "urban ecology" is not only about dense cities but also deals with large urbanized regions and the variety of ecosystem types or habitats within them.

Demography: Referring to both people and biota, the term is defined as the age, sex, size structure of a population. In the case of humans, social identities and indicators of social status are also included in demography.

Exotic species: Species introduced, either on purpose or accidentally, into a region distant from that in which they originally evolved. Some exotic species can become invasive, displacing species that have been a component of local biodiversity for centuries or millennia.

Geomorphology: The lay of the land, including its geology, elevation, steepness, and directions of slopes. Although these features are often profoundly modified by urbanization and urban spread, they are also part of the enduring context of urban systems.

Georeference: The process of linking a data record to a specific spot on the Earth's surface, via either an address or latitude and longitude.

Hydrology: The science of the patterns and controls on water flow both above and below the earth's surface.

Introduced or invasive species: Exotic species.

Island biogeography: An ecological specialty that studies the patterns of species distributions on literal or habitat islands, and the controls on those patterns. Originally developed to understand the patterns on literal islands in oceanic settings, the field has been extended to island-like habitats on land. Mountaintops, isolated lakes, and green spaces within urban mosaics are examples of nonoceanic subjects of island biogeography.

Isopod: A group of crustaceans. Small, terrestrial isopods are commonly called pillbugs or woodlice, although they are neither bugs nor lice. They are a part of the terrestrial decomposition chain.

Legacy: The persistent effect of a prior ecological or social condition that affects current ecosystem structure or function.

LTER: Long-Term Ecological Research. A program of the U.S. National Science Foundation that aims to understand ecological processes that are slow or rare in time but which nevertheless can be fundamental to ecosystem structure and function. The abbreviation has been adopted by similar programs in other countries.

Metacommunity: A community of communities, that is, a group of ecological communities separated in space yet sometimes exchanging species. Individual communities in the collection can sometimes disappear, and new ones become established.

Metapopulation: Similar to a metacommunity, this concept refers to a collection of populations of a species that are separate from one another but which may exchange individuals and genetic information. The dynamics of individual populations is influenced by at least some of the others in the collection. Individual populations may vanish or be established somewhat independently.

Paradigm: The set of background assumptions and ways of approaching explanations that characterize fields of research or study. These deep assumptions and approaches are rarely explicitly examined in the day-to-day working of a field. Paradigms may shift when entirely new ways of looking at a problem come into play, either due to new technology or massive infusion of new types of data.

Patch dynamics: An ecological or social model of spatial organization that views systems as being composed of distinct patches, but which change and interact with each other. Interactions may shift the identity or content of patches, and patches may cease to exist due to external or internal drivers. An entire suite or mosaic of interacting patches can change in content, connectivity, configuration, or function.

Path dependency: The idea that particular outcomes of processes over time depend on the order of events. Outcomes may depend on local or regional conditions that affect the order of events.

Production: The biological conversion of solar energy into biologically usable form, primarily by photosynthesis.

Redlining: The practice institutionalized in the 1930s by the federal Home Owners' Loan Corporation that rated neighborhoods by creditworthiness. The process was prejudicial and usually rated neighborhoods of African Americans or recent European immigrants as unworthy of mortgage or other bank loans, constraining investment, renovation, and maintenance in such neighborhoods. Such neighborhoods were indicated by red on maps.

SETS: Social-Ecological-Technological Systems is a concept that describes ecosystems as containing these three kinds of interacting components. Although it is similar to the human ecosystem concept, it emphasizes the role of technology of all kinds in urban ecosystems.

Succession: The process of change over time in ecological communities. Although it was originally thought to proceed by a deterministic series of transitions, leading to a stable community, research has judged the process to be much more open-ended and probabilistic.

Urban–rural gradient: An approach to understanding contrasts among systems experiencing or representing different degrees of urban development, compared to systems in rural areas. An urban–rural gradient is an analytic result of comparing samples along spatial transects or collected across extensive regional landscapes.

Vector: An organism that carries a disease organism; for example, mosquitoes are vectors of disease-generating viruses or other parasites, such as malaria.

BOX 1.2. THE ORIGINAL GUIDING QUESTIONS OF BES

1. How do the spatial structures of ecological, physical, and socioeconomic factors in the metropolis affect ecosystem function?
2. What are the fluxes of energy and matter in urban ecosystems, and
   how do they change over the long term?
3. How can urban residents develop and use an understanding of the
   metropolis as an ecological system to improve the quality of their
   environment and their daily lives?

---

addressing structure and change, ecosystem processes, and the influence
of ecological knowledge in a social-ecological system.

Structure

Science has long had a core concern with the structure and functioning of
systems, be they atoms, molecules, organisms, the biosphere, or the solar
system. In urban systems, structure includes the existence and layout of
buildings and infrastructure as well as the distribution and composition of
planted and volunteer vegetation and the spatial arrangement of people and
the institutions they constitute. Such institutions range from households to
firms to neighborhoods to social groups. Institutions can be highly structured or ad hoc and can be persistent or fleeting. Governments are among
the most persistent institutions, while households change as families grow,
children age, number of members expands or contracts, and perhaps the
family ultimately dissolves or moves. Businesses may last weeks or decades. Some neighborhoods retain their nicknames and spatial identity even
after all original residents have left or new groups predominate.

One of the central characteristics of urban systems is their dense and
often abrupt heterogeneity. While large suburban tract developments or
regional shopping malls may occupy dozens or hundreds of acres, urban
theorists have long remarked on the fine-grained mosaic of neighborhoods
and districts in cities. Even in suburbs and exurbs, the kind of development
close to transportation networks differs from the development and land
covers away from major roads and rail lines. Such heterogeneity provides
a ready link with mainstream ecology and the geosciences given their con-

cern with diversity of organisms and soils, for example, and with landscape ecology given its focus on spatial pattern. Because the spatial heterogeneity of a CSE includes constructed, social, and biophysical elements, patch dynamics served as a core concept in unifying various disciplines in BES. These urban patterns are layered on top of the natural physical template of the landscape, which remains evident and functional to some degree. Watersheds are an obvious manifestation of the physical template and have long been useful for integrating physical and biological properties in natural ecosystems. Furthermore, watersheds can foster integration with social sciences through the formation of watershed associations or other institutions that function as nodes for human–environment interactions and for environmental justice.

Interest in structure leads quickly to interest in its reciprocal relationship with various processes. Most immediately, a long-term perspective raises questions about how the various structures develop and change through time. Historical records, maps, paleoecology, remote sensing, and direct observation are rich sources of data on structural change in urban systems. The interaction between structure in each of the three focal realms of BES—biological, physical, and social—raises questions about how functions in each of those different realms interact with structure. Numerous examples of structure-function relationships appear in part 2 of this book.

Process

The second guiding question used to establish BES focused on the movement of matter and energy in the urban ecosystem. The watershed approach is useful beyond its structural application described above. It has also served BES as a functional unit. In this role it is the basis for comparison of urban ecosystems with the less human-dominated ecosystems that traditionally have been studied more broadly by ecologists. Furthermore, watersheds serve as a research platform for cycles of question generation, hypothesis testing, and model development. This approach, which allows for the quantification of inputs and outputs of water and nutrients in hydrologically defined basins, has been central to ecosystem ecology over the past fifty years, allowing for holistic evaluation of ecosystem structure and function over areas that differ in important characteristics. In fact, the powerful, productive examples of watershed-based research in other LTER projects, such as those at Hubbard Brook in New Hampshire, Coweeta in the southern

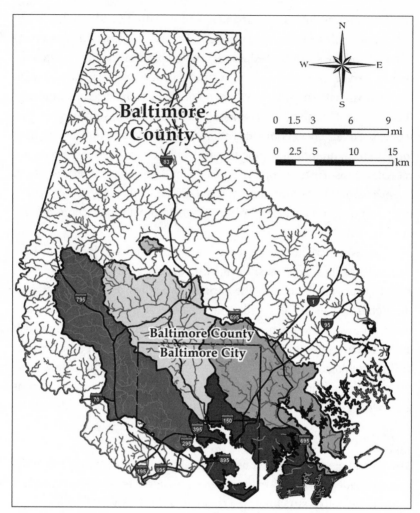

FIGURE 1.1 The principal watersheds of Baltimore City and Baltimore County. The watershed to the west in dark gray is the Gwynns Falls, with Jones Falls shown in light gray in the center, and the Back River, encompassing Herring Run, in medium gray to the east. The watershed to the southeast is the direct harbor drainage, shown in black. Interstate highways are shown for orientation. All the large watersheds shown drain into the Chesapeake Bay, with the Gwynns Falls and Jones Falls emptying first into Baltimore Harbor, an arm of the Patapsco River, and the Herring Run emptying first into the Back River estuary.

Appalachians, and the H. J. Andrews Forest in the Cascade Range of Oregon, stimulated our use of the concept. The watershed studies in Baltimore examine stream flow and water quality in a forested reference catchment and at gauging stations along the Gwynns Falls stream (figure 1.1). These measurements are coupled to long-term measurements in terrestrial permanent plots that provide data on processes occurring in the forest, grassland, and riparian components of the watersheds. Sampling campaigns have characterized the distribution of contaminants, such as soil lead, and the social factors, such as building age and zoning variances, that influence this distribution.

Understanding

The flow of information was purposefully omitted from Question 2, which focused on other social fluxes as well as biophysical fluxes. That intentional omission was due to the crucial importance of information flows in CSE systems and the need to highlight them explicitly. Therefore, Question 3 recognizes that BES is a part of the Baltimore ecosystem and that it may impact the promulgation of environmental information and subsequent decision making by officials and residents. Hence Question 3 asks whether ecological understanding is used by residents and, if so, does that change their quality of life? This question suggests three approaches: social surveys; engagement with communities, managers, and decision makers; and education.

Social surveys seek to document the characteristics, knowledge, perspectives, and values held by residents across the full spectrum of class, race, wealth, and education in the Baltimore region. In addition to surveying households, organizational connections are assessed. Attention to community and decision-maker engagement seeks not only to share our scientific insights with people who might use it, but also to learn from residents and leaders the needs and opportunities for gathering new environmental knowledge in the Baltimore ecosystem. Regular, reciprocal interactions with residents and decision leaders have emphasized listening on the part of BES researchers, not just one-way presentation of scientific findings. In several cases this mutual learning and sharing led to joint projects, such as planning, management, and assessment of stormwater fluxes in a thinning neighborhood in the center city, the assessment and planning for increasing tree canopy cover throughout the city, and understanding the environmental effects of streambank restoration. The final channel for understanding is through formal and informal education. Early efforts in education focused on building a constituency within state and local education systems and on developing curricula for schools and after-school and summer programs. Based on the successes and challenges of these programmatic efforts, the BES education team subsequently emphasized research on teaching and learning in the Baltimore region (chapter 17).

The three fundamental questions motivating BES are related to the core research areas required by the LTER program. But our questions also reflect the evolution of urban ecology as a field, which prompted the National Science Foundation (NSF) to require additional areas for long-term research in urban areas.

BES AND THE LTER CORE RESEARCH AREAS

The three original guiding questions of BES can be parsed among the traditional core areas of the LTER program at NSF. The following descriptions of each core area are quoted from the LTER website. Examples of the studies addressing each core area are listed below, with reference to later chapters.

> 1. **Primary Production**—Plant growth in most ecosystems forms the base or "primary" component of the food web. The amount and type of plant growth in an ecosystem helps to determine the amount and kind of animals (or "secondary" productivity) that can survive there.

BES measures parameters that support understanding of the growth of dominant plants in terrestrial and stream environments (chapters 8, 9, 11, 12, and 13). Woody plant biomass and change over time are assessed via extensive and intensive sampling. Woody plant production is measured every five years using 195 randomly located plots. Eight intensively measured permanent plots combine assessment of vegetation and soil biogeochemistry. An ecological-hydrological model of watersheds (chapter 9) simulates coupled ecosystem primary production, hydrology, and nutrient cycling. The primary production by microbial organisms coating surfaces in streams and the effects of pharmaceutical contaminants on that production are under investigation.

2. **Population Studies**—A population is a group of organisms
of the same species. Like canaries in the coal mine, changes in
populations of organisms can be important indicators of environmental change.

Population studies and biodiversity assessments are a component of BES
research (chapters 11 and 13). Populations of birds, squirrels, and soil invertebrates, such as earthworms and isopods, are measured in both long-term
and short-term studies. Aquatic invertebrates are quantified in streams and
in constructed stormwater detention ponds. Plant populations are assessed
in randomly distributed permanent plots, which are assessed in all land
uses. Population studies also include research in vacant lots and residential
yards and of the habitat and food web relationships of introduced mosquitoes that can transmit human diseases.

3. **Movement of Organic Matter**—The entire ecosystem relies
on the recycling of organic matter (and the nutrients it contains),
including dead plants, animals, and other organisms. Decomposition of organic matter and its movement through the ecosystem
is an important component of the food web.

Soil organic matter is assessed in Baltimore soils, including forests and
lawns (chapters 8 and 12). Organic matter dynamics are also included in our
studies of stream ecosystems. Several of the focal groups in the biotic population studies are important in decomposition of organic matter in soil.

4. **Movement of Inorganic Matter**—Nitrogen, phosphorus, and
other mineral nutrients are cycled through the ecosystem by way
of decay and disturbances such as fire and flood. In excessive
quantities nitrogen and other nutrients can have far-reaching and
harmful effects on the environment.

Inorganic nutrients and nutrient pollutants are regularly measured in BES
watershed and stream research (chapters 8, 9, and 12). Nitrate, phosphate,
and particulates are key pollutants in metropolitan streams and downstream
in the Chesapeake Bay. Nitrate and chloride are also drinking water pollutants. Nutrient processing data are collected in the permanent plots, in
streams, and in riparian zones. Decades of road salt application have increased the chloride concentration in Baltimore region streams, including

those draining into the region's reservoirs. Heavy metal contamination of soils has implications for public health and soil nutrient processing (chapter 8).

5. **Disturbance Patterns**—Disturbances often shape ecosystems by periodically reorganizing structure, allowing for significant changes in plant and animal populations and communities.

Disturbances, detected as pulsed structural alterations in ecosystems and landscapes of the Baltimore region, appear in the geomorphology of streams, the alteration of forest cover, and the mortality of trees in permanent plots and coarse-scale vegetation surveys (chapters 9 and 11). More subtle press disturbances include invasion of novel exotic species and differential alteration of forest regeneration along urban–rural contrasts (chapter 13). Disturbances also take the form of social presses and pulses, such as shifting economic investment and disinvestment, migration of racial groups and social classes, and policy interventions such as redlining or altered regulations in response to the disruptions (chapters 3, 4, 6, and 7). An integrated research program such as BES must account for both biophysical and social disturbances.

Demographic and Social Phenomena: New Core Areas for Urban LTERs
BES has also focused on social, demographic, and land-change data beyond the five core areas identified in 1980, when the LTER Network was established. The 1997 request for proposals for urban LTER by NSF included three additional requirements. To quote from the 1997 request:

In addition to the traditional LTER core areas, an Urban LTER will:
- examine the human impact on land use and land-cover change in urban systems and relate these effects to ecosystem dynamics,
- monitor the effects of human-environmental interactions in urban systems, develop appropriate tools (such as GIS) for data collection and analysis of socio-economic and ecosystem data, and develop integrated approaches to linking human and natural systems in an urban ecosystem environment, and
- integrate research with local K–12 educational systems.

The first two requirements are mandates for research. Consequently, urban LTERs necessarily have *seven* core areas, which include the five traditional

ones that the LTER program has required since 1980. The third new requirement mandates a channel for engaging with the local community and
institutions.

Fortunately, human systems possess a plethora of long-term records
that are relevant to ecosystem structure and function. The decennial census, municipal records of construction, maintenance, or demolition of real
property and infrastructure, the minutes of formal and informal civic bodies, and newspaper accounts are among the data sources available for understanding the social structures and processes in metropolitan Baltimore
(chapters 3, 4, 5, and 15). Legal and administrative information, such as
zoning and environmentally relevant variances, add depth to understanding of social trends and heterogeneity. De jure and de facto segregation are
significant in the social structure of the Baltimore region (chapter 3) and
hence have environmental implications, in terms both of inequity and of
environmental perceptions and actions.

A significant contribution of BES to urban ecological research is the
triennial social survey, carried out via telephone questionnaire. At each iteration the survey has included more than three thousand respondents whose
specific locations are known but protected for privacy. The thirty-question
survey documents people's environmental perceptions, attitudes, and activities. When combined with marketing data, the survey yields a highly re-
solved and spatially registered model of social cohesion, social capital, and
lifestyle differentiation. Additional surveys of types, rates, and social determinants of lawn and garden fertilization and management have directly
addressed links with biophysical processes (chapter 12).

The three foundational questions about structure, flux, and the feedback of understanding frame the synthesis presented in this book. They
point to an emerging conceptual context for BES, which may be useful for
other urban ecological research projects and their applications. These concepts can be labeled as a "school" of urban ecology to indicate conceptual
cohesion and practical value. We do not claim this school is a complete or
mature theory of urban ecology. Surely contemporary urban ecology will
continue to evolve (chapters 2, 14, and 15).

EVOLVING APPROACH OF THE BALTIMORE SCHOOL

The added core areas of the 1997 LTER request for proposals necessitated
moving beyond the bioecological comfort zone and engaging the ecological

structure and functioning of all kinds of habitats and areas within complex urban systems (figure 1.2). Furthermore, we were anxious to solidify the interactions between the social and biophysical scientists in our newly acquainted group. To facilitate these interactions, BES adopted the goal of pursuing an ecology *of* the city as a complement to ecology *in* the city. It is important to recognize that both research approaches—ecology of and ecology in—are necessary and indeed complementary for a full understanding of the ecology of social-ecological systems. This conception has proven widely useful well beyond Baltimore.

A major aid to our efforts to integrate biophysical and social sciences was the human ecosystem framework, developed by a team of social ecologists, to incorporate social structures and interactions within the scope of the ecosystem. While the approach was labeled the "human ecosystem model," we have used it as a framework rather than a formal model to remind us of the kinds of social structures and social processes we would have to consider in constructing hypotheses, models, and explanations for ecology *of* the city. We made the distinction between model and framework to avoid the temptation to include all variables in every model we might construct. Our strategy would become one of erecting multiple, intersecting models that embodied different kinds of structures and interactions rather than to construct one gigantic, inclusive model.

A second integrating concept employed in BES was the watershed approach. It is commonly agreed that watersheds integrate inputs over their entire extents, integrate terrestrial and aquatic processes, link biotic and abiotic components and processes, and thus allow evaluation of whole-system function and response to disturbance and environmental change. We hypothesized that the watershed approach could serve as a tool for further integration in urban areas, allowing for evaluation of human structures and the flows of matter and energy in our assessments of whole-system function. The watershed could also be used as an empirical platform for integration across disciplines. In addition, the watershed idea had been used as a social and political tool in Baltimore since the 1970s, providing a ready bridge between science and policy (chapter 15).

The Baltimore School addresses spatial heterogeneity as a key feature. Since the middle of the twentieth century ecology had become increasingly aware of the need to understand spatial heterogeneity. Gone are the days when ecosystems could be considered uniformly mixed, equilibrium sys-

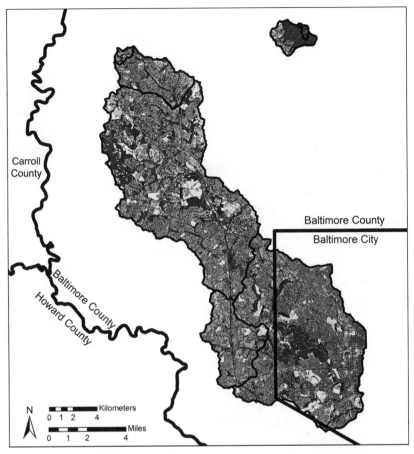

FIGURE 1.2 The eleven land use/land cover categories and their distribution in the large Gwynns Falls watershed and the smaller Baisman Run watershed to the north. Along Gwynns Falls there are four gauging stations on the main stem of the stream, and on four subwatersheds representing different land uses. On the main stem of the Baisman Run watershed, water flow and quality are measured to represent large lot, exurban development, while the Pond Branch subwatershed represents a forested reference catchment.

tems or populations could be assumed to be uniformly distributed across their ranges. Landscape ecology, metacommunity ecology, and metapopulation ecology, emerging from developments in island biogeography and population genetics, made spatial heterogeneity a core topic of ecological research and application. This awareness prepared ecologists to act on the fine-scale heterogeneity so often encountered in urban systems, where turning a corner might reveal a new patch defined by contrasts in biotic, social, or physical structures or, more likely, all three. The functional significance of such a patchwork was an irresistible research topic, resonating as it did with both urban reality and new thinking in the ecological and geosciences (chapter 8). Spatial heterogeneity, the changes in heterogeneous patchworks, and the fluxes across patch boundaries and entire patchy mosaics are the purview of the theory of hierarchical patch dynamics. This concept has provided a bridge for comparison with other urban areas, notably the Central Arizona–Phoenix LTER, as well as a bridge to the theory and practice of urban design, which itself is acutely tuned to spatial heterogeneity.

Spatial heterogeneity is important not only in contemporary time. Different kinds of biophysical and social heterogeneities can interact with each other and with external disturbances in a very dynamic way through time. Therefore, an additional important assumption of the Baltimore School is that historical legacies and path dependencies as well as temporal lags are important in urban ecological processes. The spatial distribution of physical environmental conditions, soils, and biota can influence ecological and social conditions today. The constructed environment itself is a legacy in many urban systems. Certainly this is the case in Baltimore, which has many built historical layers. The persistent template of the old market roads is a powerful legacy, but one which clashes with the newer pattern of expressways. Both of these legacies have altered hydrology and the boundaries between neighborhoods and also conflict with the legacy of the partially implemented Olmsted Brothers' parks and parkways plan of 1904. Among the social legacies is the legal segregation of African Americans during the Jim Crow era, including the redlining by the federal Home Owners' Loan Corporation. Such social legacies have environmental and linked social consequences today.

In dealing with heterogeneity, BES researchers have paid considerable attention to the shortcomings of classic land use/land cover theory in the context of an integrated conception of urban ecosystems (chapter 14). The traditional "industry standard" classification, developed in 1976, conceived of natural, agricultural, and urban land uses as mutually exclusive. This system and its derivatives thus violated our basic assumption that the Baltimore ecosystem was an integrated social-biophysical phenomenon.

The human component of an urban ecosystem, or indeed of any inhabited ecosystem, is complex. Humans are present as individual organisms, as parts of social networks (chapter 5), and as behavioral, economic, and political agents (chapters 3, 4, and 7) that affect and are affected by the rest of the ecosystem. It has been useful in BES to adopt the sociological concept of human institutions as tools that structure society to provide various functions. Institutions are social norms that govern people's interactions. Institutions manifest as formal and informal structures or organizations. They can exist at various spatial and temporal scales and have different spatial contexts. Considering humans as components of ecosystems contrasts with the traditional bioecological approach to humans as external, usually negative, impacts on biological systems.

The idea that humans and their diverse formal and informal institutions are components of ecosystems suggests that Greater Baltimore must be approached as an integrated ecosystem, one consisting of biotic, physical, social, and constructed components (chapter 14). This assumption is fully compatible with the definition and use of the ecosystem concept in mainstream ecology. Indeed, the original discussion of the ecosystem concept emphasized the role of humans. The basic definition of the ecosystem is a biotic complex interacting with a physical context in a specified spatial frame. This definition readily applies to urban systems if one assumes that social features are a part of the biological component and that the constructed features are part of the physical component. However, it has proven useful to explicitly include constructed and social components in the definition, lest the terms "physical" and "biotic" appear to exclude human artifacts, technologies, and social systems.

The study of urban systems has several dimensions that may help us understand the scope and changes in urban ecological science since the middle of the twentieth century. The approaches to urban ecology over that time have included studies of material and energy budgets for entire cities (the "metabolism" of Hong Kong), studies of wildlife populations throughout cities, and studies of composition and change in the botanical assemblages—both remnant and opportunistic. The variety of studies can be viewed in an idealized conceptual space defined by contrasts between ecology *in* and ecology *of* urban areas. Studies of ecology in cities have been in the majority, and this biotically focused approach began early in the history of the discipline. In this approach, the built and social components of cities were treated largely as an external environment to the biological habitats of cities. These studies contributed greatly to the understanding of ecology as an urban phenomenon and have contributed to planning, design, and restoration in the urban realm.

We apply the term "school" to the evolving conceptual framework used by BES in order to contrast it with the older Chicago School of urban ecology. The label "urban ecology" is sometimes still used to describe the school of American sociology pioneered at the University of Chicago in the early twentieth century (chapters 2, 14, and 15). We wish to emphasize that although the sociology of the Chicago School has been abundantly critiqued, such criticism does not apply to contemporary social-ecological urban ecological research. Although the early Chicago School employed key ecological concepts, such as competition, succession, and spatial zonation of communities, it used those processes as analogies to explain the patterns of distribution and change occurring in the city of Chicago during a stunningly rapid phase of the city's growth.

There were two problems with the application of these ecological analogies to social dynamics in Chicago. First, the specific mechanisms of distribution and change were neglected, and the spatial distributions were rather idealized. Subsequent sociological critiques sought mechanisms based on behavior and heterogeneity of and within social groups and institutions. The second problem was that even as the Chicago sociologists were adopting ecological analogies as explanations in the 1920s, the ecological concepts themselves were controversial and changing. Although the Chicago sociologists used the treatises and books of leading ecologists of the time in their teaching, they apparently did not interact substantively either with those ecologists or with iconoclasts in the ecological profession. For example, the nature and causes of succession were being hotly debated as early as 1917, introducing ideas that individual organisms, heterogeneous populations, and gradations between biotic communities were important phenomena. Therefore, the Chicago School is a useful historical touchstone in the development of urban ecology, but it does not represent the contemporary ideas and empirical generalizations that are emerging from social-ecological approaches to urban systems (chapter 14).

An additional assumption of the Chicago School was that rural life was the ideal, and cities fell far from that state of grace. As a result, early sociologists emphasized urban pathologies. More recent environmental scholars have also assumed that cities were a blight on the land. In contrast, BES researchers did not assume that cities and urban systems should necessarily be viewed in negative terms. Certainly there are costs of contagion and density in cities. But there are also efficiencies of scale, energy, and materials savings and benefits of innovation and interaction that accrue to urban systems. Our concern has been not to judge cities but to understand how they are structured, how they work, and how they change as integrated social-ecological systems.

These assumptions suggest a general conceptual shape of the Baltimore School of urban ecology. The scope of this school can be summarized in four broad propositions:

1. The ecology of cities addresses the complete mosaic of land uses and management in metropolitan systems, not only the conspicuous green spaces.
2. The urban mosaic is complex in space, scale, and time. Spatial heterogeneity, a multiplicity of interacting spatial scales, and temporal legacies and path dependencies or contingencies are crucial to the structure and function of urban systems.
3. The ecology of cities is an integrative science, combining social and biophysical frameworks, theories, and methodologies. Technology, infrastructure, and design are also key perspectives to be linked in urban ecology.
4. We assume that advances in understanding of ecology in and of cities can contribute to both the basic science and the formal and informal decision making in urban systems.

These assumptions are also the precursors of principles that may be of general utility in urban ecology (chapter 14). The empirical findings presented throughout this volume and the applications in education, design, and policy are shaped by these assumptions.

CONNECTING WITH COMMUNITY

Working in an urban system obligated us to conduct ourselves and our program in ways that were respectful and potentially useful. In other words, to the complementary research approaches of ecology *in* and ecology *of* the city we had to add an ecology *for* the city. This goal of connecting with communities is consistent with the requirement that all NSF-funded research consider the "broader impacts" of research beyond pure scientific understanding.

We have learned that the most effective way to engage with communities, policy makers, and managers is simply via respectful dialogue. Analysis of the flow (or lack of flow) of information from science to society has led to a fundamental restructuring of the way scientists engage with communities: moving from a one-way delivery of scientific information to a more active engagement with audiences that focuses on presenting information in contexts useful to them and to "framing" in ways that are relevant to different groups. Our interactions with residents and leaders in Baltimore became much easier, more productive, and even more fun when we learned to listen and to converse rather than to reach out and deliver (chapter 15).

An important early venue for communication was the regular meetings of the Revitalizing Baltimore (RB) Technical Committee, a group of managers, researchers, and community leaders convened by the Parks & People Foundation between 1994 and 2006. This group of experts, community leaders, activists, and researchers gathered to share information about practical projects for neighborhood revitalization, local environmental improvement, environmental restoration, government agency enterprises, education activities, and career preparation in underserved neighborhoods. Through the RB Technical Committee we learned of the concerns of managers and residents, began to see where our data and models could be helpful, and learned where our scientific thinking was out of sync with actual changes occurring on the ground. For example, while we initially were thinking about suburban sprawl, managers and developers had moved ahead to densification and redevelopment of old urban and suburban areas. We also discovered opportunities to embed social-ecological research within ongoing or planned activities. Community engagement has also ensured that relevant advances in understanding Baltimore as an integrated social-ecological system are shaped by communities, civic institutions, and government agencies at various levels and are subsequently available in social decision making as well as in guiding future research. In other words, a participatory research approach, one which respects community concerns and knowledge as well as the proprietary nature of individual- and community-based information, has become a significant part of BES. This evolution reflects the fruits of community engagement.

ROADMAP OF THE BOOK

This book has a simple roadmap. Part 1 covers the goals and the larger historical context of the project (chapter 2). Part 2 constitutes the bulk of the book and summarizes the empirical and modeling work of the program over its first three phases, spanning 1997–2016. This empirical section ends with a chapter expressing the insights in terms of nascent principles of urban ecology. Finally, part 3 shows how the program applies to and benefits from its work in application, urban design, and conducting and understanding education. It is our hope that the empirical results of the Baltimore School will prove to be useful well beyond our metropolitan region.

Acknowledgments   The authors of this chapter are founding members of BES who have been active in the program since its inception. Among those who have not continued to be active, we wish to especially acknowledge Professor William R. Burch Jr. for his seminal role in establishing BES I. We also thank Wayne Zipperer and Robert Costanza for their early contributions to the project.

CHAPTER TWO

Urban Ecological Science in America

THE LONG MARCH TO CROSS-DISCIPLINARY RESEARCH

Sharon E. Kingsland

IN BRIEF

- Efforts in the interwar years to link ecology and social science opened an important dialogue, but it fell short of producing a coherent discipline of human ecology.
- Ecosystem ecology created a new framework for human ecology but reinforced equilibrium viewpoints.
- Urban ecology emerged as a priority around 1970, prompting experiments in cross-disciplinary exchange between ecology and other fields.
- Critiques of equilibrium viewpoints, more realistic models of ecosystem function, and focus on "resilience" of systems were key conceptual advances in the pursuit of urban ecology.
- Human ecology often involved long-term historical perspectives and less judgmental attitudes about human nature.

INTRODUCTION

Urban ecological science in the United States grew out of a long cross-disciplinary conversation between ecology and social science that began in the early twentieth century, when these disciplines were being defined. At the meeting ground of ecology and social science rested "human ecology," which started out as an eclectic, ill-defined enterprise that did not add up to

24

a coherent discipline. Nevertheless, it was deemed to be a subject worthy of pursuit, and those interested in its development promoted early exchanges across disciplinary borders. But ecologists were at the same time wary of opening up their subject to too many difficult problems, especially in the early decades of the twentieth century, when ecologists were concerned about whether ecology could survive as an independent discipline. Human ecology immediately led to problems of great complexity. It was easy to set these aside in order to focus on topics in botany and zoology, which were already difficult enough.

After the Second World War, new environmental concerns such as human population growth energized the conversation between ecological and social sciences. The shock of urban riots in the 1960s roused an intense interest in urban ecology and urban planning, but the problems of interdisciplinary interaction remained formidable. In the meantime, new ecological concepts, especially the ecosystem viewpoint, suggested new ways to integrate ecological and social sciences, while the emerging concepts of ecosystem services and ecosystem resilience suggested new ways to frame and explain ecological problems. With the continued expansion of ecosystem science in the 1970s and 1980s it became possible to get inside the black box of the ecosystem and understand the patterns and processes guiding ecosystem development. By the 1990s it was possible to transfer successful models of ecosystem structure and function to the urban environment and to move toward a more fully integrated socioecological analysis. Urban ecological science started to coalesce and settle into its niche only in the 1990s, but it had a long prehistory. Here I explore some of the key stages of this evolving conversation. The long march has been a quest to solve three difficult problems: understanding the complexity of ecological systems and environmental problems; overcoming the barriers to cross-disciplinary collaboration; and seeing humans as components of ecosystems without adopting an overly judgmental attitude toward human behavior.

EMERGING DISCIPLINES: OPENING A CONVERSATION

Ecology achieved disciplinary coherence in the United States around 1920. The founding of the Ecological Society of America (ESA) in late 1915 had helped to unite plant and animal ecology, and in 1920 the society launched its new journal, *Ecology*. The early years of the journal reveal tentative discussion of how broad ecology should be and whether it should include human ecology. ESA's second president, the geographer Ellsworth Huntington, identified human ecology as a key area needing development and requiring the input of geographers as well as medical professionals. Barrington Moore, ESA's president in 1920, considered ecology not as a narrow specialty but as a way of thinking that could be superimposed on other sciences. Stephen Forbes's presidential address to the society in 1921 emphasized the importance of humanizing ecology, by which he meant paying more attention to applied problems and forging closer links between ecologists and economic biologists. A few early articles dealt with human problems, and a symposium titled "The Relation of General Ecology to Human Ecology," held at ESA's annual meeting in 1923, focused on ecological questions relating to agriculture, horticulture, and forestry, with an anthropological contribution on humans in prehistoric America. But these broad, inclusive visions of ecology were not realized, and on the whole ecological writing focused on botanical and zoological problems, excluding humans.

The reasons are easy to understand. Ecological problems are so complex that understanding how nature was organized—or even proving that it truly exhibited organization—was a difficult task. Ecology had emerged in a climate of skepticism in the early twentieth century. As one of its founders, Charles C. Adams, remarked, there had been an "abundance of opposition to the new subject" among older biologists because it seemed implausible that nature was ordered in the way that ecologists thought it was. Given the need for ecologists to establish their disciplinary identity within biology, it was premature to move toward the higher order of complexity represented by problems of human adaptation and social development in relation to climate and environment.

While Adams understood the initial reservations, he thought that by the 1930s ecologists needed to take the plunge into human ecology. He was encouraged by the interest in ecological approaches that he saw developing within the social sciences and in urban planning. Books such as Lewis Mumford's *Technics and Civilization* (1934) and *The Culture of Cities* (1938) struck him as strongly ecological in orientation, while the Chicago School of sociology was building its reputation on the basis of an "ecological" approach to problems of city growth. Geographers and anthropologists were starting to adopt ecological perspectives, and Adams thought that ecologists too were in a good position to contribute to this growing multidisciplinary conversation.

Ecologists needed to be part of the conversation because social scientists were using ecological concepts, but they were not actually doing biological research. The Chicago sociology program was a good example of how human ecology could be construed as "ecological" by analogy only, while the research questions were grounded in sociology. Chicago's research program originated in Robert E. Park's proposal in 1915 that the city was an ideal place to investigate human behavior. That idea was innovative given the dominance of rural sociology at that time, but Park's foundational proposal was not yet cast in ecological terms. His idea was that in great cities the whole spectrum of human types was displayed, and one could observe behaviors that would normally be suppressed in smaller communities. Thus for Park the city should be seen as a "laboratory or clinic in which human nature and social processes may be most conveniently and profitably studied."

Coinciding with Park's provocative proposal to his fellow sociologists, a model of how to study community development scientifically presented itself in the field of ecology. Henry C. Cowles at the University of Chicago had been expounding the concept of ecological succession in his courses for several years, and the key theoretical monograph on the subject, Frederic Clements's magnum opus, *Plant Succession: An Analysis of the Development of Vegetation*, was published in 1916. This book marked a major shift in ecologists' attention from individual species and physiological problems relating to adaptation to the study of whole communities and their development over time. Placing the structure and development of the plant community at the center of ecology entailed drawing attention to the interspecies relations that gave the community its character and drove its development along certain paths, with climate being the ultimate determinant of its final, "climax" form. At the climax stage, Clements argued, certain species dominated the community and rendered it stable. Clements argued even that the ecological community was a type of "complex-organism" with a predictable developmental path.

Park accepted the Clementsian idea that the community formed a type of superorganism that passed through a life history of juvenile, adult, and senile phases. He was also impressed by the Darwinian idea that competitive and cooperative interdependencies among species gave rise to a kind of balance, such that the numbers and distributions of species were regulated. His interest in ecological thought, as he explained in 1936, stemmed from the way it described how competition "exercises control over the relations of individual and species within the communal habitat." Under the leadership of Park as well as that of Ernest Burgess, Roderick McKenzie, and Louis Wirth, the Chicago School began to define urban sociology as the study of succession, driven by competition between urban populations. Within the city, too, struggle and competition between different populations, institutions, and businesses set in motion successional changes that ended in the formation of dominant forms. The main difference between the ecological and the cultural community was that the cultural process was more complicated, and it could exhibit sudden change, perhaps as a result of new inventions. The core questions involved how equilibrium was achieved, what forces disrupted equilibrium, and how new equilibria were established.

These ecological concepts functioned purely on the level of analogy, on the grounds that there were similarities in the social life of plants and animals and in human beings. Sociologists used ecological concepts to argue that certain urban changes were natural or followed a typical life history, in analogy to the changes of animal and plant communities. Biological analogies suggested that the process of urban growth and development was predictable, and therefore results derived from the study of one city could be translated to another city. This analogical thinking was subject to debate and to empirical tests in the 1930s. Although sociologists recognized the limits to these analogies, the habit of analogical thinking persisted into the 1950s.

The Chicago School's understanding of human ecology did not include biological problems. Ironically, since in the city the "environment" referred to the built environment, the natural environment fell outside the definition of human ecology or sociology. Thus the sociologist Amos Hawley complained in the 1940s that the conception of human ecology remained crude and ambiguous. He put the blame on the failure to connect human ecology closely enough with general or biological ecology. He thought that sociologists like Park recognized the value of biological concepts but had not gone much beyond borrowing biological terminology, leading some sociologists to oppose the usefulness of such analogies. In addition, sociological approaches placed undue emphasis on competition and spatial relations. The latter emphasis meant that human ecology was very descriptive but lacked causal analysis. It involved compiling inventories of characteristics of community life and plotting their distributions on maps. Hawley maintained that these views were too limited or mistaken and that there was a need to take seriously the link between the biological and the social. For Hawley, however, this meant focusing on population-level analysis, that is, studies of the size, composition, and rate of growth or decline of the populations representing the various functions within the community. Hawley was also unable to integrate or synthesize ecological and sociological approaches, and as a result human ecology did not become a separate discipline.

A more serious effort to link the biological and the social emerged at this time from a different area of sociology, regional sociology. In this field an important bridge builder between sociology and ecology was Howard Washington Odum, whose two sons, Eugene P. and Howard T. Odum, became prominent ecosystem ecologists in the postwar period. To understand the developing conversation that would contribute to urban ecological science several decades later, we have to include not just urban sociology but also regional sociology, and in this case the synergies operating between father and sons. Howard W. Odum thought of ecology not as standing outside of sociological study but as intrinsic to it, and he took inspiration from his son Eugene's early interests in ecology. In *American Regionalism* (1938), Odum and his coauthor, Harry Moore, drew on Eugene Odum's unpublished writing to explain how the ecological perspective applied to regionalism. They drew attention to the areas of thought where human ecology was being advanced and where multidisciplinary approaches helped define and foster human ecology. Valuing the holistic and synthetic approaches of ecology, they argued that ecological ideas of succession and the ecological emphasis on balance or equilibrium had a lot to offer the regional approach to sociology. They were impressed especially by the ecological viewpoint advocated by Radhakamal Mukerjee, a professor of economics and sociology at Lucknow University in India, who was an early exponent of regional sociology from an ecological perspective. He developed these ideas in his 1926 book *Regional Sociology* and in a series of articles published in the 1930s.

The importance of adopting an ecological approach and developing human ecology was reinforced by the conservation and agricultural problems that confronted Americans during the Dust Bowl of the 1930s. In Odum's analysis of American social problems in 1939, he recognized that through human ecology—the interpretation of "men and institutions in terms of their relation to the living environment"—we could learn how to adjust to and master the geographic limitations of the environment. Odum also perceived the ecological approach to be focused primarily on biomes, or as starting with the division of the landscape into major biomes, relatively homogeneous areas where vegetation types were determined by climate. This practical landscape division would work well for human ecology, he thought, and it dovetailed with Odum's preference for regional sociology. His son Eugene would use this idea in his later textbook *Fundamentals of Ecology* (1953). The elder Odum was optimistic about developing human ecology, and he did not mean a form of ecology in which the biology was missing but rather a true joining of biological ecology with social science and geography. The postwar decades, during which his two sons would rise to prominence in ecological science, would provide further incentives to explore those disciplinary boundary regions.

THE ECOSYSTEM CONCEPT: EXPANDING THE VISION OF ECOLOGY

Well before the intense public discussions that culminated in the first Earth Day in 1970, conservationists, natural scientists, and social scientists were worried about how to solve the new environmental problems that emerged in the postwar period. Growing problems included fallout contamination from nuclear tests, air pollution, and pesticide residues in foods and in the environment. The expectation that the world's population would reach five to six billion by century's end prompted debate about how to increase food supply to meet the coming demand. As the Yale ecologist Paul Sears concluded, "We are an explosion. For the first time in earth history, a single species has become dominant, and we are it." These comments were made in a session devoted to "Perspectives in Human Ecology" at ESA's annual meeting in 1954. In the published version of his talk, Sears noted that one of the great challenges of the modern period was urban encroachment on rural land as the American population expanded. Who was better equipped than the ecologist, he asked, to take on the analysis of such problems?

Sears also pointed out that solutions had to conform to the cultural expectations and patterns of the community. The ecological expert must not simply read the landscape but also be capable of analyzing the human community and "understand the forces at work within it." Where proposed solutions did not match cultural values, progress was stymied. "It is significant," he observed, "that Milwaukee, with its German cultural heritage, has been a pioneer in abating pollution and utilizing waste." Edgar Anderson, a botanist from the Missouri Botanical Garden, also participated in this session. In the 1950s he started taking his botany classes to the dump heaps, alleys, and vacant lots of St. Louis to stimulate the students to think creatively about city living. It was easier in the city environment to analyze the interplay of forces that produced the natural balances ecologists valued elsewhere and to see how those balances were upset. Anderson was also open to dialogue with social scientists.

Biologists and social scientists with common interests in human ecology tended to adopt a long historical perspective, as Sears noted. This sensitivity to the history of human interaction with the environment was much in evidence in a famous conference held at Princeton in 1955 entitled "Man's Role in Changing the Face of the Earth." The conference, organized by the cultural geographer Carl Sauer, not only explored modern human impact on the environment but also took the view that humans had had significant impact on the world for a very long time, even as far back as the Ice Age. And if humans truly had a long-term impact on nature, then it made sense to study human-dominated environments. As the title of this conference suggests, the 1950s were a watershed decade when Americans awakened to the fact that human impact on the Earth was far greater than had been appreciated. As the conference participants explored these ideas, they started to grapple with the meaning of the term "equilibrium" and to suggest the possibility that the appearance of equilibrium was illusory: perhaps, they realized, we saw equilibrium because we wanted to see it. Nature appeared to be constantly changing, which meant that equilibrium states were constantly shifting or existed only in the short term. The longer the time frame adopted, the more nature appeared to be in disequilibrium. This problem of how to understand the concept of equilibrium would be debated in greater depth over the next three decades.

As awareness of the extent to which humans were transforming the Earth grew in the postwar decades, so also the need to expand ecology into the urban setting and to rethink the design of cities became a focus of attention by the mid-1960s. The postwar changes in American cities marked by suburbanization, the decline of inner cities, and the heightened racial tensions that erupted in urban riots in African American communities, emerged as the most important and potentially most difficult environmental problems of the time. Just as the Dust Bowl had in an earlier generation, these crises raised the specter that humans were dangerously out of balance with their environments.

After the war, the cross-talk between biologists and social scientists intensified, but now there was a new concept to frame the discussion: the ecosystem. The British ecologist Arthur Tansley had suggested using the term "ecosystem" in 1935 as a way to link organisms with the surrounding air, water, and soil. But the term did not catch on until the postwar period. By the 1950s the ecosystem concept focused attention on energy flow and transfer of materials through systems, and it offered a new perspective on the older ideas of balance, equilibrium, and the reaction of systems to disturbance. The concept of the ecosystem paralleled the rise of systems thinking in other fields, such as cybernetics, the science that studied feedback and control in any kind of system. Eugene P. Odum and Howard T. Odum were largely responsible for showing that the ecosystem was a system in this cybernetic sense, and each placed the ecosystem concept at the center of ecological theory.

Biologists were quicker than social scientists to see the advantage of applying the ecosystem concept to human ecology and to urban studies. Francis Evans, at the Institute of Human Biology, University of Michigan, saw that the ecosystem concept could be used to overcome the objection that human systems were fundamentally different from natural systems because of the way humans controlled their environments. Despite these obvious differences, human communities involved the circulation of materials and energy, just as in natural systems. He proposed that the ecosystem viewpoint provided a good entry into the analysis of urban areas and could be used to study the interaction of biological and social factors in the "evolution of man." The ecosystem concept offered a unifying framework that could help to synthesize biological and social sciences, thereby going beyond the analogical thinking that had dominated earlier social science.

But the very concept of the ecosystem was itself an analogy, and it was tempting to define the ecosystem using metaphorical language. The Odum brothers expressed the idea of ecosystem regulation through two metaphors. Eugene Odum preferred to think of the ecosystem as analogous to an organism and having a measurable metabolism. H. T. Odum (figure 2.1) thought of the system as a machine that could operate at different levels of efficiency and produce different levels of "output power," depending on

FIGURE 2.1 Howard T. Odum, along with his brother Eugene Odum, developed the concept of ecosystem services, which helped to link natural and social systems. (Howard T. Odum Papers, Special and Area Studies Collections, George A. Smathers Libraries, University of Florida, Gainesville, Florida)

how energy was being used to maintain the system. Just as sociologists had earlier drawn on biological analogies to develop human ecology, so ecologists relied on analogies and metaphors to explain how nature worked and how human activities affected ecosystem functions. Such analogies reinforced the idea that nature progressed to a steady-state climax, an idea that both Odums accepted. Humans therefore needed to mimic nature to make sure they too progressed toward an equilibrium state. As Eugene Odum explained, humans somehow had to pass successfully though the current rapid-growth stage and reach a more controlled growth-and-equilibrium stage to which they were not yet adapted. As the ecosystem became a central ecological concept, questions about what fostered or maintained equilibrium remained central.

The Odums' approach to ecology lent itself to a broader regional approach that might involve several ecosystems linked together, including natural systems, cities, and agricultural systems. In Eugene Odum's landmark textbook, *Fundamentals of Ecology* (1953) (to which H. T. Odum contributed), the final chapter discussed applications of ecology to human society. The discussion drew on the social science literature and mainly on the Chicago School but also recognized that Howard W. Odum's concept of regionalism offered a useful way to integrate humans and nature. Climatic zones, zonal groups, and biomes were logical natural bases for regional classifications, as Howard W. Odum had also recognized in the 1930s. Eugene Odum understood ecology to be relevant to urban studies and city planning. He ended his text with the comment that mastering our destiny required "a lot of diligent study of man and nature (as a unit, not separately)." By the time the book's third edition was published in 1971, the section on human ecology drew attention to the "almost frantic search for common ground" that was occurring in the 1960s between various disciplines and between workers in basic and applied fields, although no synthesis or consensus was yet in sight. One of the major developments to come out of this search for common ground was to explore the relationship between architecture, urban planning, and ecology, as exemplified by Ian McHarg's groundbreaking book, *Design with Nature* (1967), an extended discourse on how urban and landscape planning could benefit from an ecological method.

Just as ecologists spoke of the metabolism of ecosystems, the problems of clean air and clean water that were mounting by the 1960s led engineers to develop the concept of the "metabolism of cities," which echoed the perspective of ecosystem ecology but was more focused on urban infrastructure. As explained by Abel Wolman, a sanitary engineer at Johns Hopkins University, the city's metabolism referred to all the materials and commodities needed to sustain the city's inhabitants. Wolman's particular concern was with three "metabolic problems" that had become acute, but whose solution rested largely in the hands of local governments: clean water, sewage disposal, and air pollution. Thinking about the metabolic cycle meant solving the problem of disposing of wastes and residues with minimal nuisance or hazard. In developing the concept of urban metabolism, he also drew attention to how societies decided what level of nuisance was acceptable and when to invest in large-scale remediation. The metabolic metaphor was useful in drawing attention to the relationship of the city to its hinterland and to the problem of how a city's metabolism changes over time as it grows. An example of a study adopting an urban metabolism approach that combined physiological and community ecology was Stephen Boyden's study of Hong Kong, which was adopted by UNESCO as part of its Man and the Biosphere (MAB) program, launched in 1970. Boyden's report for the MAB program noted the urgent need for more research of this kind. Urban geographers have found the metabolism analogy helpful for analyzing city growth; it has been applied to the study of Baltimore's history.

The concept of city metabolism was closely related to the concept of "ecosystem services," which began to be articulated in the 1970s, although it was not popularized until the 1990s. The Odum brothers also led the way in developing this concept. Although it was common to depict humans in negative terms as pathogens or cancers that caused illness to ecosystems, humans were also the source of solutions and therefore had to understand why conservation of ecosystems was in their interest. As the Odums explained, humans valued cities but did not understand that cities in turn depended on their surrounding life-support systems, the natural systems that provided critical services, such as water purification, free of charge. In the absence of such essential "natural services," the cost of maintaining quality of life in a city would be prohibitive. Indeed, cities would go bankrupt if forced to pay such costs. In this way the Odums emphasized the need to see natural and built environments as connected. They understood that getting an accurate reading of how systems were functioning was still an enormous task. However, they thought it worthwhile to try to implement these new concepts in urban planning, especially in locations where public opinion and government policies already favored control of development. Walter Westman argued in 1977 that emphasis on ecosystem functioning, that is, on the flows of material and energy, would be an important way to demonstrate the significance of pollution damage to society. One offshoot of this idea was the development of the new transdisciplinary field of ecological economics, especially under the leadership of Robert Costanza, who was one of H. T. Odum's doctoral students. Costanza's landscape model based on the Patuxent River watershed in Maryland would later serve as a point of reference for BES.

With the rapid rise of the environmental movement in the 1960s, radically futuristic thinking focused on the possibility that we might be capable of engineering our way out of environmental problems. Athelstan Spilhaus, an oceanographer and meteorologist and the dean of the Institute of Technology at the University of Minnesota, had a remarkable vision for what engineering might accomplish. Having just finished an investigation into waste management for the National Academy of Sciences in 1966, Spilhaus concluded that unplanned growth was largely to blame for the current urban mess. The solution, he thought, was not just better planning but the creation of "experimental cities" which, unlike the usual planned communities or new towns, would serve as prototypes for new technologies and new ideas about the organization of cities. The closest contemporary idea to this vision was Walt Disney's "Experimental Prototype City of Tomorrow," which Spilhaus viewed with interest, although Disney's theme park (opened in 1982) was still many years in the future. By 1967 Spilhaus, with the support of the Minnesota newspaper executive Otto Silha, obtained seed funding of $330,000 from three federal agencies and several industries to try to move the project from planning to implementation stages.

Spilhaus's vision was that an experimental city, housing about a quarter million people, would contain the full complement of institutions and businesses that were normal to urban environments. This was intended to be a multidisciplinary experiment involving social science, human ecology, environmental biology, and environmental engineering. It was an experiment in creating a total system in which there would be various experiments underway involving transportation, communications, climate control, and any other technology designed to produce a healthy environment with minimal footprint. The goal would be total recycling to eliminate mismanagement of wastes, and even noise pollution would be controlled. He imagined that the building of the Experimental City, originally planned for northern Minnesota, would be overseen by a quasi-governmental, quasi-private corporation along the lines of the Communications Satellite Corporation. Spilhaus and Silha promoted the idea for several years, but it finally died in 1974. It was too ambitious, too rigid in its top-down managerial approach, and not responsive to public desires and interests.

ECOLOGISTS CONFRONT THE NEED FOR URBAN ECOLOGY: EXPERIMENTS IN INTERDISCIPLINARITY

While the grandiose visions of the Experimental City guaranteed its failure, ecologists, social scientists, and urban planners were conducting a different kind of social experiment in the hope of fostering meaningful interdisciplinary interaction. Although in some respects this experiment did not achieve its goals, it imparted valuable lessons that moved the discussion forward. The ESA, wanting to show that ecology was of service to society, was thinking of ways to translate ecological knowledge into meaningful policy. One way was through the creation of a National Institute of Ecology, and in 1968 a task force commissioned by ESA presented a proposal for such an institute, on the model of the National Center for Atmospheric Research. After a feasibility study showed interest in such an institute, an operational plan in 1970 proposed a highly ambitious program that included several divisions responsible for forecasting and planning, policy research, running a research laboratory, biome modeling and synthesis, information resources, and communication and education. The title briefly changed to the Inter-American Institute of Ecology, which was incorporated in January 1971, but then became simply the Institute of Ecology (TIE). Its organization included a director, a board of trustees, a group of founding institutional members as well as individual members who made up an assembly. The institute ended up being independent of its parent body, the ESA, a separation that contributed to its financial insecurity, although it did manage to obtain grants from government and private sources to support conferences and publications on a wide variety of current problems. The planned research laboratory was never built, and continued financial insecurity put an end to TIE in 1984.

One reason such an institute was deemed valuable was that it would encourage multidisciplinary studies of complex ecosystems that included cities. Such research was seen as beyond the capacities of single universities or regionally based groups. One goal was to make human ecology more comprehensive, so that ecologists and their colleagues could be more effective in promoting policy changes. Interest in urban ecology was growing at this time. A symposium at the meeting of the American Association for the Advancement of Science in Chicago in December 1970 drew attention to the neglect of the urban environment by biologists but noted that urban climates were under intensive scrutiny, especially in relation to air pollution, and that urban soils and hydrology were being studied. Geographers and other social scientists studied the movement and distribution of populations and the problems of the inner city. But in all these studies the biological basis of these problems was missing, and the role of the city's natural communities in the life of the city was little understood. The symposium was meant to put biology back into human ecology and also to highlight the importance of ecological principles in city planning.

In November 1971 a survey of TIE's founding members showed that urban ecosystem studies were given high priority among fifteen possible areas of research. This finding led to discussion among various national organizations and the decision to sponsor more symposia dealing with urban ecology. A joint ESA-TIE symposium in Minneapolis in 1972, attended by over 100 people, was followed by a national workshop on urban ecosystems held in the spring of 1973 in Austin, Texas. Crucial support came from NSF's new program Research Applied to National Needs. The workshop project, directed by Forest Stearns, a botanist at the University of Wisconsin–Milwaukee, involved the combined efforts of several task forces, which included 90 specialists from disciplines in the natural and social sciences. Their reports and recommendations appeared in 1974 as *The Urban Ecosystem: A Holistic Approach.* The goal was to stimulate interdisciplinary research by ecologists, urban planners, engineers, and social scientists. However, the resulting book fell short of achieving this goal. It was addressed to a general audience and therefore did not deal with any problems in detail. It could do little more than broadly define what might be involved in the analysis of the urban ecosystem, while recommending that more research be done on a variety of problems. Overall it left the impression that there were still formidable obstacles to interdisciplinary exchange across the biological, engineering, and social sciences.

In the discussion leading up to this multidisciplinary project, F. Herbert Bormann, the president of ESA and a participant in the Minneapolis symposium in 1972, appointed a Committee on Urban Ecology to consider how ecologists might productively enter this field of research. The committee's report, written by C. S. Holling, then at the Institute of Resource Ecology, University of British Columbia, and Gordon Orians, in the Department of Zoology, University of Washington, was a well-considered and cautious critique of what urban ecology could or might entail. Because urban ecology was more likely to generate policy and action than more esoteric branches of ecology, they warned that entering into this field was a step not to be taken lightly. Ecologists needed to be aware that they were still ignorant of many aspects of ecosystem functions.

Holling and Orians did think that the study of urban ecosystems was justified, on the grounds that there were many properties that urban ecosystems shared with natural ecological systems, while at the same time there were distinctive differences between urban and ecological systems. But they drew attention to several problems that ecologists would have to address when dealing with urban systems. First, ecological models of ecosystems were nonhistorical: current properties of species were taken as givens, and the goal was only to predict short-term changes in system properties. Second, ecological theories ignored spatial complexities: the natural world was not homogeneous, but the logistical problems of studying spatial heterogeneity had not yet been overcome. Third, complex feedback systems were not well understood: ecologists would have to learn from their colleagues in economics and sociology as well as from students of human behavior. Fourth, they counseled ecologists not to engage in "territorial imperialism" and to be willing to share the study of urban ecology with other disciplines, perhaps restricting ecological questions to those that had a biological component.

Their key recommendation was that in the urban environment it was not useful to think in equilibrium terms, as was the norm in ecosystem ecology and resource management. Although parts of the urban system can be in equilibrium, the equilibrium state is continuously changing and the "system itself is subjected to considerable random change." Therefore they advised that the goal of ecologists should move away from improving the efficiency of the urban system (which they saw as the expression of an equilibrium view) and instead focus on another concept, the resilience of the system. Resilience was meant to apply not just to ecological components but also to economic, physical, and social components of the system. Holling in particular was responsible for introducing and expounding on the concept of resilience in the study of ecosystem function.

The basic concept of resilience referred to the ability of a system to respond to or adjust to a disturbance while maintaining the same structure, function, and feedbacks. Holling argued that the equilibrium-centered view, because it was too static, did not give insight into the behavior of systems that were not near the equilibrium. The key point was to adopt a more realistic analysis of how systems functioned. His concern, as elaborated in his later writings, was that management strategies devised for equilibrium systems might well be antagonistic to strategies based on nonequilibrium views and that equilibrium-based strategies might actually create conditions that would cause systems to collapse. By developing the concept of resilience he hoped to create a new framework for effective management policies. This approach necessarily involved a multidisciplinary perspective on human and natural systems.

Bormann's talk at the 1972 Minneapolis meeting's roundtable on urban ecosystem studies focused on another mechanism to promote interdisciplinarity: educational reform at the university level. Bormann was on the faculty of the Yale School of Forestry, which since the late 1960s had undergone major changes, moving away from its traditional focus on forest productivity to a focus on ecological systems and contemporary environmental problems. Curricular reforms helped to advance the cause of interdisciplinarity, and Yale's faculty would provide leadership in thinking about the relationship between social sciences and ecology within an ecosystem framework. The school's dean, François Mergen, had been very responsive to the growing environmental movement of the 1960s, bringing in new faculty like Bormann and taking part in the national teach-ins that led up to the first Earth Day on April 22, 1970.

As part of these efforts to explore current environmental problems, Yale's forestry school sponsored a symposium in 1969 titled *Man and His Environment: The Ecological Limits of Optimism,* which featured lectures by Yale's faculty. Among the speakers was the sociologist William R. Burch Jr. (figure 2.2), who was later on the advisory board of TIE's urban ecosystem project. His talk was a provocative reflection on the nature of American society and the current environmental crisis; it was soon expanded into a book exploring the interpenetration of myth, social systems, and ecosystems, *Daydreams and Nightmares: A Sociological Essay on the American Environment* (1971). Burch warned academics to be wary of falling into the trap of blaming environmental problems on a flawed human nature, on single causes like overpopulation, or on various villains and conspirators. Oversimplifying the cause of the problem would not help solve it. While much talk at that time focused on the population "explosion," as though humans were cancers on the Earth, his point was that human reproduction was not a strictly biological phenomenon, overpopulation was not to be blamed on irresponsible behavior among the underclass, and the solution would not come from handing out the latest birth control technology. Always there were social, cultural, economic, and political dimensions to these problems that had to be understood. Environmental problems, in short, had broad ramifications that resisted reductionist thinking. Burch was in constant search of a middle ground between blaming humans for their destructive impacts versus relying on a totally human-centered approach that placed too much faith in technology.

Bormann's presidential address to the ESA in 1972 made a similar point: linear thinking oversimplified the depth and complexity of the current environmental problem. Not only that, but failure to connect ecological

FIGURE 2.2   William R. Burch Jr. offered cogent criticisms of the logic of environmentalism and later helped develop the Human Ecosystem Model that was the basis for the Baltimore Ecosystem Study. (With permission of William R. Burch Jr.)

and social indices had split the environmental movement, dividing those concerned with the biological environment from those concerned with the human environment. As a result, he argued, "today we are often presented with the paradox of choosing between clean air and water *or* rebuilding the cities and the human landscape. Charges of ecologic elitism are beginning to fill the air." Both Bormann and Burch argued for the need to see the whole picture by putting the biological and social dimensions together. Burch's ideas would later shape the approach of BES through his direct involvement and through the work of his doctoral student J. Morgan Grove, one of its founding members.

Bormann, on the other hand, would influence the Baltimore project through his development, with Gene Likens, of the small-watershed model of ecosystem analysis, based on their studies of a watershed in the Hubbard Brook experimental forest in New Hampshire. By defining the ecosystem as a watershed, they created a model experimental approach that proved to be highly adaptable and appropriate to Baltimore's environment. Their approach also attempted to look inside the black box of the ecosystem and to understand the smaller-scale patterns and processes occurring within the system: they referred to these dynamic processes as a "shifting mo-saic." While that idea was not entirely novel, having been prefigured in work by the British ecologist Alexander Watt in the 1940s, it signaled a general turn in ecology toward more detailed study of ecosystem processes. This would eventually expand into a more systematic critique of equilibrium viewpoints in ecology, which in turn reinforced the value of thinking of human-dominated systems in terms of resilience instead of equilibrium. The Hubbard Brook ecosystem study, although not part of the International Biological Program (IBP) that ran from 1968 to 1974, did become part of the IBP's successor, the Long-Term Ecological Research (LTER) program that started in the 1980s. By proving the value of long-term and relatively small-scale ecosystem studies, such projects as the Hubbard Brook study created a solid track record that would later justify parallel studies within urban environments.

CONCLUSION

By the 1990s the various conversations and developments in ecological and social science that had been unfolding throughout the century finally came into focus with the creation of two urban ecological studies, in Baltimore and Phoenix, as new additions to the LTER program. The commitment to long-term study in urban ecosystems made it feasible to work toward a synthetic approach in which several disciplines could operate together with-out one dominating another, although each had its own methods and ap-proaches. I have shown here that a very long history of discussion, explora-tion, trial, and error was important in achieving the eventual outcome. We have seen that a broad vision of what ecology entails has existed from the discipline's early history—a conception of ecology as embracing applied problems, dealing with the complexity and subtlety of human interactions, and having a long-term historical perspective on human impact on the en-vironment. But it has taken a long time to work out a strategy to realize this kind of vision and to engage in cross-disciplinary studies while at the same time maintaining the standards of the different disciplines and of the agencies to which one is accountable.

The relevant disciplines—ecology, geography, social science—had first to create their disciplinary identities before meaningful cross-disciplinary strategies could be devised, but the efforts to foster cross-disciplinary exchange began early. We have seen that the impetus for cross-disciplinary interaction in the ill-defined borderland called human ecology came not exclusively from urban studies but also from regional approaches in both social science and ecology. Regionalism within sociology had natural links to the study of biomes and ecosystems in ecology, and it was through the regional perspective that human ecology could be properly appreciated and not stripped of its biological content. Regionalism also promoted interest in human economic activity, reinforcing ecologists' ideas about extending ecology's range to cover economic biology. In addition, a long-term historical view of human interaction with the environment encouraged greater attention to human ecology.

New ecological ideas, especially the concepts of the ecosystem and ecosystem services, suggested better ways to bring ecological approaches into urban studies in the postwar period, while the emergence of serious environmental and social problems increasingly focused attention on cities. Prompted by the urgency of these urban problems, experiments in interdisciplinarity in the 1960s and 1970s, while not always successful, helped to open discussion about which intellectual approaches would best advance urban ecological science. The shift away from an equilibrium viewpoint along with the development of alternative concepts, such as ecosystem resilience, was meant to create a more realistic analysis of urban and other human-dominated systems. Looking inside the black box of the ecosystem created a more detailed, accurate picture of the dynamic and heterogeneous nature of these systems. Other conceptual and methodological changes within ecology produced robust ecosystem models, such as the small-watershed model, that were easily adapted to the urban environment. Finally, it took a long time to locate the middle ground between a negative view of humans as inevitable despoilers of nature and cancers on the Earth's surface and an overly optimistic view that human ingenuity and technology would be a panacea.

Acknowledgments   I am very grateful to Terry Camp at the Odum Library of the Odum School of Ecology, University of Georgia, for drawing my attention to the similarity between the sociological approach of H. W. Odum and the ecological ideas of his sons.

PART TWO WHAT HAVE WE LEARNED?

CHAPTER THREE

The History and Legacies of Mercantile, Industrial, and Sanitary Transformations

Daniel J. Bain and Geoffrey L. Buckley

IN BRIEF

- The Baltimore Ecosystem Study (BES) has many ideas about coming urban transitions, but these ideas are based on, paradoxically, relatively rich *and* relatively limited information about how previous transitions worked.
- Gathering data about historical decisions transforming urban systems is fundamental to understanding the couplings between human and natural systems.
- The transition to local-scale, systematic data recording in the early 1900s provides the first opportunity to completely probe these processes.
- The context that influences early-1900s responses remains nebulous, and continued reconstruction of earlier history is an essential research task.
- Ultimately, it is unclear that big data are more important than dark data for study of how urban systems evolve.

INTRODUCTION

The sustainable city is a planning concept, a normative vision to work toward. Like many powerful ideals, it is elastic. For example, "footprint" metrics to measure the relative sustainability of use patterns abound but vary

47

BOX 3.1. KEY TERMS

Urban transition: Originally shorthand for the switch from a predominantly rural population to a predominantly urban population. Used here to describe substantial shifts in the administration, function, etc. of urban areas.

Structural legacy effect: Persistent change in the function of landscape features due to unintended changes in structure (e.g., floodplain sedimentation) that arise from historical process.

Signal legacy effect: Changes in system function due to historical and ceased contamination events (e.g., chromium contamination in Baltimore harbor) that are diminished through time due to processes of dilution, diffusion, etc.

---

widely in complexity, scope, and definition. In contrast, the sanitary city is a deduction, an organization of empirical evidence into a generalizable conceptual model for considering the evolution of the city. Characterizing a transition (box 3.1) from the sanitary city to the sustainable city has emerged as a fundamental research theme for BES. However, transitioning from the way things are to the way things ought to be is not a clearly defined process. The examination of longer-term processes in BES is building evidence to fundamentally inform our understanding of this and future transitions.

Long-term ecological research, as often practiced, relies on an operationally defined "long" that is rather short. With some exceptions, long term is too often limited to the initiation of sampling at the site or the first existence of reliable remote sensing imagery, both generally covering a handful of decades. For the most part, transitions between urban paradigms or ecosystem states likely operate on time scales similar to or longer than the long-term studies. Further, detection of couplings between natural and human systems and characterization of their influence on urban evolution cannot be reliably done with a single transition. To expand our set of transitions for analysis without having to wait for the next several transitions, there are few choices. Examinations of emergent behavior from agent-based or other modeling frameworks are quite powerful but hard to confront with data. Alternatively, the incorporation of historical context extends the length of our research period and allows collection of transition cases and the potential for inference.

These transitions and this context are particularly important for cities. In no other system is human modification of the system as intense or heterogeneous. Moreover, historical processes are overwritten multiple times, leaving accumulations of both structural legacy effects and signal legacy effects. Yet the challenges of simultaneously characterizing all of the processes interacting in urban systems have limited most characterization in BES to the most recent transition (for example, from sediments to social change). However, work is emerging from BES that, when pulled together, allows one to examine transitions through crises, whether actual or imagined. This work will be a cornerstone in the assessment and evaluation of adaptation strategies attempting to make cities more sustainable.

THE FIRST POST-EUROPEAN TRANSITION: SWAMP TO SETTLEMENT

Although there were likely historical crises in Native American societies (European plagues, for example), limited documentation makes examination of those transition periods a substantial challenge. So BES has generally examined only periods following the initiation of European activities in the area. And the first important transition leads to a fundamental question: why does Baltimore sit on the Northwest Branch of the Patapsco River? It is not an obvious place for a city, as in the case of Pittsburgh or New Orleans. Baltimore does straddle the Falls Zone, but that was a happy accident, as the falls defined the limit of navigation, and only later, as water power emerged as vital to mechanized manufacturing, did the physiography become a positive feedback. The establishment of a second city in the Maryland colony relied heavily on political machinations in Annapolis, the seat of colonial government. And in some ways Baltimore emerged as the last city standing. Louis Gottschalk's extraordinary work described the chartering of Joppa Town in 1607, before sedimentation closed the Joppa Town harbor and the Baltimore County seat transferred to Baltimore. Therefore, Baltimore resulted from one of the first coupled natural-human crises in the colonies, the substantial, widespread pattern of erosion and sedimentation with the clearance and cultivation of the land.

The challenge in examining this early crisis is that in many cases these crises were not necessarily recognized as such. Gottschalk documented George Washington's excavation of sediment deposits from the Potomac and application to his fields as fertilizer. Early periods of Baltimore's growth dealt with the transition of marshy areas at the mouth of the Jones Falls into saleable lots. The influx of sediments and resulting elevation gains relative to harbor waters were likely a boon to early Baltimoreans. However, a mere 150 years later land use changes in the central United States, particularly unglaciated portions, set off a similar cycle of erosion and sedimentation. More recently, the legacy of floodplain sedimentation associated with the initial land use changes continues to limit potential ecosystem services such as those arising from riparian buffers (chapter 9). Ultimately, crises, when viewed in hindsight, are not necessarily useful to our understanding of the evolution and adaptation of coupled human and natural systems. Contemporary clarity in crisis definition (for example, the current recognition of unsustainable cities) is thus fundamental to a historical approach to understanding urban transitions.

Regardless of what in hindsight can be seen as the naïveté of the early European actors, the importance of these very early decisions is clear. The original property mosaics continue to influence subsequent property-holding patterns and therefore associated changes in land cover throughout the Gwynns Falls watershed. Moreover, the initial choices in property claims, unencumbered by regular systems such as the public land survey system, reveal an apparent preference for specific geophysical characteristics (flatness, soil fertility) of the landscape. While we remain unable to confirm these valuations, the influence of the initial characterization can be traced throughout the evolution of Baltimore. Datasets from this initial period are rare and generally require substantial work to organize. However, some of these early decisions are fundamental due to the heavy role of contingency. For example, once property lines were settled, there were very few opportunities to edit this mosaic. Continued attempts to clarify these decisions and then apply lessons to modern decisions remains an important and rich area of inquiry.

EARLY TRANSITIONS: TOWN TO CITY

In nascent urban systems decisions often revolve around the need to sustain growth, not quality of life. Ports generally grow out of iteration between minimization of distance to navigable waters and spatial patterns of economic production. Baltimore was closer than ports like New York or London to the Caribbean, allowing delivery of wheat before it turned rancid and enabling a purer sugar monoculture on these islands. The geometry of the Chesapeake Bay provided access to relatively inland portions of the eastern seaboard, again placing the port closer to the great carbonate valleys that served as early bread baskets. Further, the Falls Zone, previously an impediment to navigation, now provided renewable energy, with the establishment of at least ten mills along the main stem of the Gwynns Falls by 1776. This juxtaposition allowed rapid growth in Baltimore.

This period of rapid growth, probably unparalleled since, is relatively poorly documented but much clearer than periods of early settlement. This documentation arose from challenges addressing problematic couplings between humans and other natural systems. For example, water had to be captured to fight fires, control floodwaters, and minimize communicable disease. These early processes of urbanization were focused in the Jones Falls. While BES does examine the region as a whole, focused geophysical measurements largely occur in the Gwynns Falls watershed (figure 3.1), limiting comparison of this early documentation of development with the bulk of our geophysical characterization. For example, local newspapers in periods coincident with major regional floods (in 1817, 1868, 1873) do not mention damages in the Gywnns Falls until late in the 1800s (1868 is the earliest). Indeed, that first mention was only a passing remark of damages in the mill district (Dickeyville). So while the early periods of urbanization begin to generate increasing numbers of decisions, it is difficult to explore them in the Baltimore ecosystem framework without a great deal of additional archival work, not to mention luck. Nonetheless, given the development patterns in the Gwynns Falls, urbanization during this period remained minimal, and the primary changes were agricultural, as the landscape was transformed to feed the growing city and the Caribbean.

BALTIMORE AND SLAVERY: MANUMISSION AND THE CIVIL WAR

As the nation developed, political pressure to transition from an era of slavery grew in importance, dominating decision making at the federal level. Baltimore has been called "the northernmost southern city, southernmost northern city, easternmost western city, and westernmost eastern city." This geographic confusion was probably truest historically during this period given the composition and layout of states in the Union. For example, the Baltimore Company, an early ironworks located near the mouth of the Gwynns Falls, utilized a workforce that was roughly half slave labor, in con-

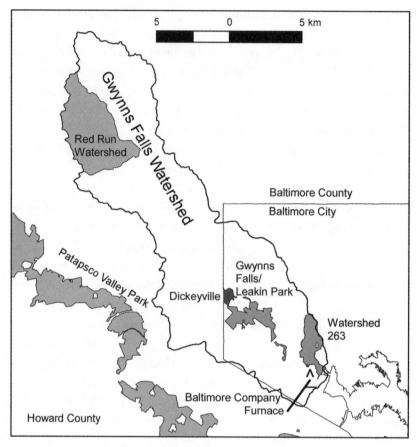

FIGURE 3.1   Gwynns Falls watershed and areas discussed in this chapter.

trast to ironworks located farther north. This ironworks utterly transformed the landscape, consuming a majority of trees in most of the lower half of the watershed for charcoal. Fundamentally, the patterns and rates of disturbance we associate with colonial and early republic periods in tidewater and piedmont regions would not have been feasible without the practice of slavery. While this understanding is clear in the abstract, the legacies arising from these changes have not been quantified either in Baltimore or in associated areas (for example, the monoculture-dominated Caribbean islands).

Further, Baltimore's role in the national transition from slavery was unique. Here, the hybridization of regional characteristics is particularly pronounced. Baltimore was the site of widespread manumission (granting slaves freedom), with 80 percent of the black population free in 1830. This was an extremely high population of free blacks for areas where slavery remained legal. This concentration engendered a wide variety of processes that continue to influence the Baltimore ecosystem. The consolidation of African American populations into segregated, concentrated areas in Baltimore lagged behind southern cities by decades. While the history of neighborhood association in sustaining segregation is part of our understanding of urban processes in Baltimore, the transitions in the antebellum period and their role in setting patterns of human populations are not necessarily considered in the more recent conceptual models. This is particularly important, as early areas of concentrated African American populations are persistent in modern Baltimore geography. Moreover, it is unclear how population geographies in the late 1800s made the process of block busting so effective in Baltimore. It is also unclear how segregation impacted tasks controlling nutrient loading and other BES focus fluxes. For example, variability in infrastructure maintenance regimes likely imparts spatially predictable contemporary nutrient legacies (such as maintenance-deferred sewer systems). The legacies of this social structure, while not unexplored by the Baltimore School, remain woefully undercharacterized. This is particularly true as many proposed paths to urban sustainability likely require spatial reorganization of people and places. If this reorganization does not face the legacies of historical events, the resulting systems could perpetuate these legacies.

BALTIMORE REACHES THE GWYNNS FALLS:
COUNTRY ESTATE TO STREETCAR SUBURB

In 1876, when the Hopkins Atlas for Baltimore County was compiled, the Gwynns Falls valley was the site of great plans but limited urban areas. The embayment at the mouth had a dense network of streets platted out, over the existing waters of the Middle Branch. The large lots at the edge of Baltimore City were in estates, often bounded by the same property lines as parcels auctioned off in the early 1800s with the breakup of the Baltimore Company. However, as annexation expanded the boundaries of Baltimore City, these estates were subdivided and grew into residential areas now found in and just outside of western Baltimore. These neighborhoods offer the first well-characterized, complete set of urban legacies, the low-hanging fruit of disentangling the couplings between human and natural systems.

For example, this is the location of Watershed 263, a storm sewershed receiving careful attention as part of BES. While scholarly work focused on Watershed 263 continues to emerge, the watershed has served as an end member in examinations of the interactions of socioeconomic status and tree canopy. It has served as an example of the potential for large-scale manipulation experiments in urban ecology. It is identified as an example of a potential hot spot of nutrient impairment and is therefore a potential management target. However, this early work does not necessarily incorporate examination of the structural legacies imparted by the history of urbanization. It is possible that changes in nutrient pools arising from early charcoaling have been swamped by the radical changes in nutrient fluxes arising from contemporary urban process, but time- and place-dependent decisions about sewer infrastructure are likely a fundamental control on the temperature of nutrient processing spots. Further, the reaction of contemporary human institutions to infrastructural crises remains tentative. While we have begun to develop a clear understanding of legacy effects and their impact on wicked problems facing urban systems, this understanding has not necessarily spurred action.

The transitions to streetcar suburb were rapid, likely close to development rates of modern subdivisions. However, these suburbs were emerging with the advent of indoor plumbing and the cascade of water use and waste water disposal that followed. That is, the lower Gwynns Falls likely carries vestigial infrastructure arising from the emergence of the sanitary city, the sum total of experiments and compromises made in environments of scarcity. As inquiry into urbanization and Baltimore continues, this juxtaposition represents a true opportunity to clarify legacy effects that arise from this transition. While we are building a decent understanding of the decisions made at the city level, we do not necessarily incorporate an understanding of smaller-scale decisions and actions, for example, variability in developer behavior in these areas. The reconstruction of this decision-making process will be laborious and often incomplete, but the insight into the accumulated results fundamental and, moreover, immediately applicable to transitions toward sustainable systems.

In addition to the early decisions in the sanitary transition preserved here, the property mosaic clearly influenced the pattern of residential development. This is particularly true in this portion of the watershed as the influences of the large holdings of early iron refining companies and the subsequent holdings in summer estates preserved the early land use decisions more completely. Yet the influence of the landscape structure on Baltimore function remains underexplored. Part of this arises from the difficulty in describing the urban landscape itself, but demonstrating the sum effect of legacies that accumulate in urban systems would mark a fundamental advance in our ability to partition and understand urban process.

THE DEPLETION MYTH AND FRED BESLEY

As Baltimore continued to grow through the early 1900s, environmental decision making based on systematic data collection began to emerge. This period was one of rapid change, with the Fire of 1904 consuming 140 acres (56.6 hectares) of the city center and allowing editing of the downtown, in the hopes of transforming Baltimore into a modern metropolis. While current measures of sustainability were not necessarily incorporated into this urban reconstruction, preoccupation about the sustainability of forests used in the rebuilding led to a strong concern about timber depletion and scientific resource management to address these challenges. These efforts are of particular interest, as BES's earliest, spatially explicit understanding of forest cover in the Gwynns Falls arose from these efforts.

As the nineteenth century drew to a close, vast expanses of forested land in Maryland were denuded. By 1890, if not earlier, Maryland could not sustain its own demands for wood products. Beginning in 1906, influential politicians and their conservation allies drafted and passed the Forest Conservation Act, bringing scientific forest management to the state. The first state forester was Fred Wilson Besley, a former student assistant of Gifford Pinchot, chief of the U.S. Forest Service. In selecting Besley, Maryland became just the third state in the union to employ a state forester, placing it at the forefront of the nascent state forestry movement in the South and among the leaders at the national level. By the time Besley stepped down in 1942 he and his colleagues had laid the foundation for a successful state forestry program.

The first project Besley undertook was an exhaustive statewide survey of forest resources. The six-year study was one of the first of its kind, allowing Besley to note, "Maryland has more detailed and accurate information concerning her forests than is known concerning the forests of any other State" (Besley 1909, 3). The resulting documents represent a fundamental shift in our ability to reconstruct and understand the legacies of historical activities. While reconstruction of earlier decisions can connect the decision with the written record, the data upon which the decisions rely are not necessarily preserved. Therefore, beginning around 1900 we are able to start inferring the connections between the data and the decision with much greater precision than was previously possible. Further, the systematic data provide a comprehensive baseline allowing the measurement of environmental change.

This Maryland focus on forest management contributed to movements to restore and protect Baltimore's urban forest. In 1904 the Board of Park Commissioners report to the mayor and City Council described the "shameful condition" of the city's trees. Ordinance No. 154 brought professional forest management to Baltimore in 1912, eight years later. The Division of Forestry struggled as a succession of city foresters came and went, and support from the mayor's office waxed and waned. One of the unexpected challenges to the division's efforts was residents who simply opposed the idea of planting trees along the city's streets. This predilection continues to challenge city arborists. However, the distribution of this attitude is patchy, as some neighborhoods eagerly support efforts to "green" their communities. Despite the spatial heterogeneity in citizen attitude, the early recognition of the importance of the urban canopy remains a cornerstone of proposed transitions to more sustainable urban systems.

URBAN REFORM AND PARK LOCATIONS

One clear benefit of forest was the access to more natural areas for urban residents. This reformist initiative during the early 1900s strongly promoted urban parks. This push has left Baltimore City and County with a legacy of large parks. However, the specific human actions during transition of these areas to parks illustrate the limitations of a naive approach to understanding decisions and resulting legacies.

The Patapsco River valley was first identified as a potential park location in 1904. Archival data indicate that politically powerful Baltimoreans joined with conservationists, public utilities, industrialists, and suburban real estate developers, as well as the Maryland State Board of Forestry, to create a reserve—one that would meet the recreational needs of urban residents, while securing a new growth of timber, enhancing the value of suburban developments, and ensuring a minimum flow to support the valley's industries and public utilities. What started with a small, 40-acre (16 ha)

donation of land in 1913 has grown tremendously, with the present-day Patapsco Valley State Park stretching approximately 32 miles (51.5 km) along
the river, encompassing more than 16,000 acres (6,500 ha). However, while
this park remains a tremendous amenity, a critical eye to the processes
growing parks is particularly important as we try to understand how to
forge sustainable futures, as these are the kinds of successes that will naturally be adopted as models in enacting this future.

Perhaps the best case to examine is that of Gwynns Falls/Leakin Park,
as it is situated squarely in the middle of the watershed and has a history
that contains several important lessons for sustainable futures. First, it is
important to note that Gwynns Falls/Leakin Park occupies some of the
most rugged landscape in the entire Gwynns Falls watershed. The selection
of the park location allowed maximization of size, as this rugged landscape
remained lightly utilized and open. So the park preserves, at least in part,
some of the least desirable land in terms of agriculture and residential development. While this is not necessarily a tragic conclusion, as the rugged
nature includes picturesque landscape elements and relative isolation, the
victory was a compromise. Further, the hard-won compromise was threatened only decades later as the emphasis on car transportation grew and
plans to route Interstate 70 through Baltimore were laid out in the late
1960s. The linear, open park landscape was a tremendous temptation to
planners seeking to minimize cost. In fact, Baltimore's celebrated Inner
Harbor was initially cleared as part of this construction process before resistance to the highway gelled around the park. Multiple citizens' groups focusing in and around the area mobilized to block construction of the expressway through the park. This resistance had to be sustained for more
than a decade, but when the funding window ended in the 1980s, the park
remained. And the Inner Harbor began its rebirth as a shopping district
and tourist attraction, starting a period of revitalization in downtown Baltimore (chapter 16).

The lessons for sustainable futures in the story of Gwynns Falls/Leakin
Park require some unraveling. First, as we seek models to forge a sustainable future, compromise can be important. Preserving prime land as park-
land would have resulted in a very different landscape (smaller and more
fragmented). Additionally, once preserved, open space requires vigilance
to keep it open, and smaller parks might not have generated the necessary
citizen response. Certainly, in the absence of citizen action the park would

be very different, as would the landscape stretching between the park and downtown. For example, BES has focused on Dead Run, which is dominated by an interstate interchange fundamentally altering the system. The construction of similar structures across the remainder of the watershed would have left an irrefutable legacy. So cultural structural legacies are likely not as permanent as physical structural legacies such as riparian burial and stream entrenchment. However, Baltimore's Inner Harbor, one of the urban renewal successes resulting from the aborted highway construction, may have appeared to be a substantial defeat at the time. In this case, the razing of buildings on a scale generally considered disastrous (for example, the 1904 fire) offered a clean slate that resulted in revitalization in the middle of the city. However, this process was an exception to strong inhibitions against the spatial reconfigurations that may be necessary for optimizing systems for sustainability. These barriers are often poorly accounted for in the best-case scenarios presented. Preservation of the park and growth of the Inner Harbor are linked in ways we cannot necessarily imagine. Both are likely the kind of thing most people imagine when they envision sustainable futures, though successes are seen largely in hindsight.

RED RUN AND THE LAKE THAT NEVER WAS

The final case to consider is that of Red Run, a small stream draining upper portions of the watershed (figure 3.1). Much of Red Run drains outcrops of serpentinite, rocks that evolve into generally poor soil. Red Run was thus another location where development lagged behind the rest of the basin. As Baltimore County planned its growth, officials aimed to construct a lake in the relatively undeveloped Red Run valley, drawing development along the concentrated growth corridor. This lake was proposed as flood control, relying on memories of Hurricane Agnes, though the role of developers holding key parcels in this planning process is unclear. After a long struggle and millions of dollars spent in planning, state and federal permissions to construct the lake were not provided.

The lake-planning process provided a rare glimpse into stream channel conditions prior to suburbanization. However, the nature of this information emphasizes the changes in our data collection process over the last century. While the lake-planning process generated reams of data, they remain generally less accessible than the H sheets Gottschalk used in his classic work to reconstruct harbor sedimentation or the forest maps Besley generated. Granted, data were collected and organized for different purposes, but the checklist-dominated assessment approach in modern land use decision making is not clearly a boon to understanding the decisions and legacies. Checklists may prevent reliance on faulty heuristics but may also impede our decision-making process due to the sprinkling of relevant data across incremental planning and permit documents. For example, "before" data were gathered from six separate sources by five sets of unique authors, all for a basin of less than twenty square kilometers. As we continue to examine systems in the detail necessary to slowly turn them to more sustainable pathways, it is imperative to ensure that these data are available and utilized. Otherwise, the potential to arrive at suboptimal answers to questions arising during the next transition increases.

In terms of structural legacies, the case here is ambiguous. Certainly, given the sediment dynamics in the region, the decision not to build a lake sidestepped major risks of rapid sedimentation and choices about dredging. Use as flood control was probably not sustainable over the foreseeable future, as storage would be consumed by sediment. And further, given the presumed nutrient loadings to such a lake, keeping it ecologically healthy would also have required substantial effort. However, discussing structural legacies that were not installed is fraught with peril. In contrast, the investigation of cultural structural legacies (for example, how did the public weigh memories of Hurricane Agnes against the ecological impacts?) remains an area where exciting and essential questions can be posed.

RECAPITULATION

Urban systems have faced transitions ever since the first town grew into a city. This next transition will be as challenging as the rest. However, if we are transitioning to things like energy sustainability, making the transition in areas tuned to water transport and water power may be simpler to do than in cities relying on cheap energy availability. Some have assigned the "legacy" moniker to cities like Baltimore and Pittsburgh, suggesting their existence relies on history and that modern economies have bypassed the importance of their location in trade. If we transition the second cities (like Owings Mills, the mall-centered area in the upper Gwynns Falls) to sustainable cities at the expense of the original city, in some cases the low-impact development will be simpler due to flexibility in open space necessary to rearrange neighborhood components. However, from a holistic perspective, development of second cities creates greater legacies in spatial distribution to overcome, mostly enabled by shortsighted incentivizing of decentralized development (see also chapter 7). The footprint at the block scale cannot be evaluated without examination of the metropolis scale.

Recent funding priorities about data have focused on so-called big data. The impulse to know everything, everywhere, all of the time is not new. But neither is the impulse to shed our mistakes and keep a forward perspective. Both impulses are problematic. At the limit, omniscience converges with perception. But the latter problem, a tendency toward hopeful forgetting, would benefit from a clear-eyed examination of "dark data," the data that remain pixilated, spread across rolls of microfilm and cabinets of musty folders. The studies woven together here demonstrate some of the power in organizing these data. Fundamentally, any transition to a new urban system will not be all milk and honey, and incorporating dark data forces this recognition during visioning.

To lay out the decision space dictating the transition from a sanitary to a sustainable city, a great deal of work remains, particularly in how historical decisions were forged. BES has illuminated the direct connections between social and geophysical forces, in some cases. However, the population of connections we can examine remains small and is therefore challenging to interpret. The attention to historical process that has begun with BES may be fundamental to our understanding of coupled human-natural systems in general. It is the only way to increase our population of place-specific observations of couplings. It is the only truly effective way to generate accurate knowledge of initial conditions. The transitions and opportunities laid out above merely scratch the surface. Fundamentally, inattention to realities underlying the successes and failures will not advance our understanding of the city. Moreover, if we are not right for the right reasons as we change urban systems, we are relying on luck to overcome important emerging challenges.

CHAPTER FOUR

Environmental Justice and Environmental History

Geoffrey L. Buckley, Christopher G. Boone, and Charles Lord

IN BRIEF

- It is prudent to exercise caution before making an environmental justice determination—appearances can be deceiving.
- Both procedural and distributive equity should be taken into consideration—together they help us see the big picture.
- Current patterns may best be understood through the lens of decades-old practices and norms.
- Current notions about what constitutes environmental justice or environmental injustice should be questioned—what we view as justice or injustice today may have been perceived very differently in the past.
- Assumptions about what is beneficial or harmful for a particular group or community should be challenged—context matters.

INTRODUCTION

Since the project's inception in 1997, ecologists and social scientists with the Baltimore Ecosystem Study (BES) have sought to understand how environmental amenities and disamenities are distributed across the urban landscape. More specifically, we have used an environmental justice frame to discover the processes—social, cultural, and political forces—that produce the patterns we see today. In some cases, the results of this research support the findings of scholars working in other cities; in other cases, BES investigators have identified patterns and processes that appear to be relatively uncommon. Taken together, this body of work has helped us to recognize the legacy effects of past decisions and to push the boundaries of our environmental justice inquiries.

What are environmental amenities and disamenities? What is environmental injustice? At first glance, these seem like straightforward questions. Amenities are "goods," such as parks and street trees, which we value and appreciate in the urban environment, while disamenities are "bads," like polluting industries, which we try to avoid. When it can be clearly demonstrated that an amenity or disamenity is distributed inequitably, such that one group of residents is benefiting at the expense of another, or if a group is excluded from participating in the decision-making process, then an environmental injustice determination may be relatively easy to make. However, it is not always this simple. As BES research over the years has shown, what constitutes a good or a bad—or an environmental injustice—can vary across time and space and from one group of residents to another. Sometimes what appears on the surface to be an environmental injustice may in the end prove not to be an injustice at all. Even more confusing, a just distribution of an amenity today—parks and playgrounds, for instance—may mask an injustice perpetrated in the past. Thus it is always good practice to exercise caution before concluding that a particular section of the city or segment of the population is suffering from an environmental injustice or, conversely, that it is benefiting from an equitable distribution of something we value. Equally, it is unwise to use patterns from a single snapshot in time to infer processes of environmental justice. In this chapter we offer examples from roughly twenty years of BES research that demonstrate the complex nature of environmental justice research as well as the importance of linking pattern with process.

CONTAMINATION AT SWANN PARK

In April 2007 Baltimore mayor Sheila Dixon announced the temporary closing of Swann Park, located in a predominantly white, working-class industrial neighborhood of South Baltimore. For fifty-six years Swann Park shared a fence with a pesticide manufacturing plant operated by Allied Chemical. According to the 2007 press release, soil samples showed elevated levels of arsenic at the park, ranging from 23 to 2,200 parts per million (ppm). Archival research conducted in support of BES revealed that this was not the first time the park had been closed due to health concerns. In 1976 city officials closed the park to inspect for kepone contamination. While no kepone was detected, soil sampling turned up a small amount of arsenic. Within days workers removed several inches of soil along the property boundary with Allied Chemical and installed new sod. Convinced the problem was solved, city officials reopened the park. Soon thereafter Allied Chemical closed the plant, and the city acquired the property. Fast-forward thirty years and Swann Park was closed again, this time due to high levels of arsenic in the soil. But what was the source of the supposedly new arsenic?

The short answer is that the arsenic was always there. In 2007 the Honeywell Corporation, which merged with Allied Chemical in 1999, discovered documents in its files concerning the 1976 kepone investigation. The documents indicated that Allied Chemical had conducted soil samples on its own dating back to at least 1970. The test results confirmed arsenic levels as high as 10,000 ppm—much higher than what was detected in 1976. However, prior to passage of the Resource Conservation and Recovery Act, the Comprehensive Environmental Response, Compensation, and Liability Act, and the Emergency Planning and Community Right-To-Know Act, and with no state or federal regulations concerning soil-based arsenic contamination on the books, Allied Chemical was not required to disclose— nor did they choose to release—their findings. The task force assembled to investigate the case in 1976 determined that the presence of arsenic in the soil was due to pesticide application following the installation of new sod. Remarkably, the task force did not require any follow-up monitoring.

If there is a lesson to be learned from the Swann Park case it is that not all amenities are created equal. What looks like an amenity on the map may, in reality, look quite different on the ground. Future studies need to address not just the equitable distribution of park amenities but also the quality of those amenities. The events that played out at Swann Park should also cause us to broaden our conception of environmental justice and to consider the impact of decisions on the larger community. For over thirty years Baltimoreans from all parts of the city converged on Swann Park to engage in recreational activities and participate in league sports. Unknown to them, they risked exposure to high levels of arsenic in the soil. In this case, risk was not limited to residents living near the park but spread over a much wider area.

URBAN TREES—AMENITY OR NUISANCE?

For many, trees in an urban environment are amenities that offer a wealth of benefits, from beauty and shade to increased property values and energy savings. To others they are disamenities that cost money, require maintenance, and attract pests. During the 1940s and 1950s Baltimore's Division of Forestry learned firsthand that not everybody likes trees, or, as the city forester Charles Young put it in 1946: "I was under the impression that all people like to have trees planted in front of their houses until I started planting trees in front of houses" (quoted in Buckley 2010, 171). In 1965, after years of neglect, government officials allocated $326,000 to plant eight thousand street trees per year along the city's streets. At the time, a survey conducted by the city forester Fred Graves indicated that several sections of the city lacked trees but possessed space for planting. Noting that the area was practically devoid of trees, Graves identified East Baltimore, in particular, as a section of the city ripe for a major planting effort. Despite high cost estimates—funding the program in this section alone would cost more than $385,000—the city initiated tree planting in East Baltimore two years later. It was then that city forestry workers encountered East Baltimore's "tree rebels," residents who claimed to prefer "clean, uncluttered concrete" to street trees. More important, they refused to give city officials permission to plant trees in front of their homes. A *Baltimore Sun* account from 1967 captured the spirit and determination behind the anti-tree movement: "'They'll never put a tree in front of my house,' Louis Averella, of 530 North Linwood Avenue, vowed. 'I think trees are a nuisance in the city,' Mr. Averella, president of the Greater Baltimore Democratic Club, said. 'Trees belong in the country, not the city. If anyone wants a tree in front of his home, let him go and live in the country'" (quoted in Buckley 2010, 172).

Nearly fifty years after residents impeded a major tree-planting project and despite a significant demographic shift in the population, from predominantly white working-class to majority African American, large portions of East Baltimore still lack trees. Fieldwork in two of these neighborhoods, Madison–East End and Berea, suggest that while there is ample space available to increase canopy cover, residents remain skeptical about the city's tree-planting efforts. Indeed, interviews with local residents reveal mixed attitudes toward trees, with approximately half supporting tree planting because of perceived benefits, such as shade and energy savings, and half resistant to the idea, citing concerns ranging from pest infestation and crime (tree pits are sometimes used to bury drugs) to gentrification and displacement. Some residents were well aware that strategically planted trees can increase the value of a home and, further, that a new tree-planting program might have the effect of pricing them out of the neighborhood where they live. Does an inequitable distribution of trees signal an environmental injustice? Are urban trees amenities or disamenities? Clearly, it depends on whom you ask.

Adopting a historical perspective may prove valuable for other reasons as well. Recent research shows that past investment in landscape features, such as trees and parks, is more likely to reflect the social characteristics of past residents. In other words, present-day residents inherit not only the effects of decisions others have made in the past but also the landscape features those decisions created. Taken together, we may think of these as legacy effects. Whether or not these landscape features are perceived and treated like amenities depends on a number of factors, including the condition of the landscape features in question (such as trees), the resources required to maintain these features, and the extent to which the demographic profile of a given area has changed. As cities like Baltimore seek to increase urban tree canopy they must leverage all the assistance they can from private property owners, learning from communities that already exhibit a high level of commitment to urban forestry while promoting the benefits of urban trees in neighborhoods with plantable space but few trees. Only by doing so will they ensure that urban tree canopy goals are met.

UNWANTED LAND USES

In a groundbreaking study Boone revealed that white Baltimoreans, not African Americans, are more likely to live close at hand to a Toxics Release Inventory (TRI) site as defined by the EPA. Although data collected from Detroit, Cleveland, and Buffalo support similar findings, these appear to be exceptions to the rule. That is, they contradict the majority of environmental justice studies that show that marginalized communities, including those made up largely of ethnic and racial minorities, are more likely to live near toxics release facilities. According to Boone, the anomalous pattern is a legacy of the past. It is also tied closely to shifting ideas about residential location. In the 1930s and 1940s white workers chose to live in neighborhoods situated close to the factories where they worked. Restrictive covenants and other practices that enforced segregation assured white workers that they would continue to benefit from short commuting times while the city's African American residents would be forced to live farther away from their jobs. In this case, proximity to polluting industries was viewed as an amenity, not a disamenity. In time, however, this amenity of proximity would backfire as more residents found themselves exposed to harmful industrial emissions. As the public health consequences of exposure became increasingly apparent, what was once viewed as an amenity came to be treated as a disamenity.

Environmental inequities have persisted over time in Baltimore. Analysis of Dun and Bradstreet business records of polluting industries from 1960, 1970, and 1980, along with TRI data for 1990, 2000, and 2010, shows that the spatial distribution of hazards in Baltimore has shifted over time but has always been uneven. In general, the highest concentrations of polluting industries moved away from the downtown core around the Inner Harbor to the eastern and southern peripheries. Residents who lived in the neighborhoods with the highest densities of polluting facilities in 1960, 1970, and 1980 tended to be lower-income families with low educational attainment. From 1990 on, they tended to be white residents with low educational attainment. This study encompassing multiple decades confirms other BES research that shows African Americans were unlikely to live in neighborhoods with a high concentration of polluting facilities. Revealed for the first time was the persistent correlation between neighborhoods of low educational attainment and high density of environmental hazards, a troubling finding given that education is an important resource for reducing vulnerability to harm and recognizing environmental injustice.

While it is clear that certain neighborhoods host a disproportionately high percentage of environmental disamenities, there has been no framework for understanding the role that race played in the distribution of such inequities. A central flaw in traditional environmental justice distributional analyses is that they tend to focus on a single point in time. In a first-of-its-kind study of environmental justice patterns over time, Lord and Norquist presented a method for understanding cities as complex, emergent systems. They argued that a close study of cities through the lens of emergence theory reveals and makes sense of urban patterns. Emergence theory suggests that cities grow from the choices and behaviors of their individual residents and institutions, not from the dictates of the central planners. They applied that lens to environmental justice to illuminate the distributional patterns in Baltimore and to identify the rules that created those patterns over time. Their analysis of zoning special-use permits and their correlation to race during the period 1940 to 2000 showed that race was a rule in the distribution of unwanted land uses during the twentieth century. The special-use data confirmed that the zoning system in Baltimore distributed unwanted land uses on the basis of race, not a post-siting market dynamic. Specifically, their study showed that explicit racism was imported into the ostensibly neutral zoning process. They illustrate that segregation ordinances and then, after these were struck down, de facto segregation campaigns involving property associations, the real estate board, and the building inspector limited the ability of black families to rent or buy in white neighborhoods. Further, they illustrate that the resulting segregation then injected considerations of race into the zoning process through neutral standards linked to "appropriateness." They illustrate even further that these standards, coupled with redlining, the doctrine of prior nonconforming uses, perhaps the decisions of private actors, and the resources available to actors in the zoning system, essentially brought forward the earlier explicit racism into the zoning system. This system of interacting factors can explain the inequities in the distribution of unwanted land uses in the present day. This study confirmed methodologically that it is possible to evaluate the feedback loops that operated to define the emergent patterns of cities during the twentieth century.

PARKS AND EQUITY

Among eastern cities today, Baltimore is unusual in that it offers African Americans, who now make up 63.7 percent of the population, high access to park facilities. While this should come as welcome news to resource managers and others interested in providing greater recreational access to underserved communities, the story of *how* and *when* African Americans achieved this level of access is more troubling. In 1904 Baltimore's Municipal Art Society, aspiring to create a world-class system of parks, hired the landscape architecture firm Olmsted Brothers to develop a plan for Baltimore. The plan called for protecting stream valleys, purchasing parcels of land for parks and playgrounds, linking parks together via tree-lined boulevards, and reserving large tracts of property beyond the city limits to satisfy the recreational needs of future suburban dwellers. A key tenet of the scheme was to ensure that all Baltimoreans, regardless of income, race, or ethnicity, would enjoy equitable access to park resources.

If one were to examine a map of Baltimore's parks in the 1930s and compare it to a distribution of the city's African American population, one might be tempted to conclude that blacks were already enjoying relatively high access to park facilities at this time. If we simply use proximity as our measure, then Carroll Park, located nearby, would appear to be serving their needs. In reality, however, Baltimore's black population, crowded into ghettos to the south and west of the downtown, was prohibited from participating in a wide variety of sporting activities at this neighborhood park. While they were permitted to enter the park, their movements were greatly restricted, begging the question of whether blacks actually had equitable access to park facilities. The answer is no. As Edgar Jones and Jack Levin reminded us in their 1960 report *Towards Equality: Baltimore's Progress Report*, "From birth in a colored ward to burial in a colored cemetery, Negroes lived an almost entirely separate existence bounded on all sides by racial discrimination" (quoted in Wells et al. 2008, 151). In time, African Americans did gain access to recreational facilities in Baltimore. A decades-long struggle by African Americans to be able to use the city's public golf courses, starting with the Carroll Park course in 1923, eventually opened the door to other facilities in later years. Even so, integration—whether it was ball fields and tennis courts or swimming pools and beaches—was a slow, painful process, even after the end of official segregation in 1956.

Sometimes even the best intentions of planners and decision makers are thwarted. Such was the case when it came time to select the site of a new park for Baltimore in 1922. When J. Wilson Leakin died he bequeathed his home to the City of Baltimore, requesting that the proceeds from its sale be used to establish a large city park in his name. The stock market crash threw a monkey wrench in the works, delaying the sale of the property. Progress was delayed further when members of the Board of Park Commissioners could not agree on the location of a new park. Contemporary sources indicate that several sections of town were in desperate need of recreational space. A report by the South Baltimore Improvement Association in 1938, for example, "regretted that many [children] have great distances to walk before reaching an area that is safe for play" and that only "very meager facilities" were available for the "Negro youth" (quoted in Boone et al.

2009, 779). A compromise solution to establish numerous small playgrounds throughout the city failed when the deceased benefactor's sister took the city to court, arguing that her brother's wish was to establish only one park in his name. Both the circuit court and state Supreme Court sided with her, and the playground idea was dropped.

With nowhere else to turn, the city solicited the advice of Frederick Law Olmsted Jr., whose 1904 and 1926 plans laid the groundwork for the city's current network of parks and green spaces. In a letter written in 1939 he identified the present site of Leakin Park as the best option. While acknowledging the need for parks elsewhere in the city, he justified his decision to locate the new park in the Dead Run valley, adjacent to an already existing park:

> The belated introduction of suitable parks might be a very important and valuable element in a sound general program of rehabilitation, intelligently and realistically directed toward salvaging such districts and reconverting them into really satisfactory and attractive residential communities. . . . But, in the face of widespread uncertainty about the future trends of a district, random and uncoordinated expenditures of a few hundred thousand dollars here and a few more there, intended to benefit a diminishing population, the future distribution of which no one now has a sound basis for estimating, is far too liable to be an extravagant and futile gesture. Certainly it is a very risky venture in which to invest trust funds when a safer one that is entirely appropriate is clearly available (quoted in Korth and Buckley 2006, unpaged).

Although Olmsted was committed to the idea of equitable access, even he could not resist the urge to acquire a large parcel of green space in a section of town that was already well served—at least in comparison to other districts.

As the examples above illustrate, African American movement and access were constrained throughout the nineteenth and much of the twentieth century. Exclusionary zoning, racial covenants, unfair mortgage lending practices, and other forms of discrimination and intimidation worked to ensure that blacks remained largely segregated from their white neighbors. Neighborhood improvement associations in particular proved adept at attracting amenities, such as park improvements and expansions, while deflecting unwanted land uses—and people, especially African Americans and immigrants—to other parts of the city. Only after large numbers of white Baltimoreans left the city during the first major postwar wave of sub-urbanization in the 1940s and 1950s was the black population able to spread out, occupying sections of the city formerly inhabited by whites. As this process unfolded, black residents gained access to park facilities that just a few years prior had been off limits to them. In the end, we must conclude that Baltimore's black population inherited a system of parks that was not designed with their needs in mind. If justice means just outcomes *justly achieved*, then it is difficult to interpret current high access to parks for African Americans as positive environmental justice.

THE HUMANE METROPOLIS

According to Pickett and others, opportunities to build a more sustainable and hopeful future abound. Building on the work of Platt, they describe a more "humane metropolis" as one that maintains and restores ecological services; supports initiatives that foster mental and physical health and safety; advocates efficiency; strives to be socially and environmentally just; and promotes a sense of place and community among residents. Community gardens and schoolyard greening are examples of two activities that meet these critical criteria.

In the Broadway East neighborhood of Baltimore the Duncan Street Miracle Garden has helped alleviate some of the food insecurity problems that afflict this section of the city. What factors contribute to the success of a community garden? Key ingredients include a dedicated leader, a core group of residents with knowledge of gardening, a supportive community, and the financial backing of city and community organizations. Additionally, a bottom-up or community-based approach to garden development, as opposed to a top-down strategy, is most likely to succeed.

In their study of schoolyard greening Buckley, Boone, and Grove examine the use of asphalt as a paving material in Baltimore beginning in the early twentieth century. In particular they investigate its application as a recreational surface for playgrounds and schoolyards, the origins of which can be traced to the 1940s. Despite low maintenance costs, the deployment of asphalt as a replacement for grass started to meet resistance in the 1960s. Residents in neighborhoods such as Bolton Hill complained that their chil-

dren should not have to play on asphalt playgrounds when children elsewhere were provided with grass-covered recreational surfaces. Forty years
later stringent new stormwater requirements have forged an alliance of
government agencies, nonprofit organizations, and private developers dedicated to the removal of deteriorating schoolyard asphalt in underserved
communities throughout the city. This collaboration has proven to be both
affordable and popular.

CONCLUSION

Environmental justice research in BES underscores the need for investigating processes rather than pattern alone. The considerable volume of work
from BES that examines the historic, site-specific processes of past and present environmental injustices is a significant contribution to environmental justice theory. Above all, it demonstrates that legacies of past decisions
and actions—or inactions—can powerfully shape outcomes in the present
and therefore future pathways. However, this does not mean that environmental injustice is inevitable. A commitment to sustainability is one means
of breaking harmful path dependencies—decisions or processes in the past
that eliminate or narrow our options today. Like other municipalities, the
City of Baltimore created a comprehensive plan that is designed around
sustainability principles. Successful sustainability plans are those that incorporate a shared vision of a desirable future built on justice principles. A
vigilant commitment to a just, sustainable future can begin to mend the
decades of environmental injustice in Baltimore.

Stewardship Networks and the Evolution of Environmental Governance for the Sustainable City

Michele Romolini, Shawn E. Dalton, and J. Morgan Grove

IN BRIEF

- The shift from the Sanitary City to the Sustainable City requires changes in governance approaches.
- Governance of the Sustainable City will include participation by a diversity of organizations, collaborating and coordinating through networks.
- Longitudinal analysis of networks is key to understanding them as forms of governance.
- This chapter describes changes in the environmental networks working in Baltimore's Gwynns Falls watershed (GFW) from 1999 to 2011.
- These data have been used to engage stakeholders in the active management of their networks.

INTRODUCTION

The first decade of the twenty-first century experienced a proliferation of sustainability policies in cities across the United States and the globe. It has been widely proposed that the success of these urban sustainability initiatives will require new forms of adaptive governance in which city agencies must partner with, and even cede authority to, organizations from other sectors and levels of government. Yet the resulting collaborative networks are often poorly understood. The study of collaborative networks has been a challenge for researchers at regional scales and over time. This may be especially the case in complex urban environments because of the number of diverse groups and organizations working on a patchwork of varying projects and covering multiple land types and land uses. Longitudinal studies of dynamic, urban collaborative networks could be important to inform cities as they seek adaptive management strategies to respond to the needs of a rapidly changing population and landscape. Indeed, a heightened understanding of governance network structures, functions, locations, and outcomes could improve the likelihood of their success.

In this chapter we examine long-term changes in the environmental stewardship network in Baltimore's GFW by comparing two in-depth case studies separated by 12 years. Social network data were collected from organizations conducting stewardship activities in the watershed in 1999 and 2011. During this period the Baltimore Sustainability Plan was formulated and legally enacted. The evaluation of the network over time reveals changes in organizational composition and network structure within this region. We discuss here findings from the study, strengths and limitations, future directions for continued Baltimore Ecosystem Study (BES) research, and the implications for urban sustainability policies and governance.

FOUNDATIONAL THEORY AND DEFINITIONS

Governance Shifts from Sanitary to Sustainable Cities
As discussed in chapter 3, the Sanitary City of the 1900s was characterized by concerns for making cities safe and healthy places to live. The goals of the twenty-first-century Sustainable City are likely to add further concerns for how to make cities more self-regulating, self-sufficient, and adaptive. The achievement of sustainability goals may require cities to employ new approaches to governance and management of urban systems.

Beginning in the second half of the twentieth century the United States went through major changes in both the conceptualization and practice of governance and management of natural resources. First-generation environmental governance was focused on addressing the environmental and public health concerns that stemmed from the industrial revolution. The need for strong regulation seemed clear, and the result was a generation of environmental governance characterized by centralized, command and control regulation of single media (such as air, water, land) through legislation like the Clear Air Act of 1970 and Clean Water Act of 1972 and the creation of the U.S. Environmental Protection Agency. These policies and agencies were established to regulate sources of pollution that adversely affected natural resources and public health.

The traditional emphasis on centralized, top-down government management practices became increasingly challenged by decentralized, bottom-up management in the 1980s and 1990s. Bottom-up strategies employed place-based, nongovernmental organizations to focus on local environmental problems and solutions. These approaches were successful in engaging diverse groups of stakeholders outside of government, addressing geographically and scale-appropriate issues, and often providing incentives for compliance rather than punishment for noncompliance. Examples of stakeholders include environmental nonprofits, community groups, religious organizations, and environmental consulting firms. However, bottom-up strategies alone were insufficient to address complex and rapidly changing social-ecological issues that may require a response from centralized governmental organizations in coordination with local stakeholders. As a result, twenty-first-century cities began to employ adaptive management strategies best provided by polycentric approaches to governance with an array of interacting institutions having overlapping and varying objectives, authorities, and strengths of linkages. This diversity of interests and perspectives may allow for greater adaptability to promote sustainability and resilience. Four key principles are theorized to change from Sanitary City government regimes to the network governance structures of the Sustainable City (table 5.1, figure 5.1).

The diagrams in figure 5.1 are simplified to highlight the theorized changes in overall governance structure. We acknowledge that the Sanitary City governance regimes were not perfectly hierarchical and that Sustainable City governance will retain some of the hierarchical, centralized structures characteristic of the Sanitary City. However, it is with this framing that we set forth to conduct governance research and test our hypotheses, using the network research methodologies we describe in this chapter.

Network Governance Research

Recent research has applied social network analysis techniques to examine governance and management of social-ecological systems. Relevant exam-

Table 5.1

Key principles related to shifts from traditional government of the Sanitary City to the network governance of the Sustainable City

KEY PRINCIPLES	SANITARY	SUSTAINABLE
Governance	Technical/regulatory	Polycentric/mixed
Decision making	Specialized & separate	Generalized & integrated
Stakeholders	Sectoral segregation	Multi-sectoral linkages
Management	Individuals & islands	Collectives & mosaics

ples of empirical research include using network analysis to categorize and assess stakeholder relationships in resource management, to evaluate social capital in collaborative planning, and to examine structure and effectiveness of natural resources management networks facilitated by federal programs. Historically, few studies focused on the networks of organizations that operate in urban social and ecological systems. The study of large, urban networks also presents many challenges. Critics of past network analytical approaches argue that these studies have focused only on the relationships within the network and ignored how the network functions as a

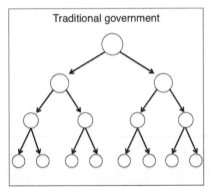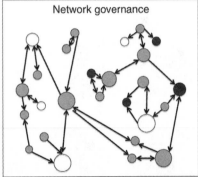

FIGURE 5.1 Simplified illustration of the difference between traditional, centralized governance approaches (*left*), and a polycentric governance structure that allows participation and interaction of multiple organizations (*right*). White circles are government organizations, and the gray and black circles are nongovernmental organizations, such civic or private organizations. (Reprinted from Muñoz-Erickson et al. 2016 under a Creative Commons license)

whole. Indeed, a review of interorganizational network studies from 1985 to 2005 found only twenty-six whole network empirical studies during that time. Further, few studies have evaluated the success of network governance. Understanding network effectiveness presents methodological challenges, such as how to combine network analysis with analyses of variations in local conditions, which is necessary to assess network governance outcomes. There is no single solution to the governance of social-ecological systems. Networks can take many forms, and not all will be effective. Evaluating the structures and functions of networks over the long term is essential to evaluating their performance and assessing their usefulness as forms of governance. This will require creative and novel methodological and analytical approaches. Finally, there is a need to study networks over time, in order to evaluate how changes in natural resources characteristics may or may not correlate with changes in network structures. Findings from these types of studies can allow decision makers and others involved in urban governance to actively manage the network using outcomes-based research.

BES researchers and others have proposed a social-ecological framework for urban stewardship network research: the STEW-MAP (Stewardship Mapping and Assessment Project) framework. The STEW-MAP framework (figure 5.2; table 5.2) is based on a set of ecological lenses: organismal, population, community, ecosystem, and landscape. Key to this framework is its utility for structuring social and ecological data, analysis, and interpretation for network analyses; facilitating comparisons among network

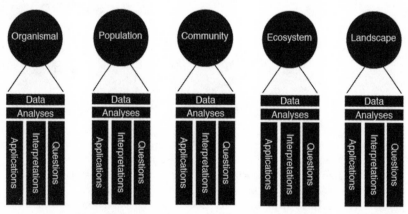

FIGURE 5.2   The five lenses of the STEW-MAP framework. (Reprinted from Romolini et al. 2016 under a Creative Commons license)

Table 5.2

Summary of the STEW-MAP framework lenses and examples of data and analytical techniques

		NAME	FOCUS OF DATA	ANALYTICAL TECHNIQUES
Lens 1	Ecological component	Organismal	Characteristics of individual organisms	Functional cycle, form, life history, speciation
	Social component	Network organization	Characteristics of individual organizations	Staff, budget, mission, history
Lens 2	Ecological component	Population	Size and composition of population of organisms	Birth rate, mortality rate, presence/absence, total number
	Social component	Network population	Size and composition of network members	Total nodes, total ties, composition
Lens 3	Ecological component	Community	Interaction between organisms, e.g., predation, parasitism, competition, etc.	Gradient analysis, site analysis, vegetation composition, habitats and niches,
	Social component	Network structure	Interaction between organizations	Centrality, centralization, density, modularity (statistical subgroups)
Lens 4	Ecological component	Ecosystem	Biotic and abiotic components of the ecosystem and their interactions	Fluxes and pathways of matter and energy, cycling (nutrient, carbon)
	Social component	Network flows	Substance of network ties, network flows	Strength of ties, change over time
Lens 5	Ecological component	Landscape	Geographical distribution of ecosystem processes	Remote sensing data, edges, cores, corridors
	Social component	Network spatial structure	Geographical distribution of social networks	Geocoding organizational locations, network relations over space

Source: From Romolini, Bixler, and Grove, 2016

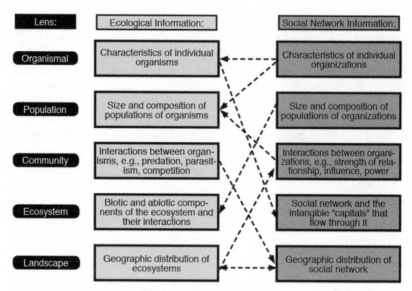

FIGURE 5.3 Descriptions of the types of social and ecological information that could be captured through each lens of the framework. Arrows represent potential interactions that form the basis of research questions. (Originally printed in Romolini et al. 2016; reproduced with permission)

systems; and connecting network analyses to decision making. Further, this framework can be used for novel combinations of social and ecological lenses (figure 5.3).

To date, BES network governance research has focused on network population (lens 2) and network structure (lens 3). Certain concepts are relevant to network structure analyses (box 5.1)

STUDIES OF NETWORK MANAGEMENT AND GOVERNANCE

A key challenge since the inception of BES has been to demonstrate the ability and value of adapting biophysically based, watershed studies from forested, rural areas to urban areas and its land use mosaics (chapter 9). This watershed approach was complemented initially by two types of socially based watershed studies. Grove developed a social patch approach using time-series census data and geographies to examine social and ecological differentiation (vegetation types) for the GFW from 1970 to 1990 (chapters 3 and 12). Dalton examined the civic governance of the Gwynns Falls in

BOX 5.1. KEY TERMS

Betweenness centrality: Describes how many times a node lies between any two other nodes that are themselves disconnected. Organizations with high betweenness centrality can be the bridging tie between the others, giving the bridging organization the ability to influence the flow of resources. Organizations with high betweenness can also diffuse information quickly and to the less-connected parts of the network. However, their position in between others can constrain these organizations to act on one decision.

Centrality: Measures the relative positions of individual nodes within the network.

Centralization: Describes the distribution of ties. High centralization exists when few organizations hold most of the ties in the network and thus can effectively diffuse information through the network. This can be positive for collective action. However, the uneven distribution of ties may result in asymmetric relations of influence and power, which can leave out the less-connected organizations. Highly centralized networks are also more vulnerable if one of the more centralized nodes is removed.

Density: Measures how many ties exist divided by the number of available ties. In general, density has a positive relationship with joint action for natural resource management. In particular, density may enhance development of knowledge and understanding through exposure to new ideas and an increased amount of information. However, very dense networks may be limited in their ability to receive new information and adapt.

Networks: Sets of nodes—in our case environmental stewardship organizations—connected by ties, or relationships. Quantitative measurements of network characteristics allow researchers to assess how they relate to natural resource management. Network analysis concepts include Centrality, Centralization, and Density, as described above.

Nodes: The structural components of networks, in our case environmental stewardship organizations.

Ties: The relationships among nodes in a network.

1999 through an examination of the network structure among public and nonprofit organizations associated with the natural resource management of the GFW. Romolini built on Dalton's work with a second survey in 2011.

Case 1. The Gwynns Falls Watershed Natural Resource Management Regime
The aim of this 1999 research project was to explore the activities and roles of the nonprofit and public sectors in the natural resource management regime in the GFW, in western Baltimore City and County. The GFW is the most heavily instrumented of all of the watersheds monitored by BES, which allows for comparison of social and biophysical long-term datasets.

This study included quantitative analysis of the relative activities and positions of organizations in the public and nonprofit sectors; and a detailed, qualitative analysis of the interactions among the Baltimore City Department of Recreation & Parks (a public agency) and the Parks & People Foundation (a nonprofit organization). A survey of nonprofit organizations and government agencies conducted in the GFW, a watershed of 66.2 square miles (circa 17,150 hectares), indicated that 154 organizations were involved in management of public natural resources in the GFW, of which 43 percent were nonprofit organizations, 40 percent were public agencies, and 17 percent were businesses, institutions, partnerships, regional entities, or other. The density of this overall network was 2 percent. More detailed social network analyses were performed on a subset of 45 organizations considered to be the core of the network. This core group displayed 13 percent density, meaning these organizations were much more tightly connected. As described above, high density can be a foundation for joint action but can hinder the ability to adapt to changing conditions. Measures of betweenness centrality were based on the number of projects on which each organization was identified as a partner, with the Parks & People Foundation shown to be the most central organization in the system, followed by the Maryland Department of Natural Resources, the Baltimore County Department of Environmental Protection & Resource Management, and Civic Works.

Thus in 1999 the top four most central organizations in the GFW were evenly distributed in the public and nonprofit sectors. While the two public agencies were responsible for environmental protection at the city and state levels, the two nonprofits had distinct missions. The Parks & People Foundation had the mission of uniting Baltimore through parks, while Civic

Works had a mission of strengthening Baltimore's communities through education, skills development, and community service.

Case 2. Environmental Stewardship Networks in Baltimore City

The intent of this 2011 research project was to better understand the activities, partnerships, and geographical extent of stewardship in Baltimore City. Building on a multi-city research program of the USDA Forest Service, this study employed survey and analysis methods comparable to research carried out in four other U.S. cities. Through descriptive statistics, social network, and spatial analyses, this study characterized the network of environmental stewardship organizations in Baltimore City in terms of the attributes of the organizational population, the relationships within and overall structure of the network, and the spatial distribution of stewardship activities across neighborhoods. The 165 responses to an organizational survey revealed that a network of 390 organizations work on environmental stewardship activities in the city. Sectoral distribution of respondents was 79 percent nonprofit, 18 percent public, and 3 percent private/other. The network displayed less than 1 percent density, which is consistent with large networks. However, a relatively high centralization of 18 percent was found, indicating that a small number of organizations held most of the ties. Measures of betweenness centrality in this network revealed the Parks & People Foundation was also the most central organization in the system, followed by the Baltimore City Office of Sustainability, Blue Water Baltimore, and the Baltimore City Department of Recreation & Parks. This high centralization can be useful to quickly diffuse information through the network, but it can be detrimental for resilience. Reliance on a small number of organizations means that the network would be vulnerable if one of these most central organizations disappeared.

In the 2011 study we advanced the research by not only characterizing the structure of the network but also comparing it to local environmental conditions. Specifically, we used spatial regression analysis to assess how the spatial distribution of land cover by neighborhood compared to network characteristics found in the environmental stewardship networks in Baltimore. This effort began to examine relationships between networks and environmental outcomes. In this case, we expected increased network activity—number of organizations, number of ties between them, and network density—to correlate with increased tree canopy. Conversely, we hypothesized that increases in network activity across neighborhoods would correlate with increases in tree cover and improvements in water quality. We did not find evidence to support the second hypothesis, but we did find that the number of stewardship organizations was positively related to tree canopy cover and that the number of ties found between stewardship organizations was negatively related to tree canopy cover. Preliminary interpretation of these results suggests that there might be two types of stewardship groups: those focusing on conservation of existing tree canopy, and those focusing on revegetating neighborhoods with low tree canopy. Revegetation is more resource intensive and may require collaboration among groups, which would explain the increased number of ties between organizations in neighborhoods with low canopy. This interpretation will need to be tested as we collect more data and have the ability to compare over time.

Longitudinal Comparative Analysis of the GFW Network

Data for the two studies were collected using different methodologies and over different geographic areas, so a reliable comparison of the two required researchers to reanalyze the data. First, data had to be trimmed to fit within the overlapping geographic area of these studies, which was the boundary of the GWF within Baltimore City. Baltimore County organizations were excluded from the 1999 data set, since the county does not have jurisdiction in the city. In the 2011 data set, organizations were excluded if they did not indicate working in one of the eighty-seven Baltimore neighborhoods that fall within the GFW. Next, organizations were only included if they had three or more ties, to conform to the core dataset of the 1999 project. This trimming of the data yielded a population of thirty-six organizations working in the GFW in 1999, and fifty-nine working there in 2011. Additional data preparation included the removal of valued ties. That is, each study asked several questions to which the respondent could nominate other organizations. Thus Respondent A could name Organization X in multiple categories, and the strength of the relationship would be considered to increase with each additional nomination. In the comparative study, organizational ties were simply valued as present or absent, since the original survey questions were asked differently.

Analyses were then conducted on the two reconfigured data sets. There were substantial changes in the twelve-year period. In general, there were twenty-three more organizations working in this city watershed in 2011 than 1999. There were twenty organizations from 1999 that still were active twelve years later. Thus sixteen organizations either stopped working in the watershed or, as in at least four known cases, ceased operations altogether; and thirty-nine organizations either moved or expanded their work to the GFW or were founded during this time. Some notable public organizations established within the twelve-year period were Baltimore City's Office of Sustainability (node #516 in figure 5.4b), located in the Department of Planning (#514) and TreeBaltimore (#378), located in the Department of Recreation & Parks (#237). These groups had such a distinct presence in stewardship activities in 2011 that they were included as separate organizations. Some prominent nonprofits established during the 12 years were the Community Greening Resource Network (#159 in figure 5.4b), Baltimore Green Space (#609), and Blue Water Baltimore (#539), a nonprofit representing the merger of Baltimore's five watershed associations (including the GFW Association, #539 in figure 5.4a). Finally, relevant to this particular volume, BES was nominated enough times to be considered a core network participant in both 1999 and 2011 and is represented by #413 in figure 5.4.

Network characteristics changed from 1999 to 2011 (table 5.3). Analysis of the sectoral distribution indicated the nonprofit sector taking a larger role and the public sector displaying a shrinking role in natural resources management in the GFW. Nonprofits increased by 7 percent, and public organizations decreased by 7 percent overall. However, public sector presence was not diminished at all levels—Baltimore City's organizational role in the network actually increased 5 percent during this time. State and federal organizations each lost 6 percent of their relative network presence. The social network analysis revealed a less dense network of organizations in 2011 than had been working in the GFW in 1999, with a 3.5 percent decrease. A large decrease of 14.5 percent in overall network centralization indicates that the ties became distributed more evenly among organizations during the twelve years between these studies.

The size of the nodes in figure 5.4 is related to the betweenness centrality of each organization. As described earlier, betweenness centrality is a measure of how many times an organization lies between two organizations that otherwise would not be connected. This is sometimes referred to as a bridging or brokering role, depending on the nature of the relationship. In natural resources governance, betweenness centrality is considered a measure of an organization's influence within the network (table 5.4). In

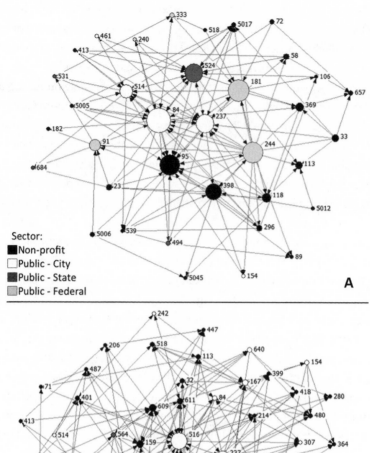

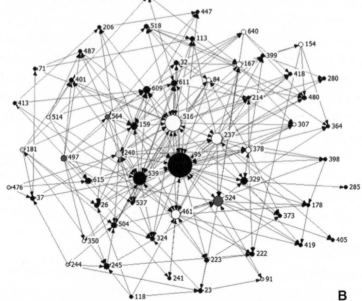

Sector:
■ Non-profit
□ Public - City
■ Public - State
■ Public - Federal

FIGURE 5.4 Changes in structure of the GFW organizational network from 1999 (A) to 2011 (B). (Reprinted from Muñoz-Erickson et al. 2016 under a Creative Commons license)

Table 5.3

Changes in the GFW organizational network from 1999 to 2011

MEASURE	1999 RESULTS	2011 RESULTS	CHANGE
Population	36 organizations	59 organizations	+ 23 organizations (+ 64%)
Nonprofit	22 (61%)	40 (68%)	+ 7%
Public	14 (39%)	19 (32%)	− 7%
City agency	5 (14%)	11 (19%)	+ 5%
State agency	4 (11%)	3 (5%)	− 6%
Federal agency	5 (14%)	5 (8%)	− 6%
Network Density	12%	8.5%	− 3.5%
Network Centralization	36%	21.5%	− 14.5%

the original 1999 study the network ties were valued based on number of projects the two organizations worked on together. The comparative analysis measured the relationship as a presence or absence, without any value attached, which had an effect on the relative positions of organizations in the network. For example, in the original 1999 study, the Parks & People Foundation had the highest betweenness centrality, while in this comparative analysis it ranks as the eighth most central. As shown, the Maryland Department of Natural Resources was quantified as the most influential in the 1999 network, while Parks & People ranked first in 2011.

There is also clearly a diminished influence of the federal organizations, with the National Oceanic & Atmospheric Administration (#244), the U.S. Environmental Protection Agency (#181), and the National Park Service (#91) as three of the top ten most active organizations in 1999, and only the USDA Forest Service (#240) representing the federal sector in the top ten most active in 2011. In contrast, local organizations gained prominence, with eight Baltimore City public sector and nonprofit organizations ranking in the top ten most active. This includes Baltimore City Public Schools (#461), which was not among the most active in 1999. Perhaps the most surprising finding is the inclusion of the nonprofit Community Law Center, which is ranked as seventh most influential.

The trends reported here support the theory of a shift in governance of the Sustainable City. The 1999 study demonstrated a more polycentric,

Table 5.4

Changes in relative influence (betweenness) of members of the GFW organizational network from 1999 to 2011

TOP TEN ORGANIZATIONS WITH HIGHEST BETWEENNESS CENTRALITY (NODE ID)	
1999	2011
1. Maryland Dept. Natural Resources (524)	1. Parks & People Foundation (95)
2. Baltimore Dept. Public Works (84)	2. Baltimore Office of Sustainability (516)
3. Friends of Gwynns Falls–Leakin Park (398)	3. Maryland Dept. Natural Resources (524)
4. Baltimore Dept. Rec & Parks (237)	4. Blue Water Baltimore (539)
5. National Oceanic & Atmospheric Admin. (244)	5. Baltimore City Public Schools (461)
6. Baltimore Dept. Planning (514)	6. Baltimore Dept. Rec & Parks (237)
7. U.S. Environmental Protection Agency (181)	7. Community Law Center (611)
8. Parks & People Foundation (95)	8. USDA Forest Service (240)
9. Maryland Save Our Streams (369)	9. TreeBaltimore (378)
10. National Park Service (91)	10. Community Greening Resource Network (159)

multi-sectoral, interconnected management regime than would be expected from traditional government in the Sanitary City. Twelve years later, in 2011, these shifts were even more evident, as the governance network was less centralized and distributed largely among local nonprofits and city agencies. The majority presence of the nonprofit sector in Baltimore supports governance theories that public agencies rely more heavily on nonprofit organizations for the service and delivery of public goods than in traditional government systems.

Local conditions must be considered when describing the mechanisms for change. One possible contributor may have been the development, adoption, and implementation of the Baltimore Sustainability Plan, which began with legislation in 2007, the establishment of a Sustainability Commission in 2008, and the adoption of the plan by the City Council in April 2009.

This two-year process was well publicized and engaged the community and representatives of organizations with various missions and purposes, and the observed activity and influence of the Baltimore Office of Sustainability are clearly quantified through this study. The 2011 increase in network participation at the local level may be partially explained by the rise of awareness about the interconnection between social, ecological, and economic issues—or, as the Baltimore Sustainability Plan describes it, "People, Planet, and Prosperity." More groups may have self-identified as doing some type of environmental work, even if their organizational missions are not focused on the environment, which may explain the influence of groups like the Community Law Center. Additionally, the Baltimore City Public Schools (BCPS) became a much more active and influential node in the network in 2011. This could be a result of efforts to involve schools in sustainability through activities such as the establishment of the Green Schools Network, the development of the 33-acre Great Kids Farm, and the BCPS system's focused efforts on planting trees on school properties. Another contributor to the central role of the schools could be the work of BES to establish environmental education in schools in Watershed 263, within the GFW (chapter 17). A deeper understanding of the reasons behind the changes in network structure and composition may be acquired through further research.

While we have made great progress in characterizing the structure of environmental stewardship networks and how those have changed over time, we recognize that there is much work still to do to assess relationships between network changes and environmental changes. As discussed, we have conducted some initial analyses with the 2011 dataset to compare networks and land cover, but this was for a point in time. As we continue the research, we have the opportunity to delve deeper into this area, thus contributing to the development of a theory of urban social-ecological governance.

CROSS-CITY COMPARISONS AND FUTURE DIRECTIONS OF BES NETWORK RESEARCH

Research on social-ecological governance is a promising area of study. The collection of organizational network data through BES provides a core dataset that can contribute to this field of study in several ways, including continued comparative, longitudinal, and applied research.

The 2011 research project is part of an ongoing national multi-city program of the USDA Forest Service, with comparable studies being conducted in Chicago, Los Angeles, New York City, San Juan, Philadelphia, and Seattle. We have been involved in a number of comparative research projects with researchers in these cities to share results and best practices, thus strengthening our methods, improving our ability to inform practice, and allowing us to develop theoretical frameworks. First, researchers in New York City, Baltimore, Chicago, and Seattle published a comprehensive guide to conduct this type of research in urban settings. In this report, we provide an overview of the time and other resource commitments required for research design, data collection, analyses, and development of research products, including technical tools such as maps, databases, and websites to be created in partnership with local stakeholders. Next, BES researchers led an effort to address the challenges of conducting governance studies that can be compared to ecological studies. Using lessons learned from Baltimore and several of the cities mentioned above, we offer a framework for data collection and analysis intended to facilitate the use of this research to evaluate networks and promote sustainable and resilient cities. Finally, in a synthesis paper of empirical social-ecological governance research in four cities, including Baltimore, we assert several propositions about urban social-ecological governance and its importance in sustainability transitions. Comparing stewardship networks across cities, now and over time, can greatly contribute to our understanding of urban social-ecological governance. These research findings, in turn, can inform the ongoing management and success of Baltimore's natural resource governance.

BES is pioneering the collection and analysis of large-scale, longitudinal network data. With two datasets separated by 12 years, BES is still in the early stages of building the capacity for the long-term analysis of governance structures. These types of analyses are still a relatively nascent methodology in network research generally, and specifically in the study of governance of social-ecological systems. While governance research is part of BES's long-term plan, with network data as a core long-term dataset, we anticipate our approach to network governance will evolve in terms of how we ask both social and social-ecological questions.

First, using the STEW-MAP Framework and its lenses, we anticipate to progress from description of governance networks (what), to explanation (how it works), to integrated modeling (predictions and scenarios of change), to assessment (what are the outcomes?). Second, we anticipate moving

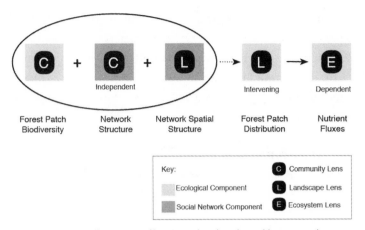

FIGURE 5.5 An illustration of how social and ecological lenses can be combined in examinations of governance networks and urban ecosystems. (Reprinted from Romolini et al. 2016 under a Creative Commons license)

toward increasing integration of social and ecological phenomena, using long-term BES data, by combining social and ecological lenses from our framework. For instance, returning to our focus on the GWF, we might investigate how the combination of forest patch biodiversity and network structure across the watershed interact with the spatial distribution of forest patches to produce different nutrient fluxes of the Gwynns Falls (figure 5.5).

Finally, as we develop network data, analyses, resources, and applications, we anticipate that some organizations may begin to actively self-manage the network based on their knowledge of the overall structures and relationships. Indeed, this research is already being used to directly engage with stakeholders and decision makers in Baltimore City. Since the 2011 data collection, BES researchers have presented to local, state, and federal stakeholders at several all-project and individual meetings. The participants in these meetings intentionally included the most central nodes in the network, in addition to more peripheral organizations. Participating organizations were shown the data, including the network figures showing their own positions in the network. Participants were asked to react to the findings and to identify their needs and interests in using the data to inform their governance practices. The most recent stage of this research is directly working with organizations to develop resources powered by the stewardship database that can facilitate their practices; for example, a directory and

associated map that allows organizations to search geographically for partners or projects. These types of resources may facilitate collaborative projects
and decision making about natural resources in the city. We are working to
make these data part of the operational informatics that city agencies, nonprofits and community groups, and businesses use to advance sustainability
activities.

ANSWERS TO GUIDING QUESTIONS

The preface of this book posed several questions, which we answer as follows:

1. What theory or set of concepts were used?

This work is guided by concepts related to network governance and
natural resources management. Particularly, we were motivated by theories
that governance approaches and structures shift from government-centric
and top-down in sanitary cities to more multi-sectoral and participatory as
cities transition toward sustainability.

2. How did BES allow that theory to be tested? Was the theory changed
or confirmed due to one or more conditions: (a) application in an urban
system; (b) use in an interdisciplinary context; (c) at different levels of complexity; or (d) because of the availability of new data. Note that the term
"urban" is used here in the broadest sense, as a city-suburb-exurban complex, spatially extensive system. If a chapter uses "urban" in the narrower
sense of downtown, center city, or dense residential development, that will
be stated explicitly.

BES provided the infrastructure for the longitudinal collection of core
data sets, which provided evidence to support the network governance
theory.

3. What was learned? How does this compare with information from
other places, for example, contrasting cities or other ecosystem types?

We discovered that there were some notable changes in stewardship
networks in the GWF over twelve years. The size of the network grew considerably, with increased participation from nonprofit organizations and
local government agencies and decreased participation from state and federal agencies. The influence of state and federal agencies also decreased
during this time, and the networks became less centralized.

4. How was the knowledge applied to one or more of the following
realms: education, public understanding, policy, or management?

The most recent stage of this research is directly working with organizations to develop resources powered by the stewardship database that can facilitate their practices; for example, a directory and associated map that allows organizations to search geographically for partners or projects. These types of resources may facilitate collaborative projects and decision making about natural resources in the city. We are working to make these data part of the operational informatics that city agencies, nonprofits and community groups, and businesses use to advance sustainability activities.

5. What is needed next? Such needs can address theory, methods, and translations, among others.

First, using the STEW-MAP Framework and its lenses, we anticipate to progress from description of governance networks (what), to explanation (how it works), to integrated modeling (predictions and scenarios of change), to assessment (what are the outcomes?). Second, we anticipate moving toward increasing integration of social and ecological phenomena, using long-term BES data, by combining social and ecological lenses from our framework.

CHAPTER SIX

Effects of Disamenities and Amenities on Housing Markets and Locational Choices

Elena G. Irwin, J. Morgan Grove, Nicholas Irwin,
H. Allen Klaiber, Charles Towe, and Austin Troy

IN BRIEF

- Amenities and disamenities are generated by spatially heterogeneous ecological features of urban and suburban landscapes, including parks and green infrastructure projects.
- The capitalization of these amenities and disamenities into housing and land prices provides a means of measuring their value as an ecosystem service.
- The capitalization of local green infrastructure projects into nearby housing values generates either amenities or disamenities depending on the type of project and whether it is located on public or private lands.
- Heterogeneous park attributes generate both amenities and disamenities. The net benefits to nearby homeowners depend critically on specific park attributes rather than just the presence or absence of a local park.
- High-resolution spatial data and novel identification techniques are necessary to test long-standing theories regarding urban spatial structure and land and housing values.
- Current and future work focuses on the development of structural models of household location choices that can be integrated with biophysical models to simulate policy scenarios.

INTRODUCTION

A range of economic, social, and institutional factors influence housing markets and the locational choices of households. These are described generally as "push" and "pull" factors, which can be classified as formal versus informal and exogenous versus endogenous. Formal push and pull factors originate from social institutions or policies, such as government-built roads or public parks; whereas informal factors emerge from social arrangements, such as exclusionary practices by neighborhood associations and other social processes. Exogenous factors, or "drivers," of the system originate from forces outside the system under study: an economic or natural disaster, for example, or federal policy. Endogenous factors are dynamic feedbacks generated by the cumulative effects of individual location and land use decisions within the system. These feedbacks often reinforce existing patterns by acting as a constraint to some households' location choice while reinforcing the location choices of others.

A novel link between social science theories of locational choice and the social-ecological context of the research of the Baltimore Ecosystem Study (BES) is to connect push and pull drivers with ecosystem services (table 6.1). For example, certain ecosystem services act as push/pull drivers by affecting locational choices, which in turn influence land and housing markets and the well-being of households. These human impacts can generate feedbacks that influence biophysical functioning and the quantity and spatial distribution of ecosystem services, which can spur further locational adjustments, institutional responses, and changes in urban spatial structure at more aggregate scales. The role of ecosystem services as push/pull drivers have changed systematically over time. For example, environmental catastrophes such as the burning Cuyahoga River in 1969 led to the Clean Air Act in 1970 and the Clean Water Act in 1972, which led to substantial reductions in the health impacts and disamenities from polluting industrial facilities, which in turn altered push/pull dynamics at neighborhood scales.

In this chapter we review theories of household location choice and the contributions that BES researchers have made to integrating these theories within the social-ecological context of Baltimore. Specifically, we focus on BES research that has linked ecological amenities and disamenities with land and housing values and other outcomes that impact the well-being of households. We first provide an overview of location and land value theories

Table 6.1

Ecosystem services associated with push/pull forces that influence household locational choices and migration flows

PUSH/PULL DRIVERS	ECOSYSTEM SERVICES
Safety (crime)	
Tax rates (e.g., fuel tax)	Climate regulation, Gas regulation
Housing cost	Raw materials
Housing quality	Raw materials, Gas regulation, Climate regulation, Waste regulation
Lot size	Aesthetics, Nutrient cycling, Habitat, Gas regulation, Climate regulation, Biological regulation, Nutrient regulation, Water regulation, Water supply, Raw materials
Proximity to work	Gas regulation, Climate regulation
Proximity to transportation	Gas regulation, Climate regulation
Public school quality	
Racial diversity	
Environmental quality and pollution	Hydrologic cycle, Water regulation, Waste regulation, Water supply, Air quality, Nutrient regulation
Proximity to open space/ recreation	Disturbance regulation, Science and education, Net primary productivity, Food, Nutrient regulation, Habitat, Water regulation, Raw materials, Recreation, Spiritual and historic
Aesthetics and residential landscaping	Net primary productivity, Gas regulation, Climate regulation, Biological regulation, Hydrologic cycle, Water regulation, Water supply, Nutrient regulation, Nutrient cycle, Habitat, Nutrient regulation, Water regulation, Raw materials, Aesthetics

that provides the theoretical framework for our research. We then review the major findings that relate to the spillover effects of green infrastructure projects and the amenities and disamenities associated with park attributes. We conclude the chapter with a discussion of current and future research directions, including the role of ecological amenities in urban redevelopment and the goal of sustainable and resilient cities.

THEORETICAL BACKGROUND

The location choices of households and the economic, social, and institutional constraints to these choices are fundamental processes underlying the spatial dynamics of urban socioecological systems. Location theory identifies fundamental economic forces that influence location, including (1) natural advantages that attract firms and households; (2) economies of concentration that enhance the productive efficiency of firms that cluster; (3) transportation and communication costs that spatially differentiate markets and their geographical extent; and (4) amenities and disamenities that explain the location of population and jobs. These include urban amenities, such as per capita cultural activities; urban disamenities, such as crime; and natural amenities, such as climate, coastlines, and open space.

Hedonic pricing models (box 6.1) offer a means to estimate the marginal value of amenities, disamenities, and other characteristics associated with a differentiated market good, such as housing. Because housing is spatially immobile, the values of amenities and disamenities are capitalized in the sales price of the home. House prices will therefore reflect the value of both the structural characteristics of the house as well as the local public goods and amenities and disamenities in the house's neighborhood. The hedonic pricing function, which posits price as a function of the quantities of a good's attributes, arises through the interactions of many buyers and sellers in the market. As a result, it describes the equilibrium between buyers and sellers in the market. The marginal effect of any attribute on the house price is called the marginal implicit price and represents the individual's willingness to pay for a one-unit increase in the attribute. These implicit prices can be empirically estimated using data on housing prices, structural characteristics of the house, and locational attributes of the house and neighborhood to specify the hedonic price function.

Underlying the hedonic framework are the demand and supply decisions of households to buy and sell houses. On the demand side, households that differ in their income or locational preferences sort themselves across different neighborhoods by choosing the location that maximizes their utility subject to their budget constraint. As households sort, the bid on different neighborhoods and prices is determined by the intersection of

---

BOX 6.1. KEY TERMS

Amenity: An enviromental feature, whether constructed or natural, that is percieved by some persons or groups to provide benefits.

Disamenity: An engineered or natural environmental feature that yields a perceived burden or risk.

Endogenous: A factor affected by other interacting factors within a system, in contrast to exogenous factors, which are not influenced by the interactions within a system.

Formal: A relationship or influence established by law or institutional structure. In contrast to informal controls, which emerge from custom or interactions within a group or community.

Green infrastructure: Infrastructure designed to perform ecological work within cities and suburbs. It consistes of several types, including the following: green streets are curb extensions or sidewalk planters planted with vegetation that absorb stormwater; green roofs are roofs with vegetation that utilizes stormwater; retention basins are ponds that retain and filter stormwater before gradually releasing it into nearby waterways.

GWR: Geographically Weighted Regression is a statistical technique to quantify spatial relationships among variables.

Hedonic model: A regression modeling technique that breaks down the contributions of various factors to the prices of homes. Preferences for different properties are revealed by purchasing behavior.

Locational choice: A theory describing the factors that drive the decisions of households and firms on where to locate across an urban region and which account for the formal and informal constraints on such decisions.

Pull factor: A factor attracting households or firms to a location, in contrast to factors which operate in other locations and encourage people or organizations to leave those locations.

---

the bid and offer curves of the households moving to different neighborhoods. Households choose a neighborhood to maximize their utility, subject to their budget constraint. Other constraints also often play a role. For example, race and religion have been significant factors over time in constraining where people can live and in biasing the allocation of amenities and disamenities. The sorting process continues until no one has the incentive to move, and the region reaches a spatial equilibrium. The hedonic pricing model is based on the equilibrium outcome of this sorting process and therefore is conditional on the underlying demand and supply relationships that in equilibrium reflect the capitalized value of local amenities and disamenities.

THE EFFECTS OF ECOLOGICAL DIS/AMENITIES

BES researchers have used the hedonic pricing model and other econometric models to investigate a number of intriguing hypotheses regarding the effect of specific ecological features on housing values and other quality-of-life outcomes. One aspect of this research focuses on the question of whether, in addition to their intended environmental effects, local green infrastructure projects generate so-called co-benefits in terms of positive amenities to nearby residents. Below we discuss research results of stream restorations and stormwater basins on housing values and the effect of trees on neighborhood crime rates. In addition, we review BES research that addresses an ongoing debate about the value of local parks and the extent to which they provide positive amenities versus disamenities in the form of neighborhood crime and congestion.

Stream Restoration

Recent EPA policy changes aimed at reducing nutrient levels in waterways have invigorated local officials' interest in expanding the scope of stream restorations to meet these new federal standards and reduce nitrogen loadings into the Chesapeake Bay. While much of the existing research regarding stream restorations has focused on the biology and ecology of the stream or the design of the restoration itself, little academic work has attempted to quantify the amenity potential of restoring damaged urban waterways. We have examined the value of stream restorations as perceived by households to answer the question of whether particular types of restorations may result in improved housing values and argue that differences in values across restoration types could be important additional criteria when deciding where to site future restorations. This work focuses on stream restorations in Montgomery and Baltimore Counties, both of which were heavily developed in the post–World War II era, a time when developers were held to very little, if any, environmental regulation. The resulting developments, more often than not, regarded streams as a nuisance and spent more time burying or channeling them than preserving them. The results were fewer streams channeling more water with greater velocity and thus eroding stream banks, increasing stream incision, reducing infiltration, and damaging aquatic health.

The difference between degraded and restored streams is visually dramatic. Our empirical strategy relies on this change in visual amenities to identify the capitalization in surrounding home prices. Using a large dataset of housing transactions located near a number of stream restorations, various empirical strategies, including household fixed effects and repeat sales models, are used to reveal the causal effect of stream restorations on nearby housing values. This allows for heterogeneity in restoration value across three types of restorations: those occurring in private lands, parks, and publicly owned lands with limited access. The data demands are intensive for this approach because (a) a home must be within fifteen hundred feet of a stream, (b) a home must be sold prior to the restoration event, and (c) the same home must be sold again after the restoration event is complete.

A number of controls are necessary for this analysis. The overall identification strategy is to select a small enough spatial and temporal window to be reasonably sure dramatic changes to the home or the neighborhood amenities have not taken place, which could confound estimates if they were to have occurred at the same time as the restoration event. Time windows of sales range from five years to nine years, and spatial windows range from five hundred to fifteen hundred feet (152.5–457 m). We further control for the time of each sale with sale year fixed effects and for any "neighborhood" specific trends in house price change using watershed fixed effects interacted with sale year fixed effects. Finally, within this overall strategy, similar homes not near a stream restoration but located near a stream are used as a control group (figure 6.1).

In contrast to the existing literature, which has largely focused on a small number of restorations and identified only a single effect of restoration, we find significant heterogeneity in restoration impacts depending on the locations and ownership regimes where restorations occur. The findings range from no impact when located in private lands, potentially reflecting the creation of easements on private lands, which was common in this

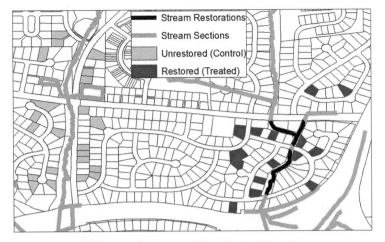

FIGURE 6.1 Location of houses within fifteen hundred feet of a stream restoration site (treated group) and houses within the same distance of a stream site that was not restored (control group) in Baltimore County, Maryland.

area, to a larger impact of upward of 2 percent for public parks and around 3 percent for other tax exempt lands. The tax exempt lands are mostly government owned, nonpark properties but also include churches, schools, and cemeteries. These findings provide new evidence that spatially targeting restorations should be a key goal of policy makers seeking to gain the greatest economic, and ecological, return. They underscore the private benefits that public investments can yield and suggest that restoration on public lands creates ancillary benefits, for example, from recreation, which restoration on private lands does not.

Stormwater Basins

The installation of gray infrastructure, such as large underground storage tanks to collect rainfall or increases in treatment plant capacity, is the traditional means by which municipalities have addressed stormwater management. To offset the high cost of installing new gray infrastructure, urban areas have increasingly incorporated green infrastructure, such as green streets, green roofs, and stormwater retention basins, into citywide comprehensive stormwater management plans. These green techniques have been noted not only to lower costs but also to provide potential co-benefits to residents from the natural amenities that are generated.

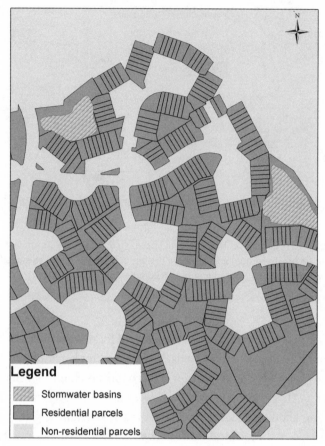

FIGURE 6.2 Example of stormwater basins located within a residential subdivision in Baltimore County, Maryland.

We have empirically identified the capitalization effect of stormwater basins on nearby housing prices in Baltimore County (figure 6.2) to investigate the potential for this particular type of green infrastructure to generate (dis)amenities that would augment or offset the intended environmental benefits of improved urban water management. Basins could provide an attractive amenity for proximate homeowners, as the grassy vegetation around the basins could be aesthetically pleasing and, when filled with water, serve as a point for wildlife observation. If this were true, then basins would capitalize positively into house prices. Basins could also be positively valued if households value the ecosystem services they provide. However, if basins are unkempt, overgrown, or unsightly, they may capitalize negatively into house prices.

Using data on the location of 2,950 stormwater basins and over 90,000 observations of housing sales between 1996 and 2007, we used an empirical approach that exploits spatial and temporal variation in the placement of basins within subdivisions to determine causally the capitalized value of stormwater basins into house prices. In order to control for unobserved spatial correlation in the data that could otherwise confound a causal estimate, a difference-in-difference regression is used that also accounts for the age of the basin. The results show that stormwater basin adjacency leads to housing prices that are consistently lower, with estimates between 13 and 14 percent, depending on model specification. For the mean house in the sample, this corresponds to a house price decrease between $28,185 and $30,579, a factor solely attributed to stormwater basin adjacency. This negative capitalization effect accentuates as the basin ages. For the case of a house adjacent to a basin that is at least seven years old, it is estimated to have a compounding negative capitalization of approximately 17 percent when compared to an identical home not adjacent to such a basin. This spillover effect is not found to have a significant effect on nearby houses that are not adjacent, suggesting that the effect is highly localized.

The chapter is the first study to date that identifies the causal effect of stormwater retention basins on housing values across a heterogeneous geographic area using revealed preference data on housing markets. The approach controls for a number of sources of bias that could arise from the presence of unobservable landscape features, lending confidence that the estimated negative effect is robust. The results show that, in the absence of a purposeful approach to amenity creation, the stormwater regulations implemented in Baltimore County have resulted in stormwater basins that confer a substantial negative impact on adjacent houses. Thus households adjacent to stormwater basins bear a disproportionate share of the cost of providing green infrastructure that is intended to generate environmental benefits for all residents of the region. While this work does not consider the full costs and benefits of stormwater basins, including their costs of construction and maintenance and the value of the ecosystem services they deliver, it does imply that any ecological benefit further downstream should be sufficiently large to offset these losses in adjacent housing values in order to provide an overall net benefit to the region.

Vegetation and Crime

Urban vegetation is known to have a wide range of benefits, including aesthetic amenities, reducing energy use for cooling, carbon sequestration, and filtering and attenuating stormwater runoff. A potential benefit of urban vegetation less often appreciated is reduction in crime risk. If this were true, it would have tremendous ramifications for Baltimore City, where robbery rates are eight times the national average and where crime is one of the leading factors behind middle-class flight to the suburbs.

We have sought to address the uncertainty surrounding the question of crime and vegetation by studying the relationship between crime and trees using high-resolution data across the entire Baltimore City and Baltimore County area. Summarizing data at the census block group level, we regressed a crime index (derived from actual geo-coded crime locations and based on street crimes, such as robbery, burglary, and shootings) against tree canopy cover and a number of covariates meant to control for confounding factors of population density, income, percent single-family homes, percent rural, race, median year of construction, amount of protected open space, and amount of agricultural land. A subsequent set of regressions broke down tree canopy cover by private land, public land including rights of way (street trees), and public land excluding rights of way. These regressions were run as both ordinary least squares regressions and as spatial lag regressions, to adjust for spatial autocorrelation in the dependent variable. Additionally, a geographically weighted regression (GWR) was run to assess spatially dependent omitted variables.

The regressions, which explained approximately 90 percent of the variance in crime, indicated that overall crime is inversely correlated with tree cover, even after controlling for all the potential confounders. The spatial lag regression, which has the most conservative estimation of coefficient values, suggests that a 10 percent increase in tree cover is associated with an 11.8 percent decrease in crime. When tree cover is distinguished by private versus public land, the magnitude of the effect goes down slightly for each, and a large difference appears between the effects of public and private trees, the former being nearly 40 percent larger. This would suggest that planting trees on public lands might yield somewhat higher crime-reduction benefits than planting on private. When public rights of way are not included in the analysis of public trees, the gap between public and private increases relative to the previous model. The magnitude of the public trees coefficient is nearly double that of private trees, suggesting that trees in non–right of way public lands such as parks and public institution lands could be the most effective components in terms of reducing crime. However, these regressions reveal association, not causation, and therefore these results may be confounded by spurious unobserved correlations. Finally, GWR analysis indicates that, particularly for private trees, there is some omitted interaction effect that conditions the relationship between private trees and crime. The resulting maps (figure 6.3) show that a few neighborhoods, including Brooklyn Park, Wagners Point, and Dundalk, are exceptions to the rule that more trees mean less crime.

In a second study we extended this analysis to look at the relationship between front yard landscaping and crime. The study analyzed the relationship between block-level crime and indicators of yard management, using a visual survey that the authors conducted of over 1,000 yards in Baltimore City and County. Over forty indicators of yard management were tested for significance that looked at the presence and condition of trees, shrubs, lawns, beds, irrigation systems, and hardscaping among many others. Point counts of crime within 150 meters of a property were regressed against these yard indicators, controlling for a number of potential confounders related to socioeconomics and housing type. Three types of regression were used, ordinary least squares, spatial error, and Poisson. Yard-level variables found to be negatively associated with crime included the presence of yard trees, garden hoses/sprinklers, and lawns in addition to the percentage of pervious area in a yard. Those found to be positively associated with crime included presence of litter, desiccation of the lawn, uncut grass, and number of small trees in front of or adjacent to the property. Results were consistent across all regression models.

A number of potential explanations have been proposed for why trees in landscaping may reduce crime. One is that maintained greenery makes spending time outdoors more appealing, leading to more "eyes on the street," which in turn mitigates dangerous behavior. More eyes in public spaces make it harder for criminals to go unnoticed and leads to informal surveillance networks. Additionally, outdoor encounters between residents foster social networks and cohesion. A second explanation of why vegetation might reduce crime is that it can be seen as a positive territorial marker or a "cue to care," signifying to criminals that the residents are actively involved with their surroundings, even in the absence of people on the street.

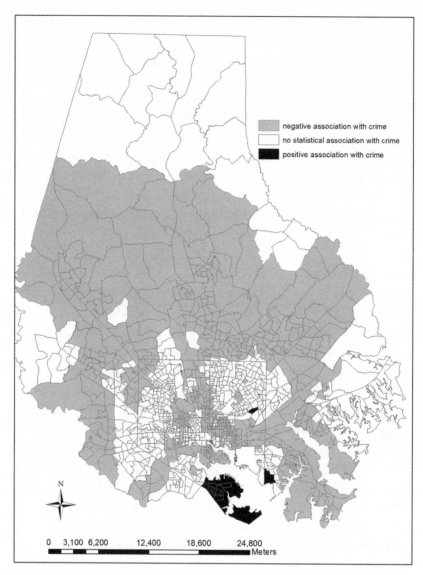

FIGURE 6.3 Spatial variation of estimated association of trees and crime in Baltimore County, Maryland, using geographically weighted regression analysis.

The presumption is that when looking for a place to commit a crime, a perpetrator would move on to a neighborhood where cues suggest weaker neighborhood organization and less social capital. This is consistent with the so-called broken window theory, which posits that neighborhoods displaying visual cues of neglect or poor maintenance experience higher crime because these cues suggest to criminals a lack of effective law enforcement, while maintained neighborhoods send the opposite cue. Finally, research suggests that green surroundings can attenuate violent behavior through psychological mechanisms. For instance, green surroundings have been found to be associated with lower levels of aggression and mental fatigue in inner-city residents, while they have also been found to be linked to cognitive forms of self-discipline among youth, such as impulse inhibition and delay of gratification.

As for the question of the few locations where GWR indicates that crime is positively associated with tree cover, a possible explanation relates to the fact that in these neighborhoods there is a considerable interface between residential and industrial lands in which lower, early successional, and apparently unmanaged stands of trees tend to grow. Such unmanaged vegetation not only offers potential concealment to criminals but also may embolden them by signaling reduced cues to care associated with neglect and fewer eyes on the street.

The combined evidence of these studies, then, suggests that trees and landscaping have a strong inverse association with crime. Determining whether this relationship is causal requires additional research, although some research has established some degree of causality, such as in the case of the greening of vacant lots in Philadelphia and the subsequent drop in crime. If causality can be more firmly established, this will show that urban vegetation can act as a powerful crime deterrent if planted and managed in way that maximizes the positive (cues to care, eyes on the street, and so forth) and minimizes the negative (concealment, signs of neglect).

Maintaining and Improving Public Spaces
Urban parks are typically considered to be an amenity that can strongly influence location choice decisions. There is considerable diversity of findings in the literature about how price effects vary the characteristics of the park or the surrounding context. For instance, some research reveals a vegetation "scarcity effect"; that is, property values increase as the proportion of trees on a property relative to that in the immediate neighborhood (visible from the property) goes up. This suggests that having many small green spaces distributed throughout an urban area might be more beneficial than having a few large parks. Other research finds that urban parks (where

more than 50 percent of the park is manicured or landscaped) are valued negatively for nearby properties, whereas natural parks (where more than 50 percent is preserved in natural vegetation) had no positive effect on property values, and linear parks had a negative effect on property values when close by, although that effect became positive when between one quarter and one half a mile distant.

Differences in price effects have also been found with respect to ownership status of the open space. Multiple kinds of open space, including public parks and private land in conservation easements, have been found to increase property values, suggesting that it's primarily the lack of development that is the most desirable attribute. Other research reveals a greater premium for private forests and a lesser premium for institutional, conserved forests.

BES researchers hypothesize that many of the inconsistencies in the results of previous studies are a consequence of unobserved heterogeneity in local park amenities. Combining data on single-family residential transactions assembled for Baltimore County, spanning the years 2000 through 2007, with detailed data provided by the Baltimore County Parks and Recreation Department on park renovation activities and costs, we have estimated a series of hedonic models to explore household valuation of local park renovations. Potential unobservable factors that may be correlated with local park amenities in traditional cross-sectional analysis are controlled by exploiting time variability in renovations to estimate property fixed effects models. Further, the large sample of homes and relatively long time period of sales in the Baltimore metro region limits concerns about sample selection in focusing the study on homes with multiple sales.

This research shows that a cross-sectional hedonic model with an aggregate indicator for renovations that does not distinguish among types of amenities is insufficient to determine the effects of renovations on nearby home sale prices. In contrast, using property fixed effects models, we find that homes within one mile of a park have a positive willingness to pay for playground replacements and field renovations, while lighting and court renovations are associated with a negative willingness to pay. These willingness to pay values decrease in magnitude as the number of years since renovation increases, and all renovations, with the exception of playground replacements, become insignificant four years after the renovation is complete.

This study is the first to use an extensive list of renovations to specific park attributes to recover willingness to pay measures that vary across those renovation types. Through estimating the value of park components separately, the approach reveals that homeowners have robust, significant willingness to pay, both positive and negative, for park amenity renovations. In comparison, estimating renovations as a single, aggregated value results in an estimate that is not significantly different from zero. This finding may partially explain the differing values associated with local parks seen in the literature. Significant expenditures on park maintenance and renovation, like those in Baltimore County, are common in many municipalities, and this study provides some of the first evidence that the value from improving existing parks is heterogeneous across amenities and changes rapidly over time.

We have extended the investigation of heterogeneous parks attributes to the question of crime levels and whether variation in the level of crime within a park impacts property values. This is an important issue in Baltimore, where high crime rates coexist with an extensive park system. The question is whether parks in high-crime areas may actually be more of a liability than an amenity. Using hedonic analysis with an interaction term between surrounding crime rate and log-transformed distance to the nearest park, our study found that there was a significant association of the interaction effect between park proximity and crime levels in determining the way that park proximity is capitalized into property values. When crime rate is relatively low, the association between parks and property values is positive. In contrast, when crime rate is high, parks are negatively associated with property values. Near the threshold, which is estimated to be between 406 percent and 484 percent of the national average, the direction of the association becomes ambiguous. As crime rates climb above this threshold, the direction of the relationship switches, and park proximity is negatively associated with home values. The result, then, is a heterogeneous mix of parks that the housing market perceives as amenities and others it sees as liabilities. High- and low-crime parks are, interestingly, fairly evenly distributed in space, rather than being clustered. This finding suggests that parks may be an amenity that can attract households to a location, but that this is not always the case. In fact, an unsafe park is a far bigger deterrent for a potential mover than is a neighborhood with no park at all.

SUMMARY AND CONCLUSIONS

BES researchers have identified a number of findings regarding the role of amenities and disamenities in land and housing markets and how spatially heterogeneous ecological attributes impact human well-being in the Baltimore region. This research is possible only because of the highly detailed spatial and temporal data on housing values and housing, neighborhood, and ecological characteristics that have been assembled as part of the BES project. There will continue to be a pressing need for these data to continue work on spatial amenities and disamenities in urban, suburban, and exurban settings. A number of data and modeling challenges remain. Here we discuss five key challenges, all of which rely on long-term perspectives and continued collection of data over time and spatially disaggregated scales: casual identification, policy scenario modeling, urban revitalization, agent heterogeneity, and preferences for dis/amenities over time.

Casual Identification of Spatial Effects

Spatial data are typically positively correlated in space, reflecting the first law of geography, that "everything is related to everything else, but near things are more related than distant things." This makes identification of spatial effects challenging. The task is further complicated by the endogenous nature of some spatial effects, such as interactions, which implies that an estimation bias may arise both from the endogeneity and from unobserved spatial autocorrelation. While the endogeneity problem may be addressed using standard econometric techniques, for example, instrumental variables, it is often impossible to find instruments that are themselves not spatially correlated, and thus it is difficult to ensure that the instrument and error are not correlated in space. Other approaches to causal identification include quasi-experimental designs, such as a treatment and control comparison that is used to identify the effect of stream restorations on housing values and the difference-in-difference approach used to identify the effect of stormwater basins on housing values.

Policy Simulations

Although hedonic models are useful for identifying the marginal effects of amenities and disamenities on housing markets and locational choices, they are a reduced form expression of an equilibrium outcome. Unless further structure is imposed, the explanatory variables included in a reduced form equation cannot be attributed to a specific demand or supply process but instead reflect the net effects of these variables on the equilibrium outcome. This implies that it isn't possible to use these models to represent feedback effects, such as the locational responses of households to changes in pollution levels or open space in their neighborhood, and how these feedbacks determine the housing market equilibrium. Instead, structural models of locational choice are needed to represent the household's demand for a particular location and the characteristics of that location. A structural modeling approach is necessary for counterfactual policy simulation in which the goal is to evaluate the impacts of a nonmarginal policy change on housing markets. For example, using a locational choice model of neighborhood location choices in Baltimore, we examine the potential for an impervious surface tax to reduce the amount of new impervious surface by increasing the costs of development and altering the spatial distribution of positively valued green space. The direct policy effect is the increased cost of development, given the added fee for impervious surface creation; the indirect effect operates through the locational choices of households, who respond to the changes in the spatial distribution of impervious surface, which is found to have a negative effect on their locational choices. Sorting out these direct and indirect effects is possible only by using this type of structural (versus reduced form) model. The results show that the fee has a significant, although small, effect on locational choices. For example, an impervious surface tax of $10,000 per acre within the Patapsco River watershed in Baltimore County was found to generate a reduction of up to 0.4 percent in the amount of impervious surface acres within a census tract due to the re-sorting of households (figure 6.4).

Ecological Dis/Amenities and Urban Revitalization
What is the role of urban revitalization and re-greening initiatives in increasing home values and spurring economic redevelopment of distressed neighborhoods in the central city? For example, the City of Baltimore began an ambitious urban redevelopment program in 2010 called Vacants to Value. This program is a comprehensive, city-led effort to revitalize neighborhoods and promote green and sustainable communities in areas of high vacancy through wholesale cleanup and demolition to remove the excessive vacant and abandoned housing stock and convert land into temporary open space before additional development. We report on other research in chap-

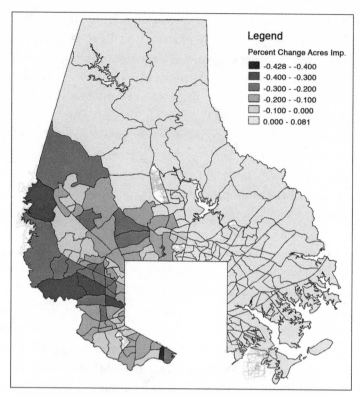

FIGURE 6.4 Predicted changes in impervious surface due to changes in household locational choices under a hypothetical impervious surface tax within the Patapsco River watershed in Baltimore County, Maryland.

ter 7 that examines the influence of this program on local housing market activities, including neighboring housing renovations, housing sales, and values.

Agent Heterogeneity

BES research reveals considerable heterogeneity in dis/amenity effects across space and types of neighborhoods and households. These findings suggest that ecological effects can generate either amenities or disamenities depending not only on the type of ecosystem service that is produced, but also depending on the composition of households within a neighborhood. For example, as illustrated by the findings reported in this chapter, the effects of parks on nearby houses may be positive or negative depending on the level of crime in a neighborhood. A better understanding of how ecological effects vary across different socioeconomic groups is critical for urban redevelopment and policy. If trees and landscaping really can lead to lower crime levels, and crime is one of the most important determinants of location choice in high-crime regions like Baltimore, then the equity significance of the spatial distribution of green infrastructure is greatly amplified. As previous research has found, Baltimore is highly variable in its spatial distribution of trees, with greater canopy cover in areas where population density is low and where single-family homeownership, property values, family sizes, and education rates are high. Furthermore, tree canopy has a legacy effect, with the greatest canopy cover found in neighborhoods built between forty and fifty years ago, all else being equal. All this suggests, then, that geographically strategic planting and maintenance could be not only an excellent strategy for fighting crime regionwide but also a relatively low-cost means of promoting spatial environmental equity and, in the long term, steering location choice decisions.

Preferences for Dis/Amenities Over Time

In this chapter we have discussed the role of dis/amenities as push and pull drivers of locational choices. Our long-term perspective offers the opportunity to examine how the relative importance of dis/amenities changes over time and varies among different social groups. Tracking these changes and how they interact with environmental conditions can help us understand long-term changes in locational choices and urban structure. At the same time, certain groups may be excluded from making locational choices or participating in the decision-making processes that determine where dis/amenities are located. Thus a long-term perspective can track changes in preferences among different social groups over time and whether they are able to exercise those preferences.

CHAPTER SEVEN

The Role of Regulations and Norms in Land Use Change

Elena G. Irwin, Geoffrey L. Buckley, Matthew Gnagey,
Nicholas Irwin, David Newburn, Erin Pierce,
Douglas Wrenn, and Wendong Zhang

IN BRIEF

- Legacy effects and past social objectives, including efforts to reinforce segregation patterns, have created path dependencies in land use dynamics and are important determinants of current land use patterns.
- Established in 1967, Baltimore County's Urban-Rural Demarcation Line (URDL) is considered a fundamental planning achievement and has provided a critical foundation for later Smart Growth initiatives.
- The primary effect of downzoning has been not to reduce development but to shift the type and location of development; in some cases this has led to greater infill, but in other cases to less dense, more scattered development that has extended the urban footprint.
- Environmental regulations that focus on protection of a single resource can generate unintended effects that result in a trade-off between enhancing one ecosystem service while degrading others.
- Urban redevelopment programs that demolish vacant and blighted housing appear to spark local housing market activity but are effective only if multiple demolitions occur in the same neighborhood.

INTRODUCTION

Human uses of land produce large social benefits in the form of food, fiber, shelter, and other essential goods and services, but they also generate a range of environmental impacts, including carbon emissions, soil and water degradation, alterations of habitat and hydrologic cycles, and loss of biodiversity. The scale of land use impacts has increased dramatically over time with growing global population and development. Many scientists believe that current global land use practices are undermining the Earth's long-term ability to sustain food production, freshwater and forest resources, and other provisioning ecosystem services. While these concerns are global, land use decisions occur in local settings in response to local, regional, and global factors. Thus, achieving more sustainable land use practices relies on policies that can effectively manage land use and land change processes at local and multiple scales. Because the impacts vary across space, an understanding of the spatial pattern of land use and land change at local scales is also important.

The framework for modeling land use change in the Baltimore Ecosystem Study (BES) derives from an interdisciplinary set of theories that address locational choice, urbanization, neighborhood change, social norms, and the evolution of regulation. The approach is innovative not only because of this diverse set of theories that ground the research but also because of the emphasis that is placed on long timeseries and clear linkages to biogeophysical processes. In this chapter we review these economic, social, and geographical theories and the contributions BES researchers have made to studying the evolution of urban-suburban-exurban land use dynamics within the social-ecological context of Baltimore. Specifically, we focus on BES research that has traced the evolution of urban land use change from a pre-zoning era through the current time period, in which zoning and environmental regulations are the primary tools of land use planners and sustainability managers.

BES research relies on long timeseries data that provide detailed historical accounts of land use and regulation changes over time. Archival data, including newspaper accounts, Board of Park Commissioners reports, neighborhood association meeting minutes and promotional materials, planning documents, and government records, such as the maps and notes generated by the federal Home Owners' Loan Corporation (box 7.1) in the 1930s, have allowed BES researchers to uncover at least some of the processes that influenced land use decisions in the past. In some cases the legacy effects of past decisions are still evident in today's urban landscape. Spatial data sets have proven invaluable as well. For example, much of our

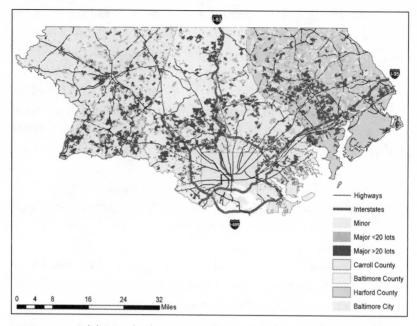

FIGURE 7.1   Subdivision development in Baltimore, Carroll, and Harford Counties by decade, 1960–2007.

research uses residential subdivision histories for Baltimore, Carroll, and Harford Counties as well as subdivision plat image files from the Maryland State Archives and county tax assessors' GIS databases. We have recreated these data from 1960 onward. Across Carroll, Harford, and Baltimore Counties a total of 7,370 subdivisions containing a total of 208,131 land parcels were developed between 1960 and 2008, leading to a mix of clustered and scattered patterns of residential development (figure 7.1). We have also linked other data to these parcels, including information on the timing of approvals by the county of subdivision plans; the timing and location of agricultural preservation; zoning; distances to nearby towns and large cities; surrounding land uses and other amenities; soil quality and slope data. As we emphasize throughout our discussion, these and other historical, highly detailed data that we highlight here are critical to understanding the relationships among individual choices, community norms, heterogeneous regulations, and spatial spillovers.

In what follows we first provide an overview of land use theories that constitute the theoretical framework for our research. We then review the

---

BOX 7.1. KEY TERMS

CWA: The federal Clean Water Act, established in 1972.

Downzoning: A planning strategy reducing the density of development in particular areas.

Endogenous: Factors operating within an area that are influenced by other factors in that same area.

Home Owners' Loan Corporation (HOLC): A federally chartered body to refinance mortgages compromised by the Great Depression. Established in 1933, it classified the loan-worthiness of neighborhoods in major cities across the United States, including criteria now judged to be prejudicial based on race and national origin.

Legacy: The effect of past decisions or prior conditions on current system state and processes.

Locational choice: The study of the mechanisms shaping where households and firms locate within an urban region. The term "choice" in this theoretical realm also includes how the available locations may be constrained by formal and informal means.

Redlining: The HOLC-mapped neighborhoods and other areas in cities during the 1930s using red shading to indicate those areas judged to be unworthy of mortgage lending. Green and yellow (and sometimes, in various cities, other colors) indicated good versus more risky areas for lending.

Spillover: A spatially correlated influence. May be positive or negative.

URDL: Urban-Rural Demarcation Line, established by Baltimore County in 1967 to set a boundary of intense urban development and promote smart growth.

V2V: Baltimore City's Vacants to Value program, aimed at revitalizing neighborhoods experiencing high levels of building abandonment.

---

major findings that relate to the social, economic, and environmental factors that have influenced the evolution of community norms and zoning regulation and, in turn, the influence of zoning and other land development policies on economic and environmental outcomes in the region. The chapter concludes with a discussion of current and future research directions.

THEORETICAL BACKGROUND

Theories of land use and location provide the backbone for land change models. The theoretical foundations of land economics were laid long ago with the concept of land rents that are generated by differences in land quality and that are capitalized into output prices. In addition, because transporting people and goods is costly, distance from economically valuable locations, such as a central business district or a high-amenity coastline, also plays a fundamental role in determining the value of location. The theory of land rent provides the foundation for the spatial equilibrium framework that underlies theories of urban spatial models, including the notion that heterogeneous households reveal their demand for public goods by "voting with their feet" (chapters 4 and 15). Households and firms have higher demand for land and housing in more desirable areas, which bids up land and housing prices and bids down wages in those locations. More recent theories build on these foundations by considering other sources of spatial differentiation, intertemporal decision making, and the endogenous relationships among many of the factors that influence land use and location outcomes. For example, the standard urban economic model of land development posits a fundamental relationship among economic growth, distance to urban centers, and urban land rents, which generates predictions regarding the location, timing, and pattern of urban development.

Heterogeneity in land, in land use, or in expectations is hypothesized to generate scattered or leapfrog development, a pattern in which vacant land is skipped over while land farther from the city center is developed first. For example, when there are multiple types of land use that differ in their expected returns (high versus low density residential land use is one instance), developers have the incentive to withhold land closer to the city from development at lower density for lower returns in anticipation of future development at higher density for higher returns. This can generate temporary leapfrog development that will decrease over time as the region fills up. Other explanations of leapfrog development include the presence of positive or negative open space or social spillovers and heterogeneous production costs, all of which can lead to varying patterns of clustered versus scattered development.

Economic models of land change begin with a model of the underlying microeconomic behavior (utility or profit maximization, for instance) that determines demand and supply relationships. Fundamental to models of land markets is the price mechanism, which determines individual choices and, in turn, is determined by the cumulative choices of individuals within a given market area. The concept of a price equilibrium is used to ensure that individual choices and aggregate outcomes are consistent with each other. Although equilibrium can be defined in various ways, the condition of market clearing, meaning that prices adjust such that markets clear (that is, excess demand and excess supply are zero in all factor and output markets), is standard. Equilibrium may be static, in which agents are myopic and prices and land use patterns are unchanging, or dynamic, in which agents are typically forward looking and prices and land uses are changing over time subject to a market-clearing condition. A common misperception is that economic equilibrium necessarily implies a static condition, which is not the case. For example, in a dynamic model of landowners, the forward-looking expectations of landowners over future costs and returns influence their land use decision today. Economic models of land use and land change differ in how equilibrium in the relevant input and output markets is defined. In local land and housing markets, prices are distinguished by space and depend not only on the quantity of land in alternative uses but also on the spatial distribution of land uses.

Land use regulations play a critical role in determining the outcome of housing markets, including the spatial pattern of housing and the use of land resources in the production of residential development. Spatially explicit parcel-level models of residential land use change are useful for analyzing the effect of zoning regulations on the rate of development and residential density. Some empirical evidence shows that minimum lot size zoning may actually exacerbate sprawl because when zoning is binding, homeowners are required to consume larger lots than desired. In addition, land-preservation programs may have unintended effects that can exacerbate scattered development patterns.

An empirical challenge in identifying the effect of land use zoning is that zoning itself is not a random process and, indeed, co-evolved with land use and location patterns over a longer period of time. This implies that zoning is endogenous to land use conversion, and therefore estimating the effect of zoning may be susceptible to selection bias. In this case, additional econometric techniques are needed to control for these sources of bias to identify the causal effect of zoning. Techniques that employ quasi-experimental designs or instrumental variables provide a potential means of identifying causality.

THE CO-EVOLUTION OF LAND USE AND REGULATION
IN THE BALTIMORE REGION

BES researchers have sought not only to identify the causal effects of zoning on land use outcomes, but also to explicitly study the co-evolution of land use and zoning over a long timeframe. Here we synthesize BES research on the influence of neighborhood associations prior to passage of zoning legislation, and on the evolution of the region's first growth management regulation, the urban-rural demarcation line (URDL) in Baltimore County. We then summarize BES research on the subsequent effects of this and other zoning regulations on land use patterns, focusing on how these regulations have influenced the location and density of new residential subdivision development.

Pre-Zoning Evolution of the City: The Role of Neighborhood Associations
The 1904 fire that destroyed much of Baltimore's downtown served as an important catalyst for change. Indeed, as BES researchers have shown, it was during the first two decades of the twentieth century that the city took steps to install a modern sewer system, improve its transportation infrastructure, commission a plan intended to create a world-class system of parks, and inaugurate a new urban forestry program. In time, Baltimore would also adopt strict zoning regulations.

Prior to the passage of Baltimore's first zoning ordinance, in 1923, and the significant amendments that were enacted in 1931, Baltimoreans contended with land use regulations that were uneven and poorly enforced. To deal with this uncertainty, concerned residents banded together to advocate for the needs of their neighborhoods and to protect themselves from a variety of perceived threats. Organized into neighborhood improvement associations, of which there were some seventy at the beginning of the twentieth century, these citizen groups exerted a great deal of influence when it came to city planning and development.

Homeowners associations and other neighborhood groups have long played an important part when it comes to shaping patterns of settlement and land use in urban areas. As Marcia England reminds us, neighborhoods

are often viewed by residents as "sites to be protected from outside interests," and when they believe "spatial boundaries are threatened" they are
inclined to act. In addition to pressuring city officials into providing much-
needed infrastructure and services, such as paved roads, modern sewers,
streetlamps, and telephone lines, BES research shows that Baltimore's
neighborhood improvement associations proved adept at attracting such
amenities as parks and street trees into their districts, while deflecting
unwanted land uses to other parts of the city. Several groups also figured
prominently in the effort to pass the nation's first segregation ordinance
and, later, to promote and pass zoning legislation. BES researchers rely on
numerous historical data sets to uncover the role that improvement associations played in guiding city development during the pre-zoning era.

As these data show, perhaps no neighborhood organization wielded
greater power than the Peabody Heights Improvement Association. The
association was formed when the neighborhood was absorbed by the city
after the annexation of 1888. Its founding members sought to protect what
they had created—a residential enclave far removed from the hustle and
bustle of commercial and industrial activity. The mission of the group focused on five key areas. First, the group promoted street-tree planting and
neighborhood beautification. Notable among their accomplishments was
mayoral approval of Ordinance No. 154, in 1912, which established a Division of Forestry under the direction of a trained forester to manage the city's
trees. Second, members of the association took every opportunity to expand
and enhance green space within its borders, most notably Wyman Park. A
third issue that resonated with the Peabody Heights membership was air
pollution abatement. Given the location of Peabody Heights—a few blocks
from Baltimore's busy Penn Station—it is not surprising that smoke from
coal-powered steam locomotives was a concern. The fourth issue that worried residents was commercial development. Wishing to remain a strictly
residential quarter, the association worked closely with the city's inspector
of buildings to prevent businesses from locating in the neighborhood. Finally, like other improvement associations at the time, the Peabody Heights
group used a variety of means to ensure that African Americans, Jews, and
recent immigrants could not buy or rent homes in the district.

While neighborhood groups did, on occasion, work together to beau-
tify the city, for example, by assisting the Women's Civic League with its
beautification campaigns, it was issues like segregation that often spurred

them into action. For instance, BES researchers use entries from the Peabody Heights Association's meeting minutes as evidence to indicate that the group actively supported the city's efforts to pass a segregation ordinance. When the Supreme Court struck down such ordinances in 1917, neighborhoods like Peabody Heights, Mount Royal, and Mount Washington used restrictive covenants to preserve segregation. In the 1930s the Home Owners' Loan Corporation and various lending institutions reinforced patterns of segregation and disinvestment via redlining and other practices. Another issue that drove neighborhood associations to cooperate with one another was the threat of business encroachment. This issue in particular led members of the Peabody Heights Association, in 1929, to support a new comprehensive zoning ordinance that would create residential, commercial, and industrial use districts in the city.

Prior to the passage of comprehensive zoning, neighborhood improvement associations played an important role in city planning and development, influencing where investments were made and where certain land use activities took place. As the meeting minutes of the Peabody Heights Association as well as other historical data show, however, many of these groups also had social objectives in mind as they advanced their respective agendas.

Evolution of Zoning: The Urban-Rural Demarcation Line in Baltimore County
The wartime economy of the first half of the twentieth century sparked an economic and population boom in the City of Baltimore and adjacent suburbs. As early as 1937, ten years before Maryland would pass enabling legislation to give counties the authority to plan, the State Planning Commission sought to contain growth by introducing garden suburbs contained within a web of greenbelts. Although the concept was never deployed, it served as a foundation for later efforts, most notably in 1960 and 1962. Finally, in 1967, inspired by the environmental design and planning firm Wallace and McHarg, whose Plan for the Valleys responded to a period of massive manufacturing redistribution and rapid suburban growth, Baltimore County established an urban growth boundary. The boundary drew a line on the ground separating urban land uses from rural land uses (see also chapter 16). One of the first in the nation, the Urban-Rural Demarcation Line (URDL) resulted in almost 90 percent of the population living on just one-third of the county's land area (figure 7.2).

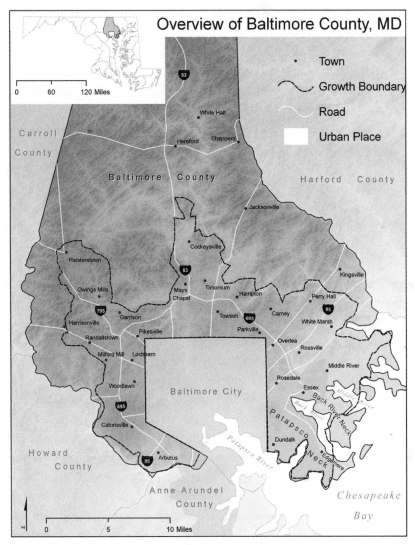

FIGURE 7.2 The Urban-Rural Demarcation Line (URDL = Growth Boundary in the legend), established in 1967, which limits the extent of public water and sewer services and divides urban and rural land uses in Baltimore County. (U.S. Geological Survey)

Baltimore County considers this tool to be one of its fundamental planning achievements—one which laid the groundwork for later "Smart Growth" efforts that are credited with preserving over fifty-five thousand acres of productive farmland and ecologically significant forest, planting almost ten thousand trees, and promoting compact development—but it has had negative consequences as well. In particular, concentrated environmental externalities in the form of urban runoff contribute significantly to the pollution of Chesapeake Bay. Over the years, the county has added teeth to the URDL by improving zoning regulations and implementing other growth management schemes, such as the Agricultural Preservation Program for farmland preservation in the 1970s, the Critical Areas Program for aquatic resources protection in the 1980s, and the Priority Funding and Rural Legacy Areas programs of the 1990s.

The Effects of Downzoning

After implementing the URDL in 1967, Baltimore County further restricted growth in the non-URDL areas by adopting resource conservation zoning areas in the Comprehensive Plan that became effective in late 1976. Prior to 1976, the zoning allowed subdivisions at one housing lot per acre throughout the entire rural area covering two-thirds of the county. After 1976 the downzoning policy in Baltimore County created three main resource conservation (RC) zoning types. The most dramatic was the creation of agricultural preservation (RC2) zoning that downzoned to allow one housing lot per fifty acres and covered about half of the rural area. Watershed protection (RC4) zoning allows one housing lot per five acres and was designated to protect those watersheds and major rivers and streams associated with three regional reservoirs (Liberty, Loch Raven, and Prettyboy), which supply water to 1.8 million residents in the Baltimore metropolitan area. Rural residential (RC5) zoning allows one housing lot per two acres and serves as the baseline-zoning category.

The effectiveness of downzoning depends on regional growth pressures and the degree to which future development patterns are affected by the downzoning. We use BES long-term data on housing sales and residential subdivision development over a forty-five-year time period, from 1960 to 2005, to examine the impacts of downzoning in three suburban counties—Baltimore, Harford, and Carroll Counties—on the pattern of leapfrog development. All three counties implemented a significant downzoning policy between 1976 and 1978. These policies, which impacted about 75 percent of the developable land in the metro region, converted land that was previously zoned to accommodate one house per acre to several new zoning classes ranging from one house per three acres to one house per fifty acres.

We developed a new measure to calculate the amount of leapfrog devel-

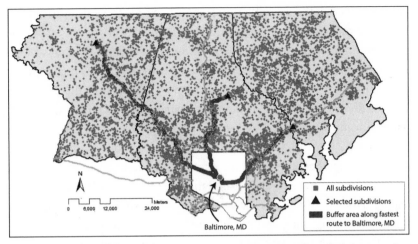

FIGURE 7.3   New measure of leapfrog development that is specific to the location and timing of each new subdivision. The amount of leapfrog development associated with a specific subdivision (black triangles) is measured as the percentage of developable vacant land that is more accessible to the city center than the subdivision itself and located within a buffer along the most expedient commuting route to Baltimore City.

opment that resulted from the creation of each new subdivision at a particular time and location in the study region. The amount of leapfrog development associated with a specific subdivision is measured as the percentage of developable vacant land that is more accessible to the city center than the subdivision itself and located within a given buffer along the most expedient commuting route to the outer boundary of Baltimore City. The leapfrog measure is expressed in percentage terms relative to the total amount of developable land that is either developed or vacant within each subdivision-specific buffer and varies between zero (no remaining developable land) to one (all land is developable).

The analysis reveals a pattern of leapfrog and infill development that is consistent with urban economic theory and that underscores the influence of zoning on the evolution of these patterns (figure 7.3). We find that the relative amount of leapfrog development is high but declines over time from 80 percent in 1960 to 36 percent in 2005. In other words, about 80 percent of developable land deemed more accessible than existing subdivisions was undeveloped in 1960. This amount declined by more than 50 percent over our forty-five-year study period at an annual rate of approximately 1 percent. In comparing this observed pattern to the unconstrained hypothetical pattern predicted by the basic intertemporal urban growth model, we find that it closely matches the predictions, but only in the early years before the downzoning policy. After the downzoning, the observed amount of leap-frog development is significantly less than the unconstrained predicted pattern. Empirical results from a first difference model show that after controlling for distance to urban centers, the spatial pattern of infill development is significantly influenced by local variations in the maximum allowable development density. Specifically, the downzoning policies enacted in the late 1970s significantly slowed the rate of infill development in more rural areas of the metro area and increased the rate of infill development in areas closer to the urban centers. These results are consistent with the hypothesis that downzoning, agricultural preservation, and other commonly used policies that restrict development introduce a substantial reduction in the relative returns to land development in these areas.

Evidence by other BES researchers suggests that the downzoning policy adopted in 1976 in Baltimore County did not have a significant effect on the probability of development but did strongly affect the density of development. A long timeseries on spatially disaggregated data was used to characterize the evolution of residential development in the pre-zoning era (1967–76) versus post-zoning era (1977–86). The average treatment effects show a reduction in the density of development of 54 percent and 60 percent, respectively, in agricultural and watershed protection zoning areas. These results indicate that the 1976 downzoning policy did not reduce the rate of acreage developed to low-density exurban development, but it did reduce the number of households living in those developed areas. An important reason for the low effectiveness of reducing the likelihood of development is the minor exemption rule. As a political compromise in the 1976 downzoning process, parcels in the agricultural zoning areas with two to one hundred acres are still allowed to be split into two housing lots to create a minor subdivision.

The Effect of Regulatory Delay and Spatially Heterogeneous Zoning
Regulatory delays in the time to complete a development project create so-called implicit costs that indirectly increase costs by extending the time required to tie up capital and delaying revenue generation. Although the explicit costs of regulation, including impact fees and required infrastructure improvements, can substantially increase development costs, real op-

tions theory suggests that increases in implicit costs can have an even larger impact on the timing, density, and location of development. We use the historical subdivision data from Carroll County with data on the timing of individual subdivision plan approvals by the local county planning authority to investigate the influence of regulatory delay on development timing, intensity, and patterns.

The most substantial difference in subdivision approval times is between major subdivisions, which are four lots or greater and typically located in higher-density suburban areas, and minor subdivisions, which are two to three lots and primarily located in the agriculturally zoned areas of the county. We hypothesize that "time is money" and that the substantial difference in the implicit regulatory costs arising from the approval time necessary for a major versus minor subdivision causes developers to substitute away from major subdivisions and build more minor developments. This hypothesis is tested by constructing a dynamic variable that predicts an ex-ante expected approval time for each undeveloped parcel and for each time period of the model, 1995–2007. With this variable, a sample selection model of land development is estimated in which the landowner chooses the optimal density of development conditional on the discrete choice to subdivide the parcel. The regulatory delay hypothesis is confirmed by the econometric model results. Specifically, we find that the regulation-induced implicit costs reduce the probability of subdivision development on any given parcel and have resulted in a substantial increase in the likelihood of exurban development relative to higher-density development.

We conclude that spatially heterogeneous costs generated by differences in the regulation of differently sized residential subdivisions have contributed to greater scattering of residential land development in exurban areas. Previous empirical studies focused on the role of demand-side amenities and disamenities and the role of these local land use spillovers in generating scattered exurban development. This work provides a new explanation of scattered residential development based on how developers respond to regulatory delay and spatially heterogeneous zoning.

The Effect of Wetlands Protection on Land Development

Environmental regulations frequently target land development activities due to their negative impacts on water quality, biodiversity, and other ecosystem services. However, these regulations have varying levels of efficacy as well as frequent unintended consequences. These offsetting effects can reduce the net benefits of an environmental policy, which, given the irreversibility of most development projects, can have long-lasting consequences. We consider a wetlands protection policy to examine the potential unintended consequences on land development in Harford County. Specifically, Section 404 of the Clean Water Act (CWA) regulates dredging and filling of water features to minimize degradation of wetlands. In 1985 the definition of the "waters of the United States" under Section 404 was expanded in an effort to specifically reduce environmental damage caused by new residential developments. The expansion of the CWA's jurisdiction was the result of a 1985 United States Supreme Court ruling, *United States v. Riverside Bayview*, 474 U.S. 121 (hereafter RBH). The ruling redefined the "waters of the United States," increasing the jurisdiction of the U.S. Army Corps of Engineers in the enforcement of the CWA. Interviews with environmental and development planners from Harford County during this time period provide evidence of a shock to the subdivision approval process as a result of the county's implementation of the ruling. The county implemented these CWA regulations by limiting modification to streams and nontidal wetlands, requiring new developments that impacted these water features to gain permit approval through the corps.

We treat the implementation of the RBH ruling as an exogenous effect and estimate its impact on the timing and density of subdivision development in Harford County. Using parcel-level data on new residential subdivision development from 1980 through 1990, we use a difference-in-difference estimator with a duration analysis to identify changes in the rate of development between the five-year period prior to the Supreme Court decision (1980–84) and a five-year period (1986–90) following the ruling. The results demonstrate that the regulation significantly delayed the development of parcels with water features. The rate of development decreased by 33 percent for affected parcels. However, the density of development also decreased for new developments, suggesting that the regulation may have also contributed to an increase in the urban footprint by lowering overall density of new development.

Urban Land Redevelopment: Baltimore's Vacants to Value Program
Between 1960 and 2010 the population of Baltimore City shrank nearly 34 percent despite a 60 percent increase in the population of the surrounding

metro region. Given the durability of housing, this has resulted in a tremendous number of vacant housing and lots—an estimated fourteen thousand vacant properties at last city count. As a former industrial city with a still functioning major seaport, Baltimore continues to feel the aftereffects of its history as an industrial hub with approximately 4 percent of all available land in the city in brownfield status.

To address the problems of urban vacancy and spark urban renewal, the City of Baltimore began a unique urban renewal program in 2010 called Vacants to Value (V2V). V2V is a comprehensive, city-led effort to revitalize neighborhoods that specifically targets areas of high vacancy through increased code enforcement, providing homebuyer incentives, expediting the sale of city-owned properties, and promoting green, sustainable communities. The centerpiece of the program is the use of targeted demolition of vacant properties in distressed neighborhoods with high vacancy and crime (figure 7.4). Previous literature has revealed the disamenity effects that vacant and blighted houses can have on nearby properties, leading to suboptimal outcomes in the market. However, whether or not the removal of such housing effects positively impacts local housing markets is an empirical question.

BES researchers have provided the first empirical evidence of the effects of publicly funded demolitions on neighborhood housing activity by investigating the outcomes of the V2V program in the City of Baltimore. We take advantage of the program's unique structure to build a quasi-experimental design with treated groups—neighborhoods funded one time by the program and neighborhoods funded multiple times by the program—and a control group—unfunded but shortlisted neighborhoods. In 2013 V2V created a shortlist of possible demolition sites in targeted neighborhoods based on Baltimore's 2011 Housing Market Topology, a large-scale housing study created by and for the city to help identify and strategically match limited public resources in neighborhoods based on vacancy, occupancy rates, and housing market activity in 2009 and 2010. From the initial shortlist, selected projects were funded each year during the study period, 2013–15. Funding decisions were made by city officials using the same criteria as above, and only projects on the initial shortlist were funded. Some neighborhoods had multiple projects funded while others had none despite the presence of multiple shortlisted projects during the study period.

We find that neighborhoods targeted by the program have housing

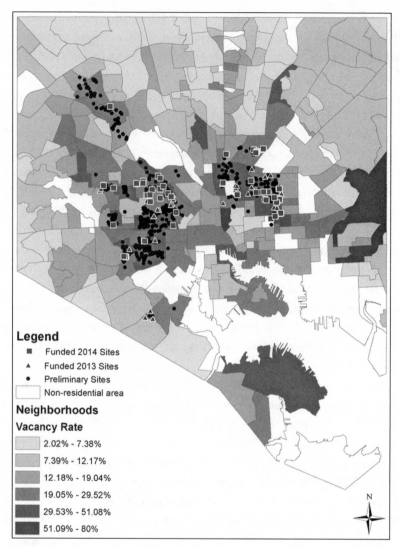

FIGURE 7.4 The City of Baltimore's Vacants to Value (V2V) preliminary and
funded sites (2013–15) and neighborhood vacancy rates. (Based on data from the
American Community Survey and the City of Baltimore Department of Planning,
2006–10)

renovations that are 82.7 percent higher and that total rental housing sales
are 31 percent higher, but only if the neighborhood had multiple V2V-
funded sites during the study period. No significant effect is found in the
neighborhoods that had only a single funded site. To put these findings in

context, the increase in renovations corresponds to an additional twenty-one neighborhood renovations while the increase in housing sales corresponds to fifty additional housing sales over the study's two-year time period. We also find that high levels of crime can diminish the effects of the V2V program on neighborhood housing markets.

SUMMARY AND CONCLUSIONS

Urban ecosystems are complex systems composed of many interacting areas—from the built-up and redeveloped city to suburban areas to sparsely settled exurban regions connected by commuting and economic flows—and many interacting social, economic, and biophysical processes. Like any component parts of such a complex system, land use and land regulations cannot be understood in isolation. Land use regulation generates both direct and indirect effects as well as intended and unintended consequences. Local land use regulations are not a random occurrence but instead evolve from the social, political, and environmental realities of a community or region. In turn, these regulations can have both intended and unintended impacts on environmental, social, and economic outcomes that impact individual and community well-being and the sustainability of the region. Identifying these effects requires long timeseries of spatially disaggregated data that can account for individual choices, neighborhood change, heterogeneous regulations, and spatial spillovers. BES researchers have unpacked several of these complex interactions among community norms, local regulations, land use change, and land and housing market outcomes. This research has focused on the evolution of regulations in both the pre-zoning and zoning phases of the Baltimore region as well as the effects of historical and more recent regulations and policies on land development and land use patterns. Several synthetic findings emerge from this research:

- The legacy effects of past decisions are clearly manifested in today's urban landscape. Prior to the passage of comprehensive zoning, neighborhood improvement associations seeking to reinforce segregation played an important role in influencing investments that determined where certain land use activities took place. Many of these investments are extremely durable, for example, infrastructure and local parks, and generate amenities and disamenities that continue to influence household location and housing values today.

- In addition to preserving open space and guiding development, the establishment of the URDL in Baltimore County had unintended consequences, namely, the concentration of urban runoff that ultimately drains to Chesapeake Bay.

- Autonomy in local land use regulations, but economic interdependence via regional labor and housing markets, creates unintended consequences in terms of land development spillovers. For example, downzoning in one area reduces the supply of new development in that area and leads to increased demand and development in lesser-regulated adjacent areas. These spillovers may occur within the same county, for example, as the result of spatial heterogeneity in zoning of minor versus major subdivisions, or across counties as the result of uncoordinated local policies. The primary effect of downzoning and other local growth management regulations has been to shift the type and location of development across the region. While downzoning in Baltimore County led to localized growth spillovers and increased the likelihood of low-density development in these neighboring areas, the combined effect of all downzoning policies across multiple counties appears to have worked in the intended direction by reducing the overall amount of leapfrog development across the region and increasing the overall amount of infill development.

- Regulations that focus on protection of a single resource can generate unintended effects that result in a trade-off between enhancing one ecosystem service while degrading others. For example, we find that the wetlands protection policy enacted under Section 404 of the Clean Water Act resulted not only in a significant delay in development on the affected parcels but also in a reduction in their density of development. Thus while the regulation was successful at limiting modification to streams and nontidal wetlands, it also fostered a lower density of development. These offsetting effects reduce the net benefits of an environmental policy, which, given the irreversibility of most development projects, can have long-lasting consequences. As the research on downzoning has shown, a reduction in the amount or density of development in one area often results in displacing development to other, as-yet-undeveloped areas.

- Preliminary work based on analysis of the V2V program in the City of Baltimore indicates that targeted demolitions may be an effective renewal strategy in urban neighborhoods with excessive housing supply and urban blight. This suggests that public policy can achieve the intended spillover effects that generate positive multiplier effects that can magnify across broader spatial scales.

- Because of the complexity of the many spatial processes that underlie land use/land cover change and the inevitable limitations of available data in terms of measuring these processes, identifying causal effects of land use change is extremely challenging. Techniques that employ quasi-experimental designs or instrumental variables provide a potential means of drawing causal inferences and can be extremely useful in isolating the effects of a spatially varying policy or heterogeneous landscape feature on land use change. This also underscores the importance of long timeseries of spatially disaggregated data that can account for individual choices, neighborhood change, heterogeneous regulations, and spatial spillovers.

Our current and future work continues to examine the implications of spatially heterogeneous zoning and urban redevelopment on land use change within the city and across city-suburban-exurban gradients. In addition, we are working with other BES researchers to develop integrated models of land use change, nutrient flows, and water quality to model policy scenarios. The goal of this work is to develop spatial land change models that account for market conditions and human-biophysical linkages to generate predictions of policy impacts on land use and ecosystem services. Such an approach is necessary for moving beyond the spatial heterogeneity that characterizes human and biophysical components to an integrative understanding of how these spatially heterogeneous processes interact with each other across multiple spatial and temporal scales. Understanding how such interactions influence the dynamics of urban systems is critical to achieving resilient urban futures.

CHAPTER EIGHT

Human Influences on Urban Soil Development

Richard V. Pouyat, Katalin Szlavecz, and Ian D. Yesilonis

IN BRIEF

- Although soils in urban and urbanizing landscapes are predominantly modified by human activity, they provide, like natural soils, various ecosystem services.
- Initial research addressed characteristics that distinguish "anthropogenic," or urban, soils from soils evolving under natural, or nonurban, conditions, how these characteristics vary spatially, and whether it is possible to generalize the development and characteristics of urban soils across urban and human settlements at global scales.
- Research on soils in the Baltimore Ecosystem Study (BES) and comparisons with results from other cities indicate that properties of surface soils and their assemblages of soil organisms can vary widely, making it difficult to define or describe a typical urban soil or urban soil community.
- Although characteristics of urban soils can vary widely, global comparisons of urban soils have shown greater similarity in their characteristics than the native soils they replaced.
- Urban soils have exhibited a surprising capacity to support plant growth and soil biota, and as a result they often have relatively high rates of biological activity.

INTRODUCTION

Soil provides vital, life-sustaining ecosystem services but is often under-appreciated as a natural resource. These services include retaining and supplying nutrients, serving as a growth medium for plants, capturing contaminants, and contributing to the hydrologic cycle through absorption, storage, and supply of water. Urban soils in particular are overlooked in this regard because they are assumed to be highly altered and thus lack the capacity to provide the ecosystem services of native, unaltered soils. Research in the previous three decades, however, has shown that urban soils are alive and may harbor a rich diversity of microorganisms and invertebrates, and they have the capacity to provide many of the same ecosystem services as nonurban soils. Nevertheless, soil conditions in urban areas generally correspond to a range of human or anthropogenic effects from relatively low influence, such as those found in a remnant forest patch to those that are highly disturbed or derived from human-created materials, such as concrete. In this chapter we explore the array of potential human modifications of soil, particularly those from urban uses, and consider human activities as a factor of soil formation and change. Specifically, we use results of BES as a case study in the relatively new discipline of anthropedology (box 8.1) to investigate the effects of human activity on soil development. Additionally, we present a conceptual framework to address questions related to the spatial patterning at multiple scales of chemical, physical, and biological characteristics of soils developing in urban landscapes. These questions include the following:

1. What chemical, physical, and ecological characteristics distinguish urban soils from soils evolving under nonurban conditions, and how do these characteristics vary spatially?
2. How do these characteristics relate to urban (anthropogenic) versus natural, or nonurban, soil-forming factors?
3. Can we generalize the development and characteristics of urban soils across urban and human settlements at regional and global scales?

A CONCEPTUAL FRAMEWORK OF ANTHROPOGENIC SOIL FORMATION

Soils are considered to be a collection of organized bodies with characteristic horizons that develop by pedogenic, or soil forming, processes over time. The changes in soil characteristics occurring from pedogenic processes are

BOX 8.1. KEY TERMS

Anderson level: A classification system of land use and land cover created from satellite and aircraft remote sensing data (Anderson et al. 1976).

Anthropedology: Science of human–soil relations (Richter et al. 2011)

Anthropogenic: Created by humans.

Beta diversity: Change in species composition over relatively short distances; the degree of similarity of species composition among distinct communities

Biotic homogenization: Process in which the community similarity of distinct locations increases over time

Coefficient of variation: A measure of relative variability, consisting of the ratio of the statistical standard deviation to the average.

Ecological community: An assemblage of species within a defined area that interact with each other.

Ecotope: The smallest, relatively homogeneous ecologically distinct landscape unit. The anthropogenic ecotope mapping and classification system (AEM) identifies ecotopes in anthropogenic landscapes (Ellis et al. 2006)

Heavy metals: Chemical elements that have a high density and are toxic to human health, such as lead (Pb), copper (Cu), zinc (Zn), and arsenic (As).

Heterogeneity, spatial or temporal: Spatial heterogeneity is a property generally ascribed to a landscape, soil types, or plant or animal species. In the case of urban soils, it refers to the uneven distribution of various soil characteristics. Temporal heterogeneity can be defined as the difference of the spatial characteristics of soils through time.

Horizon: A relatively distinct layer within a soil, differing from those above or below in composition of minerals, organic matter, and organisms as well as texture (see soil texture).

Metacommunity: Set of local communities linked together by dispersal, colonization, and interaction of multiple species.

Nitrification: Process of transforming ammonia into nitrite and then nitrate. Plants take up nitrate more easily than ammonia. Nitrate is more mobile than ammonia and can be easily leached out of soil.

Nitrogen mineralization: Conversion of organic nitrogen to inorganic forms that plants can take up.

Parent material: Underlying geological material; for example, granite bedrock or deposits from methods of transport such as ice, water, gravity, and wind that eventually form soils.

Soil texture: The percentage of different-sized particles such as sand, silt, and clay that comprise soil. The combination of these soil mineral fractions creates a soil texture class such as silty loam, sandy loam, or clay.

Stratified random: A method of sampling by strata such as land use or land cover. This type of sampling lowers overall variation and allows comparison between different strata, such as between turfgrass lawns and forests.

Synanthropic species: Organisms associated with humans. Synanthropic species are well adapted to habitats altered by people and often move around with people.

determined by a combination of state factors that include climate (cl), organisms (o), parent material (pm), relief (r), and time (t), where the changes in characteristics of any given soil, S, are the function $S = f(cl, o, pm, r, t)$ (box 8.2, equation 8.1). Human modifications of soil necessitate a separate anthropogenic factor because humans, unlike other organisms, are conscious manipulators of soils. Additionally, human effects on soils manifest at broader spatial and temporal scales than those of other organisms, due

BOX 8.2. HUMANS AS FACTORS OF SOIL FORMATION

Soils are considered as a collection of organized bodies with characteristic horizons that develop by pedogenic processes over time. For all soils, the changes in characteristics occurring from pedogenic processes are determined by a combination of state factors that include climate (cl), organisms (o), parent material (pm), relief (r), and time (t), where the changes in characteristics of any given soil, S, are the function $S = f(cl, o, pm, r, t)$. The state factor approach has been adopted to examine relationships among soils modified by human activity and helps to conceptualize the role of human activities in soil formation. Specifically, some soil scientists have proposed that human effects can be incorporated into the factor approach in two ways: (1) humans considered as other living organisms, or alternatively (2) humans considered independently of other organisms. The second approach necessitates the inclusion of a sixth, or anthropogenic factor (a), such that

$$S = f(a, cl, o, pm, r, t). \qquad \text{(Equation 8.1)}$$

When the factor approach was initially proposed in 1941, each factor was conceived to act independently of the others. Later, however, it was recognized that factors of soil formation can also be interdependent, that is, interact with each other, as is the case of soils developing on the same slope but with differing parent materials. In a similar fashion, the anthropogenic factor can act both dependently and independently of the other factors (see figure 8.1). For example, when soil is disturbed during urban development, the human impact on that soil occurred independently of the other soil-forming factors. In this case the temporal scale of the anthropogenic effect is much shorter than the time frame in which most natural pedogenic processes operate. Here, the material constituting the natural soil (**S** in Equation 8.1) predating the "new" disturbed soil (**S2**) is considered the "new" parent material, so that

$$S2 = f(a, cl, o, S, r, t). \qquad \text{(Equation 8.2)}$$

Essentially, there is a new time zero from which soil formation takes place.

Factors of soil formation are interdependent particularly in cases where human effects occur on time scales similar to natural soil formation (usually hundreds to thousands of years). Under such conditions, the anthropogenic factor is more likely to interact with natural soil formation. In this case the soil is not physically disturbed, although various properties may be altered through human-caused changes that interact with soil formation. These types of human effects result in a modified, yet undisturbed soil where the parent material (pm) remains constant, such that

$$S' = f(a, cl, o, pm, r, t). \qquad \text{(Equation 8.3)}$$

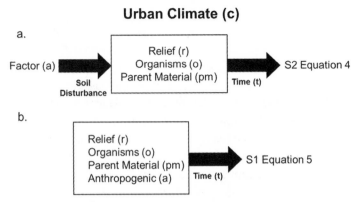

FIGURE 8.1   Schematic representation of two approaches in which to incorporate anthropogenic influences using Jenny's (1941) factor approach of soil formation: (a) human factor, *a*, is considered independently of soil-forming factors, or (b) *a* factor is dependent on other factors.

primarily to agricultural practices, the spread of urban areas, the extraction of natural resources, and the long-range transport of pollutants.

Human influences on soil are complex, with many interactions occurring between human and natural processes of soil formation (box 8.3). The inherent complexity of human interaction with soil development requires an organized representation, or conceptual framework, of the underlying relationships. Such a framework provides a simplified approach with which to explore the various ramifications of human activities, and thus urban land uses, on soil development.

BOX 8.3. ANTHROPOGENIC PROCESSES AND DISTURBANCES OF SOIL

In this chapter, a soil "modification" is any measurable human-, or anthropogenic-, induced change of one or more natural soil chemical, physical, or ecological characteristics. Pedogenic processes are classically defined by pedologists as the soil-forming processes of weathering, organic matter breakdown, translocation, and accumulation. Some soil scientists regard these as "narrow" processes in comparison to "gross" processes of soil formation, such as the podzolization of forest soils, are a combination of narrow processes working over long periods of time. In the case of anthropogenic modifications, gross processes would include cultivation or urbanization, while narrow processes of human-caused pedogenesis, or "metapedogenic" processes, could be divided into separate and distinct processes and events as (1) anthro-pedoturbation or mixing; (2) compaction; (3) addition of chemicals, materials, and organisms; (4) removals of chemicals, materials, and organisms; and (5) additions of energy, mostly in the form of heat (see table 8.1). These narrow processes are unique to human activity—e.g., anthro- is distinguishable from faunal-pedoturbation (for example, gopher holes)—and become apparent when the anthropogenic factor dominates in soil formation, which by definition results in an anthropogenic or urban soil.

The human-caused soil modifications described above may occur over varying time scales, ranging from a "pulse" to a "press" disturbance or event (see table 8.1). An ecological or soil disturbance is an event that leads to a pronounced change in an ecosystem or soil profile. Disturbances often act quickly and with great effect and result in a pronounced alteration of the physical structure of a soil. A pulse disturbance is human caused and leads to an immediate change in the physical structure of a soil in a time frame in which a recovery of that soil to its previous state cannot occur. A soil undergoes a press disturbance when a continuous or chronic human effect results in a modification of soil processes over time scales approximating natural soil formation. Human-caused effects resulting from press disturbances are typically chemical (e.g., calcium depletion due to acid deposition) or biological (e.g., human-caused changes in plant cover or introductions of invasive soil organisms) in nature. In this case there may or may not be an immediate change in measurable soil properties depending on the nature of the response, such as the differences between a threshold and continuous response. Implicit in these definitions is that with pulse disturbances there is at least some time for recovery under conditions in which natural pedogenic processes dominate, as opposed to a soil undergoing a press disturbance, in which recovery is negated until the disturbance has been alleviated.

Conceptual Framework

A conceptual framework incorporates the classic state factor approach and the potential for human alterations of soil formation (figure 8.2). In the framework, the soil-forming factors of *cl, r,* and *o* are the driving variables of soil formation. The framework is sequential, with different time scales represented by relatively long time spans of > 100 yr, short time spans of < 1 yr, and intermediate time spans of > 30 yr and having conditional end points, or sets of measureable characteristics, of either an unmodified, or native, soil (**S**), when human influences are inconsequential, or anthropogenic soils (**S1** and **S2**), when human modifications are dominant.

In the framework, pedogenesis begins when the parent material (***pm***) undergoes natural soil formation, while metapedogenesis begins when humans become a dominant factor in soil genesis (Pathway A). When a soil is altered or material is imported by human activity (Pathways B and C), soil formation starts at a new time zero when the resultant altered or transported material becomes the new parent material (box 8.2, equation 8.2). These new parent materials are referred to as human-altered and human-transported materials, with the latter often enriched in artifacts and other waste materials. In an urban landscape, pedogenic processes work in combination with human-caused press disturbances, which occur at time scales coincident with natural soil formation. Thus within an urban context, the two processes become interdependent. In this scenario, a press disturbance results in a different trajectory for soil formation over time spans of longer or intermediate duration and eventually an anthropogenic soil (**S1**) is established (Pathway D).

To illustrate, when a site is graded to build a structure, soil is disturbed as material is moved from one location to another, and thus the impact on that soil largely occurs independently of the other soil-forming factors. As revealed in the conceptual framework, the temporal scale of the urban alteration is much shorter than the time frame in which most natural pedogenic processes operate (figure 8.1). Thus there is a new time zero from which pedogenesis or metapedogenesis takes place, much in the same way that the till remaining from a retreating glacier is considered parent material or, to use an urban example, transported material used to fill in a low-lying area.

By contrast, factors of soil formation are interdependent particularly in cases where urban modifications occur on time scales similar to soil formation (usually hundreds to thousands of years). Under such conditions,

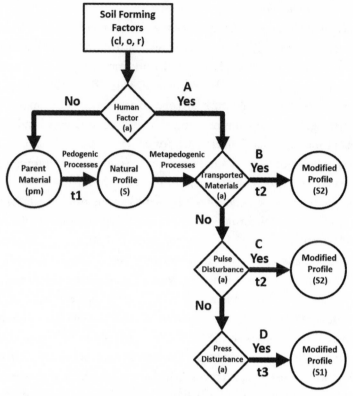

FIGURE 8.2 Conceptual framework modified from Pouyat and Effland (1999) of soil genesis incorporating the anthropogenic factor of soil formation. The soil-forming factors are represented by the symbols **pm** (parent material), **cl** (climate), **r** (relief), **o** (biota), and **a** (humans). Time scales are represented by the symbols **t1** (> 100 yr.), **t2** (< 1 yr.), and **t3** (> 30 < 100). The framework output variables consist of a natural profile (**S**) or modified profiles (**S1** and **S2**) resulting from human-altered (pulse disturbances) and human-transported materials (HAHT) or urban environmental changes (press disturbances).

interactions with other factors are more likely. Here the profile of the soil remains largely intact, although various chemical, ecological, and, to a lesser degree, physical characteristics may be altered through interactions between urban environmental factors and pedogenic processes. Examples of when the anthropogenic factor (*a*) is interdependent with other soil-forming factors include changes in water table depth, changes in soil temperature regimes, and atmospheric deposition of various elements and compounds that typically occur in urban landscapes. Even when a soil in an urban landscape is morphologically similar to a preexisting condition, the soil may be functionally altered, for example, when surface compaction inhibits the ability of an otherwise intact soil to infiltrate water. Under these circumstances, the soil parent material (*pm*) remains constant, but for a soil with an altered function, S' (box 8.2, equation 8.3).

An Approach to the Study of Urban Soils

To investigate the relative importance of individual factors, sequences of soil entities have to be identified in which a single factor varies while the other factors are held constant. For example, in a chronosequence, age *t* is the varying factor or, in a toposequence, *r* is the varying factor. Likewise, sequences can be used to investigate the effects of urban land use on soil characteristics and ecosystem processes. For example, under circumstances where the *a* factor varies over relatively short distances, while the other factors are held constant, an "anthroposequence" would result (box 8.4, equation 8.6). An anthroposequence represents a study design that investigates the effects of the *a* factor on a soil at landscape scales (tens to hundreds of km). Additionally, an anthroposequence is suitable for comparing the effects of urban land use on soils at varying temporal scales. At one extreme, sequences can be identified where the *a* factor acts interdependently with the other factors (box 8.4, equation 8.6), and at the other extreme, sequences can be identified in which the *a* factor operates independently of the other factors (equations 8.4 and 8.5).

Hypotheses

The following hypotheses were derived from the conceptual framework with respect to the spatial and temporal scales of the chemical, physical, and ecological characteristics of urban soils:

1. At local and regional scales, anthropogenic, or urban, factors overwhelm the effect of native soil-forming factors such that the spatial pattern of soil chemical, physical, and ecological characteristics will result from human activities, built environment, and deposition patterns.
2. When anthropogenic, or urban, factors overwhelm native factors, the chemical, physical, and ecological characteristics of urban soils will,

BOX 8.4. THE POTENTIAL FOR "NATURAL EXPERIMENTS" IN URBAN LANDSCAPES

Natural experiments represent an observational study in which scientists can take advantage of existing variation that is present in the landscape but not due to the intervention of the researchers. In the case of urban soils, the urban mosaic represents various natural experiments in which scientists can study anthropogenic or human impacts on soils. The advantage of a natural experiment is that observations can be done on large spatial and temporal scales, allowing scientists the opportunity to study soil responses to human activities without having to conduct the manipulations themselves.

Observations in the urban soil mosaic over relatively wide spatial–temporal scales can be used to form sequences in which a single soil factor varies while the others are kept constant. These sequences can be defined and used for investigation of urban soils:

(1) Where

$$S2 = f(a_n)cl,o,S,r,t \qquad \text{(Equation 8.4)}$$

Here, direct comparisons of different HAHT soils or management regimes associated with urban landscapes (a_1 versus a_2 and so on) are possible between soils developing with parent materials of various origins (figure 8.3b). Additionally, soil formation can be compared among similar HAHT soils along a chronosequence (figure 8.3a), where

$$S2 = f(t_n)a,cl,o,S,r \qquad \text{(Equation 8.5)}$$

In both cases, there is the opportunity to study and control factors (e.g., parent material or management regime) in the early stages of soil development that follows an urban disturbance.

(2) Where

$$S' = f(a_n)cl,o,pm,r,t \qquad \text{(Equation 8.6)}$$

Here, effects on soil formation of various urban environmental factors (a_1 versus a_2 and so on) can be studied over similar soil types, or anthroposequences (figure 8.4), such as forested soils along urban–rural environmental gradients. Such comparisons will be useful in delineating threshold responses of soil properties to press disturbances such as deposition of atmospheric pollutants, or gradual changes in soil properties due to changes in temperature regimes.

at a global scale, be more similar across metropolitan areas than the native soils typical of the region. This prediction conforms to the ecosystem convergence hypothesis and the biotic homogenization hypothesis.

THE URBAN SOIL MOSAIC: A NATURAL EXPERIMENT

As land is converted to urban uses, direct and indirect factors affect soil chemical, physical, and ecological characteristics. Direct effects include pulse and, in some cases, press events typically associated with urban soils, such as those described previously (table 8.1). For most urban landscapes these types of modifications are more pronounced during rather than after the land development process. Urban development of land typically includes clearing of existing vegetation, massive movements of soil, introduction of sealed surfaces, and building of structures. The magnitude of these initial disturbances is dependent on such local site conditions as topography, infrastructure requirements, and other site-limiting factors. On the other hand, press disturbances such as those associated with soil management practices like fertilization and irrigation and those from use (for example, trampling) are also considered direct effects. The spatial pattern of these urban-associated pulse and press events are largely the result of "parcelization," or the subdivision of land by ownership, as landscapes are developed for human settlement. The parcelization of the landscape creates distinct parcels with characteristic pulse and press events caused by humans that will affect soils over time. The net result is a mosaic of soil patches, which will vary in size and configuration dependent on human population density, development patterns, economic status, cultural heritage, and transportation networks, among other factors. As mentioned in the previous section, pulse and, in some cases, press events often lead to new soil parent material from which soil develops.

Indirect effects related to urban land use change involve changes in the abiotic and biotic environment, which can affect native soils. In our conceptual framework these press disturbances work at temporal scales in which natural soil formation processes are also at work. Urban environmental factors include the urban heat island, soil hydrophobicity, introductions of exotic plant and animal species, and atmospheric deposition of pollutants, such as nitrogen (N) and sulfur (S), heavy metals, and potentially toxic organic chemicals.

Table 8.1

Narrow processes of metapedogenesis and their relationship with pulse or press disturbance or events

Chronic is defined as a process that is repeated with regularity; episodic is defined as a process that is repeated, but not with regularity; and a single event is a process that is not likely to be repeated within a one-hundred-year period .

	FREQUENCY OF OCCURRENCE		
PROCESS	PRESS (CHRONIC)	PULSE (EPISODIC)	PULSE (SINGLE)
Mixing			
entire profile (fill)			X
surface of profile (plowing)	X	X	
Compaction			
surface (trampling)	X	X	
subsurface (settling from vibration)	X	X	
crusting (raindrop splash)			
Removals			
mineral (accelerated erosion)		X	
organic (fire loss)		X	X
chemical (accelerated leaching)	X		
water (drainage)	X		X
Additions			
mineral (sedimentation)		X	X
organic (hydrocarbons, black carbon)	X	X	
chemical (heavy metals, fertilizer, atmospheric deposition)	X	X	
water (irrigation)	X		
energy (heat)	X		

The net result is a mosaic of soil-forming conditions that are useful in comparing the effects of urban land use change on soil characteristics and the distribution of soil biota. The spatial heterogeneity of natural soil-forming factors may still underlie and constrain the effects of land use and land cover change. Additional heterogeneity is introduced from variations in human behavior and social structures that function at multiple scales. An example is the variation related to turfgrass maintenance among landowners (chapter 12). The totality of this new heterogeneity, both natural and human caused, is distinct from any relationships between landscape features and soil characteristics that have been developed for nonurban lands. This new heterogeneity, however, can be thought of as a set of natural experiments to investigate the effects of the anthropogenic factor a on soil formation at different spatial and temporal scales in urban landscapes.

The heterogeneity of the urban landscape can be a major driver in soil community assembly. Differences in site history, age, management, disturbance regimes, and the diversity of soil organisms' life histories likely result in spatially isolated populations and high species turnover (beta diversity). In BES the metacommunity concept has been proposed as a useful tool to explain species assembly and coexistence (chapter 11). This framework has produced testable hypotheses to tease out the relative importance of physicochemical and socioeconomic factors shaping urban biotic community composition. By distinguishing self-assembled and facilitated communities as two endpoints along a continuum, this framework explicitly includes human perception and behavior as drivers of urban community composition.

CASE STUDIES

The urban soil mosaic can be observed at various scales, with each scale of observation revealing a suite of soil characteristics that can be attributed to anthropogenic and natural soil-forming factors. In the following sections, we address hypotheses 1 and 2 with case studies that utilize the urban soil mosaic as a suite of natural experiments to investigate the response of soil characteristics to urban land use at multiple scales, for example, identifying sequences as in equations 8.4–8.6. We first address hypothesis 1 with studies that investigated soil responses at finer scales by comparing different categories of land use and cover, each serving as an experimental manipulation of human management activities, disturbances, site histories, and

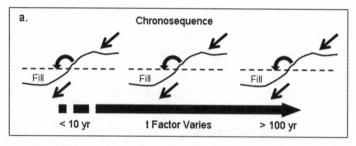

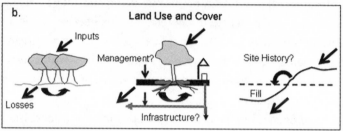

FIGURE 8.3 Schematic representations of (a) a chronosequence, where $S_2 = f(t)$ cl, o, S_1, r, a and factors of soil formation (c, pm, o, r) are held constant while time (t) since a disturbance or other human modification varies, and (b) comparisons of soils under various land use and cover types where the a factor varies as a press or pulse event, while other factors of soil formation remain constant or nearly constant.

time since a press or pulse disturbance or event (figure 8.3). We then present responses of largely undisturbed forest soils to environmental factors along urban-rural gradients at the scale of a metropolitan area (figure 8.4). Finally, to address the generality of these results across other metropolitan areas (hypothesis 2), we compare a subset of soil characteristics at regional and global scales.

City and Neighborhood Scales

Management and other press disturbances can vary significantly in urban soil mosaics, which should result in a corresponding response by the affected soils, although the response will be tempered by other anthropogenic factors and the characteristics of the native soil that remains. Therefore, at the scale of a city, neighborhood, or metropolitan area, we can spatially characterize a suite of urban soil conditions (question 1) and address the relative importance of anthropogenic factors (question 2) to natural soil-forming fac-

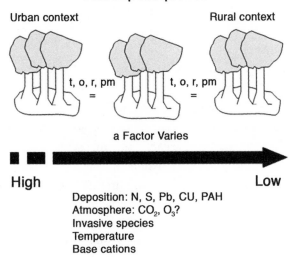

FIGURE 8.4 Schematic representation of an anthroposequence, where $S_1 = f(a_n)$ *cl, o, pm, r, t.* The primary factors
of soil formation (*c, pm, t, o, r*) are held constant while the
anthropogenic factor (*a*) varies among a single soil patch type.
The anthropogenic factor can characterize various urban
environmental influences along direct or indirect gradients,
such as a pollution, urbanization, or socioeconomic gradient.

tors, such as landscape management versus the importance of parent material or topography (that is, factor *a* versus *pm* or *r* in box 8.2, equation 8.2).

City Scale

To investigate the relationship between urban land use and surface (0–10
cm) soil characteristics at the scale of a city, 0.04-ha plots were stratified by
Anderson Level II land use and cover classes. The overall results from this
comparison suggested that measures of individual characteristics varied
widely among land use and cover classes, making it difficult to define a
typical urban soil. Moreover, base cations such as calcium and potassium fell
well within, or actually exceeded, recommended levels important for plant
growth. More interestingly, a subset of the soil characteristics (phosphorus,
or P, potassium, or K, bulk density, and pH) were differentiated by the land
use and cover classes. These differences were greatest between classes
characterized by press disturbances, that is, land management (lawns), and

the absence of press disturbances (forests). In particular, concentrations of P and K, both of which are included in most lawn fertilizers, and bulk density (a measure of soil compaction) differentiated forested plots from the turfgrass-covered plots.

Another subset of soil characteristics exhibited a different pattern. Metals such as lead (Pb), copper (Cu), and zinc (Zn) were not differentiated by the land use and cover classes, but rather were found to be related to major transportation corridors and the age of housing stock. An interesting finding related to question 2 was the continued importance of parent material that differentiated plots by physiographic province (Atlantic Coastal Plain and Piedmont) for soil texture and aluminum (Al), Mg, vanadium (V), manganese (Mn), iron (Fe), and nickel (Ni), which are important constituents of the surface rock types found in the Piedmont province. Thus observations at the city scale showed that land management practices (inferred from land use and cover) and atmospheric deposition patterns (inferred from spatial relationships with emission sources) resulted in a pattern of surface soil responses. Perhaps more interestingly, characteristics of the underlying surface geology (that is, parent material) also contributed to that pattern even in a landscape highly modified by urban development (cf. plant succession, chapter 13).

Neighborhood Scale

To investigate surface soil characteristics at finer scales, we delineated patches with higher categorical resolution at the scale of a neighborhood or subdivision and individual parcel. At a neighborhood or parcel scale, soil responses can be related to individual patches with specific site histories and activities of individual land managers. Two residential neighborhoods in the Baltimore metropolitan area, one medium (suburban) and the other relatively high (urban) in unit density, were mapped using high-resolution landscape categories. The suburban neighborhood was situated just outside the city boundary and had households with a significantly higher mean income than households in the urban neighborhood. As in the citywide analysis, eighty 0.04-ha plots were randomly stratified, but this time by ecotope classes in each neighborhood. Metal concentrations, particularly Pb and Cu, in surface soils (0–10 cm) were found to be higher in the urban than in the suburban neighborhoods. In contrast to the citywide analysis, these metals varied statistically by use and cover (ecotope classes) at a finer resolution, with the highest concentrations occurring in urban vacant and disturbed lots. Moreover, the proportion of plots with concentrations exceeding the EPA's soil Pb screening level of 400 ppm was over 16 percent in the urban neighborhood, while none of the plots in the suburban neighborhood exceeded the EPA standard for Pb.

At the parcel scale, observations of soil Pb in Baltimore City were consistent with the citywide results. Thirty percent of sixty residential properties had average Pb values that exceeded 400 ppm, the EPA screening level, while 53 percent of sites had concentrations greater than the screening level in at least one area of the yard. The highest Pb levels were found near buildings and major road networks. Elevated Pb levels were observed next to buildings regardless of building type, including brick and wood frame structures. As in the citywide results, housing age was an important predictor of soil Pb levels, with none of the structures built after the ban on lead-based paint and leaded gasoline exhibiting soil lead levels above the EPA screening limit. Finally, using the higher resolution data, an empirical model using housing age, distance from road, and distance to a building explained approximately 40 percent of the variation in surface Pb concentrations.

In addition to soil Pb variability at the parcel scale, characteristics of soil taken from planting beds and lawns from fifty residences and an adjacent forest patch in the suburban neighborhood showed interesting differences among these patch types. For example, the forest soil had almost a third more soil organic matter and was more acidic than lawn soils. Conversely, Mg, P, K, and Ca were 47 to 67 percent lower in forest compared to lawn soils even though both soils developed from similar parent materials. There was also a difference between front and back lawns. Front lawns had 26 and 10 percent higher concentrations of Ca and Mg, respectively, and a higher pH than the back lawns. The difference may imply greater application of lawn care products such as lime to sustain a healthier stand of turfgrass in the aesthetic front yard compared to the functional backyard. By contrast, at a coarser scale than this neighborhood, soil N cycling did not differ between front yards and backyards.

Metropolitan Scale and Urban to Rural Comparisons
Studies using urban-rural gradients, or urbanization gradients, suggest that soils of remnant forest patches can be altered by environmental changes occurring along the gradient. In many metropolitan areas the net effects of

multiple urban factors may be analogous to predictions of global environ-mental change, for example, increased temperatures and rising atmospheric concentrations of CO_2 or introductions of invasive soil species, both of which can have profound effects on soil C and N dynamics.

In BES we assessed earthworm communities and soil N transforma-tion rates in urban and rural forest patches of the Baltimore metropolitan area. We expected to find a mixture of native and nonnative species because the region has never been glaciated. We also expected that urban remnant forests would have higher densities of earthworms than in rural forests, as has been the case in other urban versus rural comparisons in the northeast-ern United States.

Both N-mineralization and nitrification rates were higher in the urban than in the rural stands. These results are consistent with the New York City urban-rural gradient study, where exotic earthworm invasion was a domi-nant controller of soil N dynamics. Our results in Baltimore, however, may have been confounded by differences in parent material and soil type along our urban-rural gradient.

Regional/Global Scale

Hypothesis 2, or the urban ecosystem convergence hypothesis, suggests that ecosystem and soil responses to urban land use change will converge across regional and global scales when anthropogenic factors (manage-ment, disturbance, and environmental change) dominate over natural soil-forming factors, for example, topography and parent material, or *r* and *pm* in equation 8.2. The result is that soil characteristics are more similar across urban landscapes at regional and global scales than the native soils they replaced. Similarly, the biotic homogenization hypothesis suggests that urban land use change results in local extinctions of soil fauna and that the movements of humans and products across geographical boundaries help facilitate the establishment and spread of synanthropic soil fauna. Synan-thropic species by definition thrive in urbanized landscapes, resulting in a high degree of similarity in species composition in urban areas across re-gion and continental scales. While ecological theory suggests that commu-nity similarity should decrease with distance in natural habitats, in human-dominated landscapes the greater similarity should result in a less steep distance decay relationship.

Heavy Metals

To investigate the importance of urban environmental factors (*a*) versus parent material (*pm*) in equation 8.2 on soil chemistry, fifteen soil response variables were measured in remnant deciduous forests among urbanization gradients in the Baltimore, New York, and Budapest metropolitan areas. These metropolitan areas differed in population densities, size of area, and transportation systems. In the New York City metropolitan area the forest patches were situated on surface geology of the same type and thus approximated an anthroposequence along a direct urbanization gradient of 0 to 125 km; whereas in Baltimore and Budapest the surface geology differed along a 0 to 30-km gradient and a 0 to 20-km gradient, respectively.

In all three metropolitan areas the forest soils responded to urbanization gradients, although features of each city (spatial pattern of development, forms of transportation, parent material, and site history) influenced the soil chemical response. The changes measured were most likely the result of locally derived atmospheric pollution of Pb and Cu (industrial or vehicle) and to a lesser extent Ca (built structures), which was more extreme at the urban end of the gradient but extended beyond the political boundary of each city. Consistent with the finer scale measurements mentioned previously, the concentrations of Pb and Cu were highly correlated to traffic volume and density of roads, confirming the importance of vehicle emissions with respect to heavy metal pollution. Additionally, in two of the cities (Baltimore and Budapest) the soil chemical response was confounded by differences in parent material found along those urbanization gradients.

Similar results were found in regional analyses of pollution effects on soils in Tallinn, Estonia, and the Piedmont region of northwestern Italy, where parent material composition was important. Thus the comparison of urbanization gradients across three distinct metropolitan areas suggests that characteristics of the native parent material persist in the surface of urban soils. However, at the same time the influence of urban environmental factors (deposition of Pb, Cu, and Ca) resulted in similar soil responses across the three metropolitan areas, which also occurred at distances beyond the boundaries of each city. In a similar analysis using an urbanization gradient in the Baltimore-Washington metropolitan area, but exclusively in residential soils, Pb, Cu, Zn, and arsenic (As) concentrations were one to several fold higher in urban than in rural residential yards.

Soil pH and Organic Matter

As part of the Global Urban Soil Ecological and Educational Network (GLUSEEN) and to test the Urban Ecosystem Convergence Hypothesis, soils in the Baltimore metropolitan region were compared with those of four other cities across four distinct biomes in Europe and Africa. Comparisons were made for total carbon (TC) and nitrogen (TN), available P and K, and pH among reference, remnant, turfgrass, and ruderal, or disturbed, soil types delineated in each of five cities. Specifically, the coefficient of variation (CV) of each soil property was compared among the cities for each habitat type. A higher CV suggests less similarity, while a lower CV suggests greater similarity. Therefore, if a soil property were to have a higher CV for native versus urban habitat type, this represented a convergence, or that urban soil types are more similar across biomes than native soils.

Results of the five-city comparison showed that CVs for soil pH, OC, and TN indeed exhibited a convergence across a continuum represented by the four soil habitat types: CVs ranked in the order of reference > remnant > turfgrass ≥ ruderal types. For soil characteristics highly impacted by management activities or disturbance, such as TC and TN, these results were expected, whereas for soil pH the convergence was the result of the interaction of both anthropogenic (for example, calcium carbonate from building materials) and native (for example, calcareous parent material) soil factors. By contrast, the CV for soil K and P was higher for the ruderal and turfgrass than the reference soil habitat types and thus exhibited a divergence, which was an unexpected result.

Soil Biota

While numerous studies of birds and plants have explored biotic homogenization in urbanized landscapes, limited data are available for invertebrates, and so far those results have showed inconsistent patterns. In the GLUSEEN project a comparison of soil microbial diversity in five cities revealed of convergence of archaeal and fungal communities. Since species introduction is an important component of the homogenization process, taxon level differences are likely. Soil fauna introduction usually happens via soil transportation, and some species survive extended trips better than others. Additionally, success of colonization depends on life history, dispersal, tolerance range, and competitive ability as well as local habitat characteristics and community composition. In Baltimore carrion beetle (Silphidae) assemblages were shown to be determined by habitat quality, and, due to their unique natural history, only native species occur in urban forest fragments. In contrast, cities in temperate regions, including Baltimore, are dominated by eight to ten synanthropic, peregrine earthworm species in urban soils, including urban forest fragments with varying degrees of native species presence. Species in the family Megascolecidae are players in the more recent second wave of earthworm invasion. Conducting global scale studies, resolving taxonomic difficulties, and devising a more standardized approach to quantify biotic homogenization will enable us to compare urban soil communities at large geographical scales.

CONCLUSIONS THUS FAR

Research on soils in BES and comparisons with results from other cities have shown that properties of surface soils and their assemblages of soil organisms can vary widely, making it difficult to define or describe a typical urban soil or urban soil community. The characteristics of urban soils and their biota vary depending on the history of human- and naturally-derived press and pulse disturbances, or events, and the effect of urban environmental factors. Moreover, the importance of natural soil-forming factors, for example, parent material, or *pm* in equation 8.2, continues to influence the spatial distribution of chemical and physical characteristics even when a significant proportion of the land area has been impacted by urban development.

Although urban landscapes are highly altered by development and human activities, urban soils have exhibited a surprising capacity to support plant growth and soil organisms, and as a result they often have relatively high rates of biological activity, such as nitrogen mineralization rates, and richness of species relative to the native systems they have replaced. Given this capacity, urban soils have the potential to provide various ecosystem services to inhabitants of urban areas. However, in the case where urban soils receive frequent, highly intense pulse disturbances or are exposed to multiple press events and urban environmental factors, the capacity of these soils to support a diverse soil community may be greatly reduced, such as in land fill or ruderal soils. Under such conditions, it is vitally important to develop and implement sustainable management practices that enhance ecosystem services of these highly altered urban soils.

A few systematic patterns are emerging from our investigations using natural experiments (box 8.4, equations 8.4–8.6, and figures 8.3 and 8.4) identified in the urban soil mosaic. Human-caused press and pulse events, site history, and environmental patterns all have been shown to impact the spatial response of surface soils and soil organisms at multiple observational scales of the urban landscape. Although pulse and press disturbances or events appear to have a greater effect, environmental factors have the potential for a spatially wider effect, often having influence beyond the political boundary of most urban areas. For example, Pb, Cu, and Zn contamination has been found in many, if not most, urban areas and surrounding periurban areas across the globe. Additionally, anthropogenic and, to a lesser degree, natural soil-forming factors interact and have a significant effect on soils found in urban landscapes. Specifically, soil characteristics such as soil pH, total N and C globally converged across four soil habitat types that varied in their disturbance and management regimes, while other characteristics, such as soil phosphorus and potassium, that are closely associated with parent material found in each biome did not converge.

Needs for future research include continued quantification of the spatial and temporal heterogeneity of urban soils and soil biota at multiple scales and the use of these data to develop predictive models. These models in turn can support concepts used to map and interpret urban soils. Additionally, mapping concepts and their interpretation should necessarily include assessments of the role urban soils play in the provision of ecosystem services in urban landscapes, some of which may be novel. These services should include the role soils play in green infrastructure, urban agriculture, and cultural significance.

CHAPTER NINE

Applying the Watershed Approach to Urban Ecosystems

Peter M. Groffman, Lawrence E. Band, Kenneth T. Belt,
Neil D. Bettez, Aditi Bhaskar, Edward Doheny, Jonathan M.
Duncan, Sujay S. Kaushal, Emma J. Rosi, and Claire Welty

IN BRIEF

- The watershed approach has been central in ecosystem ecology for over fifty years.
- We have used this approach to compare urban with other ecosystem types and to drive iterative cycles of question generation, hypothesis testing, and model development.
- We have learned much about the sources, sinks, and processing of nitrogen, phosphorus, chloride, pathogens, and contaminants such as metals and pharmaceutical compounds in urban watersheds.
- Results are relevant to improving the environmental performance of urban ecosystems.
- Ongoing cycles of watershed-based monitoring, experiments, modeling, and comparative studies can drive further progress in both urban and nonurban ecosystems.

INTRODUCTION

The watershed approach is perhaps the most powerful and useful methodology in the field of ecosystem ecology. The ability to quantify inputs and outputs of water, energy, nutrients, and carbon to hydrologically defined drainage basins has provided a basis for evaluating whole ecosystem function and response to disturbance and environmental change. The watershed approach also has been fundamental to the development of theory about patterns of ecosystem development in time and space and about the regulation and maintenance of ecosystem function via mechanisms of resistance and resilience. Watershed-based approaches have been applied to fundamental studies of forest, grassland, wetland, desert, agricultural, and other ecosystem types.

The watershed approach has been a central component of the Baltimore Ecosystem Study (BES) since its inception in 1997. We hypothesized that this approach would allow us to compare the relatively novel urban, suburban, and exurban ecosystems we were studying with the better-studied but less human-dominated systems in the LTER Network. Key aspects of this comparison included theories related to the ability of ecosystems to influence fluxes of water, energy, carbon, and nutrients by establishing "biotic control over the abiotic environment." These theories were developed in studies of natural ecosystems, such as the forests at the Hubbard Brook Experimental Forest in New Hampshire, where a variety of forest harvest disturbances were imposed and mechanisms of recovery were analyzed. In Baltimore we sought to determine whether these mechanisms would be maintained after forests (the natural vegetation in Baltimore) and agricultural fields were converted to human settlements. Would water and nutrient retention by biogeochemical processes (box 9.1) in these novel systems be similar to freshly disturbed, highly leaky forest ecosystems or to forests dominated by young, actively growing vegetation? Or would the behavior of urban watershed ecosystems not resemble natural ecosystems at all, for example, with minimal biotic control over the abiotic environment, stochastic patterns of biological activity, and very high nutrient losses to surrounding ecosystems?

The watershed approach also provided an opportunity to develop theories to guide analysis of the role of humans as components of ecosystems. Would human fluxes of water and nutrients be independent of or overwhelm natural fluxes? Are there coherent feedback relationships between environmental changes and human actions that function to maintain a quasi "steady-state" in urban watersheds? Finally, how do the built environment and human social processes coevolve with abiotic (geologic, geomorphic, and climatic framework) and biotic systems (plants, mammals, fish, etc.) as components of developing ecosystems?

BOX 9.1. KEY TERMS

Anaerobic: Literally, without oxygen. The opposite of aerobic. Wetland soils are often anaerobic in contrast to upland soils, which are aerobic.

Atmospheric deposition: The movement of substances such as dust, nitrogen, and sulfur from the atmosphere to the land surface as either particles or vapors.

Biogeochemistry: The study of how biological, chemical, and geological factors interact to control the fluxes of energy, water, and matter across the surface of the Earth.

Denitrification: An anaerobic form of respiration in which specialized microorganisms use nitrate as an electron acceptor and produce nitrogen gases as an end product.

Eutrophication: Excessive enrichment of ecosystems with nutrients that can lead to abundant growth and biomass of algae, which can lead to depletion of oxygen levels in water when this biomass dies and decomposes.

Hyporheic: The sediments beneath and adjacent to a stream channel. Also refers to the flow of water through these sediments.

Karst, urban: The complex system of natural and artificial flowpaths that commonly occurs beneath cities. Labeled in analogy to the subterranean crevices and channels eroded in limestone, through which water can flow. Sinkholes are a natural example of karst.

Riparian: Streamside, or the vegetation or habitats adjacent to a stream.

Stable isotope: Naturally occurring, stable (nonradioactive) forms of elements.

Watershed: The area of land that contributes water to a particular point on a stream.

ESTABLISHMENT OF LONG-TERM WATERSHED STUDIES IN BALTIMORE

Baltimore City and County are ideally suited for watershed ecosystem studies. Three principal watersheds span the urban region: the Gwynns Falls, Jones Falls, and Herring Run. These watersheds have all played prominent roles in the commercial, industrial, and residential development of the region and are of critical concern in current environmental issues related to the Chesapeake Bay (chapter 3). The watersheds also serve as a nexus for human–environment interactions, with an active watershed association that addresses environmental quality, quality of life, and environmental justice issues.

Watershed studies in BES have focused on the Gwynns Falls, which has headwaters in suburban Baltimore County, traverses older (1950s) suburban areas, enters the northwest corner of Baltimore City, and drains into Baltimore harbor just south and west of the Inner Harbor. We established four longitudinal long-term sampling stations on this stream (figure 9.1): (1) Glyndon, in the headwaters; (2) Gwynnbrook, approximately 25 percent downstream; (3) Villa Nova, at the midpoint of the watershed; and (4) Carroll Park, near the confluence with the harbor but above the tidal reach. We also established an array of additional sampling stations to provide land use contrasts in similar-sized watersheds, including (1) Pond Branch, a forested reference site; (2) Baisman Run, an exurban site with low-density residential development served by septic systems; (3) McDonogh, an agricultural watershed; and (4) a series of urban watersheds (Dead Run, Gwynns Run, Rognel Heights Storm Sewer Outfall, Maiden Choice Run).

Continuous data on stream stage and stream discharge are collected by the USGS. USGS protocols for data collection and processing are well developed and widely used, allowing for comparison of the data collected at our sites with watersheds across the world. Data from most of the gages utilized by BES are available in real time (for example, http://waterdata.usgs .gov/usa/nwis/uv?01589352). A fortunate aspect of watershed studies in urban areas is that there are multiple entities interested in the hydrologic data sets. There are therefore numerous stream and rain gages in the region and opportunities for cost sharing in the collection of hydrologic data. In Baltimore, BES has shared the cost of stream stage and discharge monitoring with Baltimore City and County and the U.S. EPA, to the benefit of all.

Weekly manual sampling of stream water for water quality analyses began at most sites in fall 1998. Automated samplers have been added at several sites to provide flow-proportional sampling along with the weekly grab sampling. Weekly analyses include nitrate, phosphate, total nitrogen, total phosphorus, chloride, sulfate, turbidity, temperature, dissolved oxygen, and pH. Cations, dissolved organic carbon and nitrogen, *E. coli,* and contaminants such as metals and pharmaceuticals have been analyzed for se-

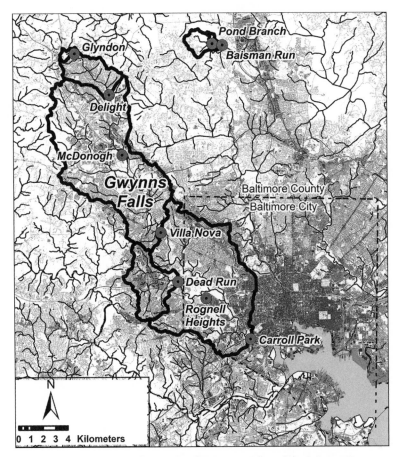

FIGURE 9.1   Stream sampling sites in the Gwynns Falls and the Baisman Run watersheds, providing reference sites for forest and agricultural land uses and for differing degrees and types of urbanization. The Gwynns Falls empties into the Patapsco River just below the Carroll Park sampling station.

lected samples. Data have been made publicly available through the BES website (http://beslter.org) within one year of collection.

THE LONG-TERM SIGNAL IN BALTIMORE WATERSHEDS

As is common in watershed studies, long-term monitoring data provide a signal of watershed ecosystem behavior that drives an iterative process of question generation, hypothesis testing, experimental work, model development, and comparative studies leading to comprehensive understanding

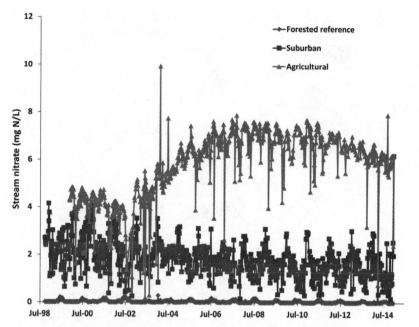

FIGURE 9.2 Nitrate concentrations in streams draining forested, suburban, and agricultural watersheds in the Baltimore metropolitan area sampled weekly from October 1998 through October 2014. (Redrawn from Groffman et al. [2004] with additional years; used by permission)

of a complex system. The signature long-term watershed graph for BES shows nitrate concentrations in streams draining forested reference, suburban, and agricultural watersheds (figure 9.2). Nitrate is important for three reasons. First, it is the most common and mobile form of reactive nitrogen in the environment; second, it is a drinking water pollutant; and, finally, it is a prime agent of eutrophication in coastal waters such as Chesapeake Bay. The long-term estimates of nitrate and total nitrogen export are compared with inputs from the atmosphere, fertilizer, food, and other sources to compute watershed input/output budgets and to calculate watershed nitrogen retention, as described below.

The long-term nitrate signals from the forested and agricultural watersheds are as expected. Forests are known to have conservative, highly retentive nitrogen cycles, leading to low concentrations of nitrate in streams draining forested watersheds. Indeed, the nitrate concentrations in the BES forested reference watershed are quite similar to those observed at the forested Hubbard Brook, Coweeta, Harvard Forest, and Andrews LTER sites. We note that, like some of the Coweeta watersheds in North Carolina, peak nitrate concentrations and loads in our forested reference watershed are in the summer, in contrast to the snowmelt-dominated Hubbard Brook site in New Hampshire.

The nitrate signal from the agricultural watershed is also not a surprise. Agricultural watersheds often have high nitrogen inputs from fertilizer, resulting in high nitrate concentrations in streams. The concentrations of nitrate observed in the BES agricultural stream are similar to those observed in other agricultural streams that have been studied throughout the eastern United States. In this small catchment, concentrations and loads are very sensitive to local agricultural practice, and the large rise in concentrations is probably related to manure application and management.

The long-term nitrate signals in forested reference and agricultural watersheds provide an established context and are a platform for detailed, process-level research for the urban and suburban watersheds that are the focus of BES. Why do suburban watersheds have nitrate concentrations that are clearly higher than concentrations in forested watersheds and lower than agricultural watersheds? What mix of nitrogen sources (for example, fertilizers, atmospheric deposition, human waste, animal waste), sinks (for example, plant uptake, wetland denitrification), and hydrologic processes produce these patterns? Can we develop models capable of depicting these patterns and predicting nitrate concentrations in the future that can then be verified by continued monitoring? As in other LTER sites, long-term watershed-scale monitoring is useful for propelling the progress of our analysis of urban watershed ecosystems.

WHAT HAVE WE LEARNED?

One of the most important results of the long-term watershed studies in BES is that nitrogen retention is much higher than we expected. Based on previous literature on urban watersheds, we expected that watersheds with as little as 10 or 20 percent impervious surface would be so hydrologically and biogeochemically disrupted that the capacity of plants and soils to retain nitrogen would be greatly reduced. We predicted that this disruption would reduce nitrogen retention in the watersheds and that more than 50 percent of the nitrogen that enters the watersheds from the atmosphere and fertilizer would be exported in the stream. Certainly, our urban, suburban,

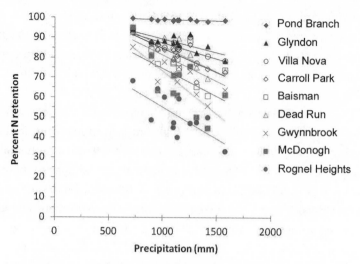

FIGURE 9.3 Watershed nitrogen retention versus precipitation for forested (Pond Branch), suburban (Glyndon, Villa Nova, Baisman, Gwynnbrook), urban (Carroll Park, Dead Run, Rognel Heights), and agricultural (McDonogh) watersheds in the Baltimore metropolitan area from 1998 to 2011. (Reprinted by permission of Springer Nature: *Ecosystems*, "Climate variation overwhelms efforts to reduce nitrogen delivery to coastal waters," by Bettez et al., © 2015)

and agricultural watersheds have lower retention (60–80 percent) than the forested watershed (~95 percent), but these results suggest that vast amounts of anthropogenically-derived N are being processed, stored, and retained in urban, suburban, and exurban watersheds (figure 9.3). This result raises a series of basic science questions about watershed hydrology and biogeochemistry as well as a series of social science questions about human activities that influence the anthropogenic fluxes. In a basic science sense, the high retention suggests there is a significant amount of the "biotic control over the abiotic environment" that has been defined in studies of natural ecosystems, such as the forests at Hubbard Brook and Coweeta. Water and nitrogen retention in these novel urban water ecosystems appears to be strongly influenced by biogeochemical processes and not completely dominated by physical and chemical processes dictated by the engineered components of the urban landscape.

The high nitrogen retention that we observed also has significant implications for management and planning efforts to improve the environmental performance of these watersheds. This performance is of particular interest in watersheds draining into the Chesapeake Bay, where there is an ambitious effort to establish and enforce total maximum daily load (TMDL) regulations for nitrogen.

The Importance of Climate

We observed marked variation in nitrogen retention both within and among watersheds that appears to be driven by precipitation (climate) and land use (urbanization) While retention was consistently high in the forested reference watershed (> 95 percent), retention in urban and agricultural watersheds ranged from 33 to 95 percent between 1999 and 2013. There were marked decreases in retention with increased precipitation and urbanization. If hydroclimatic variability increases with climate change, our results suggest that this will lead to marked increases in the temporal range of nitrate export from these watersheds, with the potential for very high loading rates from urbanized areas during wet years.

What Are the Key Sources of Nitrogen in Urban Watersheds?

The surprisingly high nitrogen retention that we observed in watershed input-output budgets raised questions about the magnitude of key sources and sinks of nitrogen in urban and suburban ecosystems. Deposition from the atmosphere had been shown to be elevated in other urban ecosystems. This was shown to also be the case in Baltimore, with estimates of total (wet and dry) deposition of 9.1, 11.1, 13.3, and 13.6 kg N ha^{-1} yr^{-1} for forested, suburban, urban, and agricultural sites, respectively.

More surprising were results for lawn fertilizer. We suspected that fertilizer inputs to lawns would be high, that retention of this input would be low, and that lawns would be hotspots of nitrate losses to groundwater and to the flux of nitrous oxide—an important greenhouse gas—to the atmosphere. Using detailed social survey and biogeochemical work, we determined that this was not the case (chapter 12). While fertilizer is a significant input, amounting to ~15 kg N ha^{-1} y^{-1} in suburban watersheds as a whole, homeowner surveys showed that only ~50 percent of lawns are fertilized. Biogeochemical data show that retention of added nitrogen is high and that leaching losses and nitrous oxide fluxes are relatively low. These results ultimately led to the surprising conclusion that lawn soils were an important location of nitrogen retention in urban watersheds. This result was interesting in a basic science sense, as our studies produced new insights into interactions between plants, soils, and microbes that produce significant biotic control of the abiotic environment that results in nitrogen retention in urban watershed ecosystems. To highlight the need to consider these controls, we began to refer to lawns as "urban grasslands" to signal to researchers and managers that these areas represent functioning ecosystems that should be evaluated with approaches similar to those taken with the less human-dominated ecosystems within the LTER Network. The high nitrogen retention observed in lawns was also interesting in applied science and management contexts because this retention is tightly linked to a specific input to an ecosystem that is actively managed, with many opportunities for changes to improve environmental performance and reduce nitrogen loads entering aquatic ecosystems.

Since lawns were not a major source of the nitrogen in streams draining urban watersheds, we were curious to know whether atmospheric N, which is usually tightly retained in watersheds, was being rapidly transmitted to streams or if there were other sources of N in urban watersheds. The most likely candidate was sewage. While none of the BES streams have deliberate discharges of wastewater (that is, there are no sewage treatment plants in BES watersheds), sanitary sewer pipes run parallel to many of the streams, and there is abundant visual and olfactory evidence of sewage leaks from these pipes. There is limited on-site sewage disposal via septic systems in the Gwynns Falls, but septic systems are found in the low-density suburban developments in the Baisman Run catchment. Stable isotope analysis of oxygen and nitrogen in stream water nitrate was used to determine whether this nitrate originated in the atmosphere or in sewage, or had been processed by soil microbes. These analyses suggested that sewage and septic effluents were the dominant source of the nitrate in our urban and suburban streams during baseflow. There was evidence of atmospheric nitrate in our most highly urbanized streams, largely during moderate storms, when impervious surfaces are flushed, while nitrate in streams from forest watersheds was dominated by nitrate that has been processed by soil microbes. Although there were clear sewage signals in most of our urban and suburban streams during baseflow, our analysis suggests that atmospheric versus wastewater sources change with hydrologic variability. These results make our nitrogen retention results even more intriguing. If a significant portion of the nitrate in our streams is coming from leaking sanitary sewage pipes and septic systems, then retention of atmospheric and fertilizer nitrogen sources is even higher in urban sites than our input-output budgets imply. As discussed below, sewage inputs to these streams suggest potential delivery to suburban and urban streams of contaminants associated with human waste. We have also demonstrated that a suite of pharmaceutical and personal care products are found in these streams, and these may affect stream ecological function.

What Are the Sinks for Nitrogen in Urban Watersheds?
Once we determined that there were significant sinks for nitrogen in urban and suburban watersheds, we asked where these might be. An immediate focus was on riparian zones, which have been shown to be significant sinks for nitrate in watersheds in many locations. Many studies have shown that this sink function depends on complex interactions between hydrologic, soil, vegetation, and microbial factors whereby nitrate moving in shallow groundwater flow from uplands is removed by plant and microbial processes in the riparian zone. There is particular interest in the ability of riparian zones to support denitrification, an anaerobic microbial process that converts nitrate into nitrogen gas.

We rapidly discovered, however, that urban riparian zones were not functioning like riparian zones in forested and agricultural landscapes. Hydrologic changes associated with increases in impervious surface lead to significant changes in the structure and function of these ecosystems. High runoff volumes and storm peaks from impervious surfaces and drainage infrastructure erode alluvial material and incise stream channels. This incision leads to lower water tables in the adjacent riparian zone, disrupting the complex interactions between hydrologic, soil, vegetation, and microbial factors that underlie riparian nitrate removal. These systematic geomorphic changes associated with urbanization were originally described decades earlier in the Baltimore region (chapters 3 and 13). Our work furthered these geomorphic observations and described how these changes convert riparian zones from sinks to sources of nitrogen in urban watersheds. This conversion is driven by the lowering of riparian water tables that increase the oxygen levels in riparian soils, decreasing denitrification and increasing nitrification, which is an aerobic process that produces nitrate. Trenching of streams to bedrock also limits hyporheic exchange. For example, we showed that the main channel of the lightly urbanized Baisman Run watershed functions largely as a flow-through system under all but the lowest flows, with the exception of the lower parts of the main stem, where alluvial material eroded from the upper watershed has accumulated.

The degradation of urban stream channels has long attracted the attention of municipal authorities. Initial responses to this degradation centered on stabilizing streams by burial and/or fortification of channels. These responses were followed by efforts to reduce the impact of urban storm flows on stream channels by construction of detention basins to hold stormwaters and dissipate their erosive energy. More recently there have been efforts to evaluate the capacity of these basins for nitrogen removal. These efforts have shown that certain types of detention basins support conditions conducive to nitrogen uptake by plants and denitrification. Indeed, some detention basins support higher potential for denitrification than natural riparian zones. There is now interest in altering the designs of these basins to optimize the balance between stormwater detention and denitrification.

Approaches to mitigating urban stormwater continue to evolve. Technologies to stabilize stream channels and provide ecological functions, such as habitat and denitrification, in the face of high flows have developed rapidly in recent decades. While these technologies do not fully "restore" degradation associated with urbanization, they are a vast improvement over stream burial and channelization and have been shown to facilitate increases in denitrification compared to degraded, untreated urban streams. Even more recently, smaller-scale methods, including green infrastructure such as green roofs and rain gardens, have been developed to treat smaller source areas of impervious surface. Evaluation of the effectiveness of such infrastructural ways of controlling stormwater and the nitrogen it carries is just beginning. The BES watershed approach that we have developed provides an opportunity to evaluate these new technological developments as they are implemented in the Baltimore region.

Frontiers in Urban Hydrology

The effects of urban land use change on patterns of runoff have been studied for decades. In Baltimore we collected baseline hydrologic data to establish differences in streamflow and nutrient regimes between watersheds in the BES and other urban watersheds that have been studied. More importantly, we have focused on principal areas of uncertainty that could be addressed with a long-term watershed approach. Key areas of focus have been the development of ecohydrological models that depict the movement

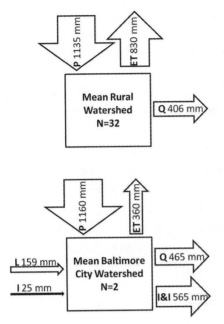

FIGURE 9.4 Comparison of annual water balance components between the mean of thirty-two rural (impervious area < 5 percent and without reservoirs) watersheds in the Baltimore region and two urban watersheds in Baltimore City (Gwynns Run and Moores Run). The width of the arrows corresponds to the magnitude of the flows (in mm/yr). Components were estimated separately (not by subtraction), so inflows do not necessarily equal outflows. P is precipitation, ET is evapotranspiration, L is water supply pipe leakage, I is irrigation, Q is stream discharge, and I&I is infiltration and inflow. (From Bhaskar and Welty [2012], reproduced with permission from *Environmental & Engineering Geoscience*, 18:1, 2012, p. 46)

and processing of water, nitrogen, and carbon across the landscape, evaluation of the importance of groundwater flow patterns and biogeochemical processes in urban watersheds, and development of new conceptual approaches to describe the complexity of urban hydrology.

A major accomplishment was the compilation of water balances of sixty-five gaged watersheds in the Baltimore metropolitan area using hydrologic data from 2001 to 2009. Both natural (precipitation, evapotranspiration, streamflow) and anthropogenic (sewer infiltration and inflow [I&I], lawn irrigation, and water-supply pipe leakage) fluxes were included in the analysis (figure 9.4). Urbanization typically alters the water balance by decreasing evapotranspiration through removal of vegetation. Indeed, our spatially distributed analysis of the urban water budget showed that in the Baltimore region the natural water balance residual, that is, precipitation minus evapotranspiration and streamflow, increases along a rural to urban gradient. However, our analysis showed that the calculated excess water of the natural water balance in urban areas is largely exported by I&I into wastewater collection pipes; for some urban watersheds this excess was greater than gaged annual streamflow. I&I is a large, poorly characterized flux, larger than lawn irrigation and water supply pipe leakage at the regional scale. If the I&I is draining water stores, consisting of non-wastewater effluent, with nitrogen concentrations similar to streamflow levels, this would comprise a significant outflow in addition to stream export and consequently lower our estimates of urban watershed nitrogen retention. This points to a significant research need that directly addresses interactions between natural and engineered drainage systems.

A focus of efforts in ecohydrological modeling has been RHESSys, a distributed ecohydrological model that operates at neighborhood to catchment scales with a computational grain of processes at the sub-parcel level. The model makes use of a hierarchical representation of the landscape, including progressively nested catchments, hillslopes, patches, and strata. Strata can include natural or built land cover patches with multiple understory and overstory species. Patch scale balances of water, carbon, and nitrogen are embedded within two-dimensional surface and subsurface hydrologic flow fields, such that upslope to downslope transport and subsidy of water and nutrients can be represented, including redirection of drainage by infrastructure, such as sewers and curbs. RHESSys allows for evaluation of changes in water, carbon, and nutrient cycles associated with actual or simulated changes in land cover and infrastructure and in individual and institutional parcel scale land management, including vegetation choice, management, irrigation, and fertilization.

A major limitation of RHESSys is that it has simplified two-dimensional shallow groundwater and treats deep groundwater flow as conceptual storage/release, when in fact the urban groundwater system is characterized by three-dimensional dynamics that are important to capture in many situations. A three-dimensional groundwater flow and transport modeling approach is a complex undertaking and requires the ability to incorporate strong heterogeneity in recharge as well as modifications to subsurface hydraulics in pipes, tunnels, fill, and foundations. While increases in impervious surface increase runoff and may decrease groundwater recharge, groundwater is a significant and dynamic component of urban water balances. We have observed marked variation in runoff response to rainfall. For example, analysis of seven rainfall events ranging from 2.3 mm to 55 mm yielded average runoff ranging from 18 to 62 percent in two adjacent urban watersheds in Baltimore City. Contrary to expectations, urban watersheds do not always efficiently convert precipitation into runoff, and a significant portion of precipitation passes through groundwater flowpaths or bypasses the stream by being shunted as I&I through wastewater pipes. Groundwater flowpaths are complicated by a heterogeneous network of surface and subsurface features and infrastructure of various ages and states of degradation that function as a sort of urban "karst."

We have modeled groundwater dynamics using the integrated hydrologic model ParFlow-CLM. ParFlow is a 3D fully coupled, distributed surface-subsurface flow model that has been integrated with the land surface model CLM to include energy processes at the land surface. The programming code is optimized for parallel computing applications so that it can be used for large-scale, high-resolution simulations. Calculation of three-dimensional flow fields enables prediction of, for example, contaminant plumes emanating from septic systems, by coupling the flow model with an appropriate transport algorithm. Work is in progress on applying ParFlow at varying degrees of spatial resolution from small watersheds (1–2 km^2), to the $5,000$-km^2 Baltimore metropolitan region and to the $166,000$-km^2 Chesapeake Bay watershed. In addition, research aims to develop a three-dimensional reactive transport code that can be coupled with ParFlow to model subsurface nitrate transport and transformation.

EMERGING ISSUES IN BALTIMORE WATERSHEDS

Establishing a platform of long-term watershed research focused on nitrogen and phosphorus has attracted interest in evaluating other contaminants of emerging concern in the BES watersheds. These include (1) elevated chloride concentrations that could be toxic to freshwater life and most land plants, (2) the presence of pharmaceuticals and personal care products that could have effects on ecological structure and function of urban aquatic ecosystems, and (3) elevated bacteria concentrations that could present threats to human health. Although interest in these contaminants emerged from practical concerns and empirical observations, they have been a productive platform for more conceptual and theoretical analyses of multifactor interactions and resilience and ecosystems.

Salinization

Routine measurements of chloride in weekly stream samples revealed several surprising patterns in chloride contamination of urban waters. The first surprise was that chloride concentrations were very high, approaching concentrations of concern for aquatic ecosystem integrity, biogeochemical functioning, and drinking water quality (figure 9.5). The second surprise was that concentrations were high in baseflow year-round. While we expected concentrations to be high in winter storm flow due to routine municipal application of road salt for snow events, the concentrations in baseflow year-round were higher than expected. These results further illustrate the importance of groundwater in our watersheds as the consistently high baseflow chloride concentrations are likely driven by recharge of groundwater from winter road runoff containing high chloride concentrations. This saline recharge contributes to baseflow during summer. Finally, we were surprised that watersheds with very low levels of development (for example, 5 percent impervious surface) had significantly elevated chloride concentrations in baseflow, although these areas typically use groundwater supply and septic systems, and increased chloride may be added in water softeners and effluent.

In addition to elevated chloride, sodium concentrations have increased in the Baltimore drinking water supply. This becomes particularly impor-

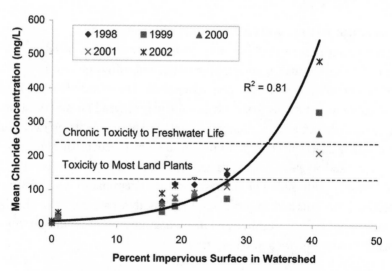

FIGURE 9.5 Relationship between impervious surface and mean annual concentration of chloride in Baltimore streams from 1998 to 2002. Dashed lines indicate thresholds for damage to some land plants and for chronic toxicity to sensitive freshwater life. (From Kaushal et al. [2005], © 2005 National Academy of Sciences; used with permission)

tant in watersheds that drain to water-supply reservoirs, suggesting a long-term concern about dissolved sodium chloride in drinking water supplies in the region. Sodium and chloride do not always follow the same pattern across BES watersheds, suggesting lingering mysteries regarding geochemical retention and sources of salinization. Furthermore, human-accelerated chemical weathering may also further exacerbate regional salinization.

Pharmaceuticals and Personal Care Products

A critical development in environmental science over the past twenty years has been recognition that thousands of anthropogenic compounds such as hormones, chemotherapy drugs, antihistamines, stimulants, antimicrobials, and cosmetic additives are present in aquatic ecosystems. An important challenge is to evaluate the effect of these compounds on the ecological structure and function of aquatic ecosystems. Watershed studies are an ideal platform for these evaluations. Studies with biofilms in streams in Baltimore and elsewhere showed significant suppression of algal production and biofilm respiration by caffeine, ciprofloxacin, and diphenhydramine and effects of diphenhydramine on bacterial community structure. Because we know that raw sewage is discharged or leaks to some BES streams, our suburban and urban streams are an excellent locale for studying the consequence of a suite of contaminants of emerging concern on ecosystem functioning. In addition, the watershed approach is an ideal platform for exploring the influence of human demographics and infrastructure degradation on concentrations of pharmaceuticals and personal care products in stream ecosystems. There is a strong need to determine whether these ubiquitous pharmaceuticals have widespread effects on urban aquatic ecosystem structure and function.

Bacteria

Fecal contamination is a common problem in urban watersheds, and Baltimore is no exception. Bacteria present in water samples taken on a weekly basis from June 2004 through June 2005 from three BES streams included twenty-six genera and seventy-eight species. In the highly contaminated Gwynns Run, *E. coli* was the most common species encountered, followed by the genera *Klebsiella* and *Aeromonas*. This was in marked contrast to the BES forested reference stream (Pond Branch), where the most common genus was *Serratia*, followed by *Yersinia* and *Aeromonas*. In the BES agricultural stream (McDonogh), *E. coli* was the most frequently isolated species, followed by the genera *Aeromonas* and *Enterobacter*. More detailed surveys using polymerase chain reaction methods found evidence of the presence of pathogenic *E. coli* in 53 percent of samples, with higher levels in urban streams compared to suburban and forested watershed streams. These results suggest that pathogenic *E. coli* are widely present in urban stream habitats and could represent an emerging health concern in the metropolitan Baltimore area.

In April 2002 the City of Baltimore reached a consent decree agreement with the Department of Justice, the EPA, and the state of Maryland to upgrade its sanitary sewer infrastructure to bring the city into compliance with the Clean Water Act. The improvements will cost approximately $940 million over fourteen years and are intended to end chronic discharges of raw sewage into local waterways through leakage and surcharging. Significant work has been carried out to reduce I&I, which can result in pressurized sanitary sewer flow and surcharges during storms. Improvements in sanitary sewer infrastructure have resulted in marked improvements in some streams. For example, repairing a major leaking sewer pipe resulted in rapid and marked decline in *E. coli* levels in Gwynns Run. A major question is whether the consent decree activities will result in widespread improvements in the bacterial quality of stream water in the metropolitan Baltimore area and whether there is a legacy effect of prior sewage releases on these ecosystems.

FUTURE PROSPECTS FOR WATERSHED-BASED RESEARCH IN BES

We expect that the watershed approach will be an enduring component of BES. In the urban setting this approach is made more complicated by the addition of engineered drainage systems in the form of sanitary and stormwater sewers, which inadvertently can act as significant output across watershed boundaries and a vector for moving human waste and associated contaminants to surface waters. However, infrastructure is central to the interaction of human/environmental systems within these ecosystems and is significantly affected and manipulated by construction and maintenance activities as well as by aging and degradation.

The watershed approach is fundamental to our ability to compare our urban ecosystems with the more natural ecosystems that dominate the LTER Network. Further, it allows us to quantify integrated ecosystem function, provides a platform for detailed question generation and hypothesis testing about relationships between ecosystem structure and function, and produces results that are useful to education, public understanding, and policy generation.

In addition to ongoing monitoring of nitrogen and phosphorus, we hope to expand preliminary studies on other constituents, for example, PPCPs, mercury (and other metals), and dissolved organic carbon. The watershed platform that we have established, with a strong USGS hydrologic backbone and ongoing sampling, will hopefully continue to attract researchers interested in a wide range of urban watershed issues.

The watershed approach should also be an important tool for evaluating the dynamic nature of urban ecosystems. We have already seen dynamic watershed responses to climate and will continue to track these into the future. A major challenge will be our ability to sample and evaluate extreme events such as tropical storms and drought, which can strongly influence the long-term performance of ecosystems.

Results from our watershed work should continue to be useful for policy makers in the region. There is great interest in understanding the nutrient performance of urban watersheds in the context of TMDL regulations and improvements mandated by the consent decree. Continued interaction between BES scientists, managers, and policy makers will help to ensure that our scientific results are useful in these efforts.

The watershed approach also has potential to improve environmental education, public understanding, and socioecological revitalization in urban ecosystems. Watershed concepts are fundamental to environmental science curricula in Maryland and elsewhere. Examples from our research have been, and will continue to be, incorporated into education materials produced by BES educators and collaborators. The watershed approach has long been central to efforts to ecologically revitalize underserved neighborhoods via stream cleanups and the activities of watershed associations. We hope to build on existing efforts to explore relationships between ecological and socioeconomic revitalization with a focus on a set of Baltimore watersheds with strong contrasts in land use, age, and infrastructure. This work will help us to develop the capacity to use the watershed approach as a nexus of science, education, management, and public engagement that will contribute to our basic science understanding of urban ecosystems and to the quality of life of the human residents of these ecosystems.

CHAPTER TEN

Urban Influences on the Atmospheres of Baltimore, Maryland, and Phoenix, Arizona

Gordon M. Heisler, Anthony Brazel, and Mary L. Cadenasso

IN BRIEF

- Using remote sensing, meteorological measurement, and modeling methods with different sources of climate data, research in the Long-Term Ecological Research (LTER) sites of Baltimore and Phoenix examined urban influences on spatial and temporal variation of air and surface temperature, total solar radiation, ultraviolet radiation, CO_2 transfer, and energy balances.
- Because the climate of Phoenix is generally hot and dry whereas that of Baltimore is relatively temperate and moist, climax vegetation around these cities differs dramatically, providing contrasts for urban heat island effects.
- Although urbanization tends to make cities in contrasting general climates more similar in physical structure and microclimate, the urban climates of Baltimore and Phoenix remain so different that the patterns of the urban heat islands differ significantly.
- The differences in general climate between Phoenix and Baltimore create a need for consideration of different strategies, including selection of plant species and use of irrigation to alleviate high temperatures in the two cities.
- Computer programs can aid in evaluating the complex of site factors

that influence air temperature, solar and thermal radiation, wind, and humidity that affect human comfort and health in urban areas.

INTRODUCTION

Baltimore, Maryland, the setting for the Baltimore Ecosystem Study LTER (BES), and Phoenix, Arizona, the focus area of the Central Arizona–Phoenix LTER (CAP), are within climate regions that could not be more different. The synoptic-scale (box 10.1)—that is, general—climates result in quite different climax vegetation in natural areas around these cities.

The process of urbanization tends to homogenize urban developed and vegetation structure among cities across the country, such that different cities have more or less similar buildings, transportation facilities, and land use patterns as well as features of managed vegetation structure, regardless of the general climate regimes. Indeed, despite the large differences in climate and their natural vegetation biomes, Baltimore and Phoenix have developed to be in many respects similar in structure. Developed- and vegetation-structural differences today may result more from historical and cultural factors than from climatic factors. However, the divergences in general climates do lead to differences in ecological function, such as contrasts in urban influences on air and surface temperatures. Furthermore, the general climate differences have led to significantly varied degrees of concern among the two cities' populations about such climate elements as temperature, solar radiation, and precipitation. The climate differences have also led to differences in social adaptations to climate, for example, the use of vegetation to ameliorate high temperatures.

The two urban LTER sites are excellent laboratories in which to integrate ideas on climate of urban areas and how climate change and local variability of climate relate to social, political, economic, and ecological processes over a long time period. The application of climate theory in such contrasting cities is facilitated by the LTER projects. For example, in both projects, one research goal was to develop and test methods to quantify the patterns, both in time and space, of urban influences on energy balances, air and surface temperatures, and solar irradiance. This is an important goal because, given the patterns and magnitude of the influences, decisions can be made on where effort and resources should be applied to remediate adverse urban influences on climate.

BOX 10.1. KEY TERMS

Climax vegetation: A plant community that, by the process of natural succession, has reached a relatively steady state.

Degree days: As used here, degree days were derived from heating and cooling degree-hours, which were the differences between the temperature for an hour (T) and 18 °C. For T > 18 °C, the difference contributed to cooling degree-hours, and if T < 18 °C, the difference contributed to heating degree-hours. Degree-hours were summed over each day to compute Heating Degree Days (HDD) and Cooling Degree Days (CDD).

Eddy correlation: A method of measuring the vertical transfer of heat, mass (water or other gas such as CO_2), or momentum (change in wind speed) with rapid response instruments that in forested or urban areas are typically deployed on a tall tower.

Energy balance: The algebraic sum of energy flows to and from a surface or into and out of a body. The energy flows can be in the form of thermal or solar radiation, conduction or convection of sensible or latent heat.

ETM+: Enhanced Thematic Mapper Plus, a sensor carried on the Landsat 7 satelite. The ETM+ sensor scans the Earth surface in eight bands of wavelength, including a 15-m resolution panchromatic band and a 60-m resolution thermal band that measures temperature of the Earth surface.

Latent heat flux: Transfer of heat to the atmosphere that involves change of phase of water (evaporation or condensation) at the earth surface.

LCZ: Local climate zone, a scheme to classify urban and rural areas at the scale of neighborhoods based on surface cover, urban structure, construction material, and human activity (Oke et al. 2017).

LiDAR: Abbreviation for light detection and ranging, a method of collecting land cover data at 1-m^2 resolution with a sensor carried by an aircraft.

Local scale: An area of climate influence with horizontal dimensions in the range of 100 m to 50 km (Oke 1987). The term is commonly used in urban climate literature for areas with dimensions of 1 to 5 km.

LST: Land Surface Temperature, the temperature of the soil or tops of buildings or vegetation as seen from above, often by sensors on satellites.

Mesoscale: An area of climate influence with horizontal dimensions of 10 to 200 km. Mesoscale computer simulation generally applies to smaller areas and shorter times than those used in general weather forecasting.

Microscale: An area of climate influence with horizontal dimensions in the range of 10^{-2} m to 1 km (Oke 1987), perhaps used most commonly for an area about the size of a typical house lot.

NDVI: Normalized Difference Vegetation Index, which shows the pattern of green vegetation on the earth by analyzing earth radiance as measured by multispectral scanners on satellite. NDVI is proportional to the difference of the red and infrared radiances divided by their sum.

SAVI: The Soil-Adjusted Vegetation Index is similar to NDVI, but it includes a correction factor (L) for the proportion of bare soil that is present in the area. If L = 0, no bare soil is present and SAVI = NDVI.

Sensible heat flux: Energy transfer to the atmosphere that changes air temperature. Contrast with latent heat flux. It is an important component of the Earth's surface energy budget.

Space conditioning: Heating and cooling systems of buildings.

Structure (urban): The material characteristics, shape, and dimensions of both human-built (buildings, transportation corridors, etc.) and vegetative objects (trees, grass, etc.) in urban areas.

Surface Urban Heat Island (SUHI): Warmer areas of LST surrounded by relatively cool LST.

Synoptic-scale: The large area of weather and climate characterization commonly used in weather prediction.

Thermal admittance, μ: The ability of a material to absorb and release heat with a given difference in temperature between the surface and the air. Wet soil and granite have very high μ, dry soil has low μ.

Urban Heat Island (UHI): The phenomenon in which cities are generally warmer than adjacent rural areas. Cities are thus often, in map view, islands of higher temperatures surrounded by cooler landscapes.

Urban Heat Island (UHI) intensity: An approximation of the maximum urban minus the minimum rural temperature.

Xeriscaping: The practice of landscaping using plants that require little water, especially in dry seasons. The term is derived from combining the scientific term for labeling plants that are adapted to dry or arid conditions—xerophytes.

This chapter summarizes the research related specifically to urban climate within the BES and CAP programs, particularly at the local and micro scales. The methods, hypotheses, and focus of these climate research programs varied because of differences in climate concerns in the two regions as well as because of differences in the institutions and personnel involved. Each program has a robust body of research focused on aspects of urban climate within its focal city. However, just as cities tend to be homogenized, so too do research methods and objectives within a discipline, especially given the ease of communication and interconnectedness in today's world, and also when there is institutional encouragement as in the LTER Network. Therefore, collaborative efforts comparing the two cities have also been accomplished.

In our collaborative work comparing our two cities, we hypothesized that differences in general climate and resulting vegetation, in addition to the human adaptation to climate, would affect urban air and surface temperature patterns, commonly known as the urban heat island (UHI) effect, and that these differences in temperature patterns between cities would be both spatial and temporal. Several UHI effects have been defined. For example, temperature differences within the air from ground level to the average height of the tops of buildings or trees, whichever are dominant, are urban canopy layer (UCL) UHIs. Urban heat islands may also be described by the temperatures of the upper surfaces of buildings, trees, streets, lawns, and so forth, as seen from above, usually in images taken from satellites or aircraft. This surface temperature pattern is sometimes called the urban skin temperature or land surface temperature (LST). We further hypothesized that the UHI patterns could be best quantified and described using combinations of remotely sensed data, on-the-ground observations, empirical and physically based models, and GIS analysis and mapping. In 1997 these observational and analytical tools and newly available data access had not been fully exploited to quantify UHI spatial and temporal patterns, even though the existence of the UHI phenomenon in cities around the world had been known for many years before the start of the BES and CAP programs. Quantification increases the possibility of using tree and water management along with urban design and planning to reduce UHIs that imperil human health and lead to environmental concerns such as high CO_2 emissions.

Contrasts in General Climate

The distinct general climates of the two urban LTER sites have been characterized in a variety of ways. The twenty-four LTER sites span the range of temperature and precipitation found on Earth for the thirty-year period from 1961 to 1990 (figure 10.1). The city of Phoenix (CAP), with about 200 mm of precipitation per year, was naturally drier than all but the Arctic Tundra site (ARC) in Alaska. With an average temperature of about 22° C, the CAP site was warmer than all but the subtropical Florida Coastal Everglades (FCE) and the tropical Luquillo Experimental Forest (LUQ) in Puerto Rico. Temperate-climate Baltimore was near the middle of the distribution, with about 1,000 mm of precipitation and an annual average temperature of about 13° C.

A system of synoptic weather type classification, or Spatial Synoptic Classification (SSC), is useful in comparing Baltimore and Phoenix. The SSC is based on categorizing the weather variables air temperature, dew point, wind, cloud cover, and air pressure for each day of a record of at least twenty years, at individual weather stations. The SSC defines eight weather types, and for each station it summarizes the frequency of each type for all

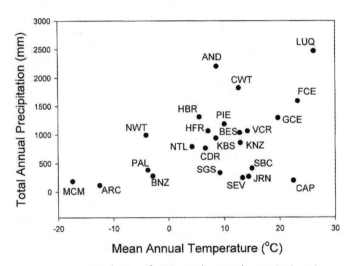

FIGURE 10.1 Distribution of LTER sites by annual mean (1961–90) temperature and precipitation, CAP in lower right, BES near center of cluster. (Reproduced from Greenland et al., 2003b, with permission of American Institute of Biological Sciences via Copyright Clearance Center)

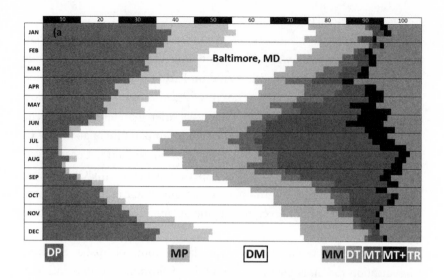

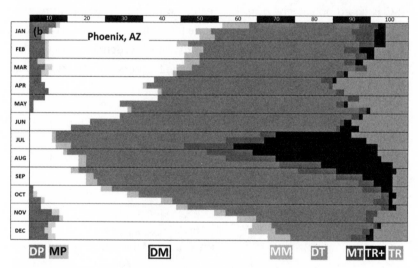

FIGURE 10.2 Synoptic weather type classification system of Sheridan (2002) for Baltimore, 1945–2017 (a), and Phoenix, 1948–2017 (b). Key to weather types at bottom of each section: DP, Dry Polar; MP, Moist Polar; DM, Dry Moderate; MM, Moist Moderate; DT, Dry Tropical; MT, Moist Tropical; MT+, Moist Tropical Plus; TR, Transition. Horizontal bars show average percentage of time in each ten-day period by weather type. (Figure provided by Scott Sheridan, using data from his website, http://sheridan .geog.kent.edu/ssc.html)

days of the record. In this scheme, Baltimore, represented by data from the Baltimore-Washington airport, has a considerable mix of dry polar, dry moderate, moist moderate, and moist tropical conditions, with the latter two dominating in summer (figure 10.2a). When the weather type moist tropical (MT) is especially extreme, the SSC system uses the name moist tropical plus (MT+). The CAP area is representative of a Sonoran desert climate of the southwest United States; dry moderate (DM) and dry tropical (DT), typical of desert climates, are the most dominant weather types, with MT+ during July, August, and September (figure 10.2b), locally known as the Arizona monsoon season. In many cities MT and especially MT+ are associated with unusually high human mortality owing to heat stress, while in other cities, including Phoenix, heat mortality is most associated with extreme DT.

Contrasts in Topography

Topography can have a major effect on local- and micro-scale climates. Important topographic features of the Baltimore region are the presence of the Chesapeake Bay and the transition from Coastal Plain to Piedmont Plateau at the fall line within the city (figure 10.3). The bay and tributary rivers have the potential to affect air temperature via the sea or bay breeze effect. The complex dissected topography of the Piedmont Plateau leads to small-scale, down-slope cold air drainage into the valleys. Such cold air flow is a major feature of air temperature patterns on clear-sky nights. The higher elevation of the Piedmont Plateau leads to generally cooler temperatures than on the Coastal Plain, but the dendritic drainage pattern probably diminishes large-scale nocturnal flow off the plateau on clear nights.

The primary topographic feature of the city of Phoenix is its location in a broad valley oriented generally east to west with mountain remnants to the north, east, and south (figure 10.4). These mountain remnants are composed of metamorphosed sedimentary and volcanic rocks or of intrusive granitic rocks. The Valley of the Sun is an extensive plain of sedimentary deposits derived from the catchments of the Salt, Verde, and Gila Rivers, which converge near Phoenix. The city is of relatively low relief, with elevation ranging from 890 to 1,099 feet (274–335 m).

RESEARCH INSIGHTS

Our LTER research in BES and CAP has produced insights on urban climate, primarily urban influences on air temperature, but also urban surface

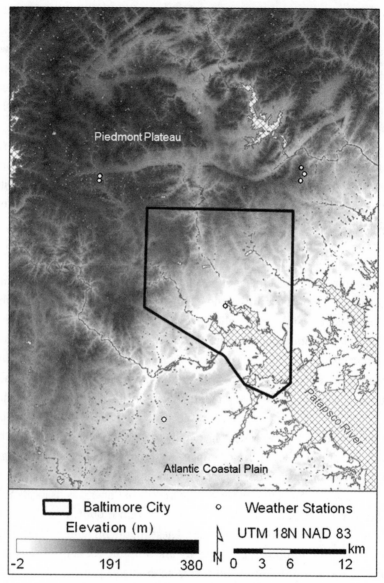

FIGURE 10.3 Elevation across the Baltimore region. (Reprinted by permission from Springer Nature: *Theoretical and Applied Climatology*, Heisler et al., © 2015)

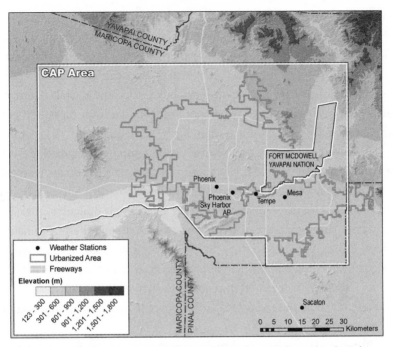

FIGURE 10.4 Phoenix, Arizona, and CAP LTER study region located in the Salt and Gila River Valleys. Note abrupt elevation increase to east and north of the city. (From Brazel et al. 2000, © Inter-Research 2000)

air temperatures, solar radiation incidence as modified by urban atmosphere and trees, and energy and mass transfer to and from urban landscapes. The goals of this research were to contribute to the body of theoretical knowledge on urban influences so that results might have direct applications for cities beyond Baltimore and Phoenix and to develop research methods that might be used for research in other cities.

Urban Air Temperatures

Current techniques to quantify urban influence on air temperatures, including our urban LTER research, generally begin with air temperature measurements, usually at a height above ground of 1.5 m (5 ft) or 2.0 m (6.3 ft). The measurements serve various purposes: to contrast temperatures between urban and rural areas or within urban areas with differing vegetative and built structure. A variety of air temperature measurement systems may be used. Fixed-location temperature sensors may be part of a national network of weather stations (the National Weather Service in the United States), a state or county network for agricultural or flood management purposes, or a measurement system established for a particular urban climate study. Mobile measurements with sensors carried by automobiles, bicycles, or even by people walking may be used to capture spatial temperature patterns over short time periods. Information from measurements may be extended spatially and temporally by modeling.

These air temperature data sources possess various strengths and weaknesses. Surface meteorological stations offer information on air temperature changes over time at discrete sites in the urban area but usually lack dense spatial coverage. Air temperature is surprisingly difficult to measure accurately, largely because of errors caused by solar and thermal radiation effects on temperature sensors. Sufficient accuracy may be achieved by adequate shielding and ventilation of sensors, and, in any case, sensor accuracy should be considered in reporting urban climate study results. All of these methods must be accompanied by some means of quantifying the urban structure.

In relating land cover to climate at various scales, the first task is usually to characterize the structure. Our urban LTERs have used the full range of available earth-surface structural data and analysis tools—digital elevation models, LiDAR data from aircraft, land use data from the United States Geological Survey), and digital aerial photographs. The list also includes an array of satellite products: Landsat thermal infrared data, Landsat TM SWIR, the Thermal Infrared Multispectral Scanner, the Enhanced Thematic Mapper Plus (ETM+), IKONOS satellite images, and the Advanced Spaceborne Thermal Emission and Reflection Radiometer (ASTER).

These land cover data sources provided the metrics for a range of indices to describe and characterize the spatial heterogeneity of the urban system that may influence urban climate. For example, we derived the pattern of impervious cover across the Baltimore region from LiDAR data (figure 10.5). Other indices included NDVI, a Soil-Adjusted Vegetation Index (SAVI), a land cover classification developed for urban systems called HERCULES, which stands for High Ecological Resolution Classification for Urban Landscapes and Environmental Systems, and Local Climate Zones (LCZs). Similar to the LCZ system, a CAP study divided Phoenix property subdivisions into 5 development zones (DZs): core, infill, agricultural fringe, desert fringe, and exurban.

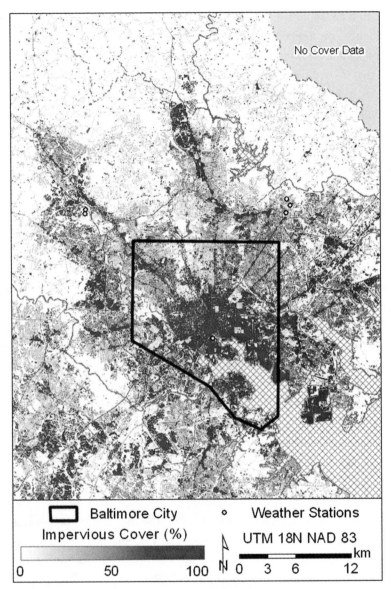

No Cover Data

Baltimore City   ○ Weather Stations

Impervious Cover (%)

UTM 18N NAD 83

km

0   50   100   N   0   3   6   12

FIGURE 10.5 Impervious cover in the Baltimore region derived from high-resolution *LiDAR* (box 10.1). Seven weather stations used in analysis by Heisler et al. (2016) indicated by small circles. (Reprinted by permission from Springer Nature: *Theoretical and Applied Climatology*, Heisler et al., © 2015)

In a collaborative BES and CAP study we examined trends in urban influences on temperature over most of the twentieth century using data from standard weather stations (figure 10.6). In comparing UHI trends we see that toward the end of the twentieth century Baltimore had a UHI intensity in July averaging about 1.5° to 2° C at midday and about 4.5° to 5° C at night. In Phoenix, May is the month most favorable to UHI formation, but a daytime "cool island" of about 1° C is likely to be present, while nighttime UHIs, averaging about 2° C in 1900, increased with city size to about 6° or 7° C in 1995. We hypothesized that the cool island resulted from greater use of energy to evaporate water from irrigated vegetation in the city compared to evaporation rates from sparsely vegetated and unirrigated desert outside the city. A different explanation, arrived at by mesoscale modeling, is that the Phoenix daytime cool island is a function of the rural desert landscape, which heats rapidly during the day. The cool island disappeared when the model was run assuming temperate vegetation in place of the rural desert. This illustrates the fact that UHI intensity, defined as urban temperature minus rural temperature, is always in part a function of the rural landscape, but it does not negate the finding that irrigation plays an important role in urban influence on air temperature in generally dry climates.

We experience uncertainty about irrigation effects on air temperatures because irrigation affects temperature in several ways. Irrigation or rain increases soil moisture, which increases thermal admittance, reducing the rate of heating or cooling at soil surfaces. Thus during the night, bare, wet soil may be equivalent to asphalt in having a slow rate of surface cooling. However, during a sunny day the wet soil will become relatively cool because the surface is slow to warm and because evaporation takes place at the surface. Further, in desert climates like Phoenix, vegetation is sparse without irrigation. During the day foliage itself keeps cool by evapotranspiration, and vegetation provides shade to high-thermal-admittance ground cover. At night vegetation cools rapidly. Thus irrigated areas in Phoenix are usually relatively cool compared to unirrigated areas, depending in part on the portion of ground covered by the irrigated vegetation.

Change in the extent of metropolitan urbanization is relevant to assessments of UHI in both cities. The rapid increase in population and size of the developed area of Phoenix has increased both the intensity of the nighttime UHI and the spatial extent of UHI influence. Over the course of the twentieth century the UHI change in the older City of Baltimore was smaller. The population of Baltimore City proper increased from about 510,000 in 1900 to nearly 1 million about 1950 and has since declined to roughly 630,000. However, urban sprawl has dispersed the human impact over a large area outside the city, so that the total metropolitan population has increased to about 2,700,000, a number not that different from the population of metropolitan Phoenix.

When contrasting urban to rural temperatures over long time periods, the stability of weather station locations and surrounding locations is a concern. Although the immediate vicinity of Baltimore's rural reference weather station in Woodstock, Maryland, remains little changed to this day, the leveling off about 1970 of the apparent annual increase of the nighttime UHI in Baltimore may be partly owing to suburban development beginning about 2.5 km from Woodstock. The Woodstock site, about 13 km west of Baltimore city limits, was discontinued in 1999. That year the downtown Baltimore site was also changed by a move of about a kilometer to a quite different micro-site on the edge of Baltimore's Inner Harbor. The rural reference for Phoenix was sufficiently distant from the city to remain distinctly rural through 1995, although in the last two decades development has caused temperature increases, so the station would not be a good rural reference for a future study. The inclusion of three other stations near Baltimore and three others in and near Phoenix bolsters the theory that urban development influences air temperature over a long time period.

Most of the world's research on urban climate has concentrated on warm-season UHIs. An LTER study of fifty-year temperature trends in Phoenix examined trends in heating and cooling degree days (HDDs and CDDs). As the city has warmed, the energy to heat buildings as indicated by the HDD index has gone down. However, the increase in the CDD index has been greater, so the total index for space conditioning of buildings has increased.

Research in CAP used a range of physically based modeling methods, whereas in BES our modeling methods were largely empirical (box 10.2).

Maximum urban–rural temperatures will often not be represented by weather stations within existing networks. In our study of UHIs over the course of the twentieth century the maximum UHI intensities of 6° to 7° C for Phoenix and 4.5° to 5° C for Baltimore represent averages over the months of May (for Phoenix) and July (for Baltimore) of daily differences

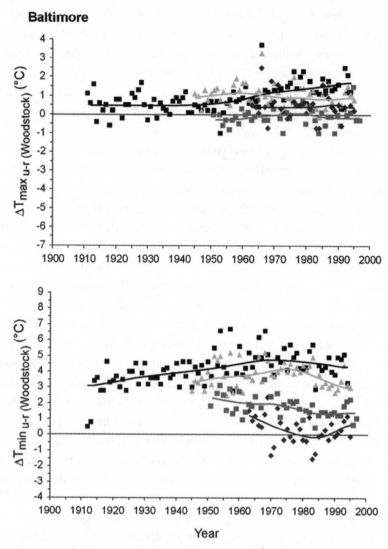

Baltimore

FIGURE 10.6 (a) Long-term July monthly averages of maximum daily urban temperature minus corresponding rural temperatures for stations in and near Baltimore (*top*) and monthly averages of minimum daily urban temperatures minus corresponding rural temperatures (*bottom*), and (b) the same for May temperatures for the Phoenix region. (From Brazel et al. 2000, © Inter-Research 2000). The rural references were at Woodstock, about thirteen miles west of Baltimore, and Sacaton, in the desert south of Phoenix. Trends over the twentieth century are shown by the data for the Baltimore WSO (Weather Service Office, panel a, black square) and the city of Phoenix station (panel b, black square). Three weather stations in or

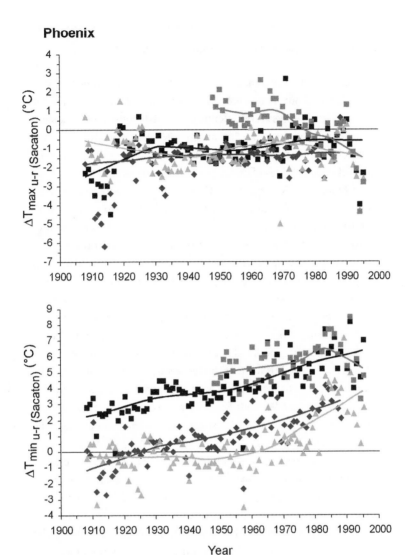

Phoenix

near each of the cities provided comparison temperature differences, shown in the respective panels: Baltimore-Washington Airport (AP; gray square); Dulles Airport near Washington, D.C. (black diamond); Washington National (gray triangle); Phoenix Sky Harbor AP (gray square); Phoenix city; Mesa, Arizona (gray diamond); and Tempe, Arizona (gray triangle). Of these, Washington National and Sky Harbor airports are within highly urbanized locations, whereas the others have suburban surroundings. Most of these data are archived in the Global Historical Climate Network (GHCN Web site, https://www.ncdc.noaa.gov/ghcnm/v3.php).

BOX 10.2. AIR TEMPERATURE MODELING METHODS

Research in CAP used a range of physically based modeling methods, whereas in BES our modeling methods were largely empirical. For the most localized temperature differences, a CAP study measured summer air temperatures in an "urban canyon" environment, an asphalt lot, and a nearby irrigated park in Tempe, Arizona. It then examined causes of different rates of cooling in the areas by using a Surface Heat Island Model (SHIM; Brazel and Crewe 2002) that incorporates thermal, morphological, and geometric features related to cooling. With the goal of finding effective urban designs to ameliorate temperatures during the summer months in Phoenix, Middel et al. (2014) simulated near-ground air temperatures for typical residential neighborhoods using the three-dimensional microclimate model ENVI-met (Chow et al. 2012). This allowed for spatial-scenario constructs under various climate change scenarios to design solutions for thermal comfort of people in neighborhoods and parks. Still another research tool used in CAP studies was the Local-Scale Urban Meteorological Parameterization Scheme (LUMPS) energy budget model of Grimmond and Oke (2002). At the mesoscale, Arizona State University researchers (Ruddell et al. 2010) used the Weather Research and Forecasting model (WRF) to locate neighborhoods where extreme heat events would cause the most hazard for residents.

To study air temperature patterns for BES we used mostly empirical computer modeling based on measurements at seven weather stations (Heisler et al. 2016). We used regression modeling to develop a prediction equation for $\Delta\hat{T}_{R\text{-}s}$, the difference in air temperature between a reference site, R, in downtown Baltimore and the six other sites, s. The weather station data included temperatures at 1.5 m above ground. Independent variables in the prediction of $\Delta\hat{T}_{R\text{-}s}$ included the difference between the downtown reference and each of the other sites in upwind tree cover, impervious cover, and water cover within 45° wedge shapes over a range of distances up to 5 km in the upwind direction. Other predictor variables included an index of atmospheric stability, topographic indices, wind speed, vapor pressure deficit, and recent antecedent precipitation. Assuming the relationships in the equation for $\Delta\hat{T}_{R\text{-}s}$ would apply to all points, p, across the Baltimore region, we used the model to estimate predicted hourly $\Delta\hat{T}_{R\text{-}p}$ across the model domain and then map $\Delta\hat{T}_{rel}$, the temperature difference between the warmest point on the map and all other points. This analysis led to predictions of the heat island intensity over the Baltimore area under a range of synoptic weather conditions. The predictions of $\Delta\hat{T}_{R\text{-}p}$ in this study were based on two-dimensional analysis of land cover. Additional accuracy might be obtained by urban structure charactization that included the vertical dimension, perhaps with a system similar to the HERCULES land cover patch system (Cadenasso et al. 2007).

between air temperatures at the times of minimum or maximum temperatures for each existing weather station used in the study. Using the method that included statistical modeling to develop maps of relative temperature differences, $\Delta\hat{T}_{rel}$, we were able to locate the predicted warmest and coolest locations in the Baltimore modeling area, with the maximum difference in ΔT, being an estimate of true maximum UHI intensity. Nights with clear skies and low wind speeds favor strong UHI formation. Under these conditions predicted maximum ΔT was 12.4° C with elevation effects included. Predicted maximum ΔT without elevation in the equation was 10.5° C, which is an estimate of the maximum Baltimore UHI owing to urban development in the absence of topographic effects. These results are for a day in September, and similar or even greater maximum UHI would be expected in other months for nights with clear skies and low wind speed.

The importance of developing efficient methods to predict the relationship between temperature and urban structure lies in the fact that potential temperature remediation actions will not be equally effective across all land use types. An example of this fact was the finding from a CAP study that measured air temperatures along car traverses through a range of land uses. Higher vegetation density tended to reduce temperatures, but not equally. Thus policies to moderate urban temperatures by tree planting can be much more effective in some land uses than in others.

Upper Urban Surface Temperatures
The land surface temperature (LST) is of prime importance to the study of urban climatology. Surface temperature is not equivalent to air temperature,

but it modulates the air temperature of the lowest layers of the urban atmosphere. Surface temperature is central to the overall urban energy balance, helps to determine the internal climates of buildings, and affects the energy exchanges that subsequently affect the thermal comfort of city dwellers. The pattern of surface temperature modifications due to urbanization, generally a heat island pattern similar in shape but not identical to atmospheric UHIs, is sometimes abbreviated as SUHI. Remote sensing provides detailed spatially and temporally consistent information on surface temperature variability within urban areas, although it is limited to discrete temporal snapshots, and surface temperatures are not necessarily an indicator of the magnitude of air temperature patterns. Upper surface temperatures are generally warmer than air temperatures at midday.

A CAP study examined the relationships between surface temperature, vegetation, and human settlement patterns in the Phoenix region using data from the Enhanced Thematic Mapper Plus (ETM+) instrument on Landsat 7 satellite for May 21, 2000, 10:56 a.m. local time, to derive surface temperature and Soil-Adjusted Vegetation Index, or SAVI. The results showed that every $10,000 increase in neighborhood annual median household income was associated with a 0.28° C decrease in surface temperature. Temperature variation within a neighborhood was negatively related to population density because tree cover is low but less variable where population density is high.

A BES study, also using surface temperatures from ETM+, examined the relationship between population characteristics and LST for 298 census block groups in the Gwynns Falls watershed. The spatial variation in land surface temperature among the block groups was greater than 16° C, suggesting that hotspots exist within the SUHI. Census block groups characterized by low income, high poverty, lower educational attainment, a larger proportion of ethnic minorities, more elderly, and at greater risk of crime had statistically higher LST. These results, along with those of the CAP study using ETM+, suggested the locations where heat prevention efforts, such as tree-planting programs, are most needed to help reduce heat exposure and moderate the SUHI effect.

Another BES study using ETM+ data investigated the effects of both the composition and configuration of land cover features on LST in Baltimore. It concluded that specific land cover features such as buildings and vegetation affect LST, but when the relative amounts of those features are held constant the spatial arrangement of those features on the landscape significantly influences LST. For example, greater nearest-neighbor distance between buildings and between individual trees reduces LST even when the percentage of area covered by buildings or trees remains constant. That is, trees scattered over an area produce lower average LST than the same number of trees in a clumped arrangement.

Urban Influences on Solar Radiation

Air temperature is merely one factor to consider in evaluating urban impacts on climate; solar radiation is another. The total solar spectrum influences the energy balance of the land surface as a whole and of individual components, including humans, thus strongly affecting human thermal comfort. The total solar spectrum also affects the energy budget of buildings, thus influencing energy used for space conditioning.

Urban buildings influence air temperatures in part by shading. A CAP study included the effects of shading by urban structures on air temperatures using a physically based computational model with 3D urban structure included as input (ENVI-met). An array of low-rise to high-rise and compact to spread-out building and landscape elements in Phoenix was defined by local climate zone (LCZ) classification to evaluate shading effects of design differences. Spatial differences in cooling were strongly related to solar radiation and local shading patterns. In midafternoon, dense urban forms can create local cool islands.

Also important is the very short-wave-length ultraviolet (UV) spectrum that influences health of animals, including people, and plant life. The urban atmosphere and urban structure influence the penetration of UV radiation to the level of pedestrians. Although we might expect that an urban atmosphere would strongly attenuate UV radiation, measurements over the course of a year at a BES solar radiation monitoring station within Baltimore City showed otherwise. There was little difference between UV measured near the center of Baltimore, in a suburban location to the south of the city, and at a rural site across the Chesapeake Bay east of Baltimore. Thus the exposure of pedestrians to UV radiation is of concern even in the city because of the decided effect of UV on human health.

Trees and built structures strongly influence UV radiation; although, because of the scattering of UV radiation by the atmosphere, the pattern of UV radiation in tree shade differs considerably from the shade pattern of the visible solar radiation spectrum. Thus people may get significantly more or less exposure to UV radiation than is intuitively apparent. Exposure is a little less than half as great within visible tree shade than just outside of visible shade. Exposure is typically about 40 percent greater with cloud-free than with average cloud conditions but still significant even with average cloud cover.

Human exposure to UV radiation may differ in different land use classes because of different amounts of tree cover. A modeling study of UV penetration to below tree canopy spaces in Baltimore suggested that in neighborhoods of high density residential housing, where tree cover is only 20 percent, UV exposure for pedestrians would typically be nearly 30 percent greater than in residential neighborhoods of mid- and low-density housing, where tree cover averages 32 percent. These results also suggested that exposure would be greatest in land uses classified as institutional because tree cover in such sites is only 13 percent on average, the lowest included in this study.

Energy and Mass Flux from Suburban Areas

Both urban LTER sites have had campaigns to measure urban energy and mass fluxes from tall towers with eddy correlation instruments that integrate the vertical fluxes from suburban areas over several kilometers. For BES, the tower is a former forest fire lookout station located on a low hill (Cub Hill) in an area that now has suburban development, mostly with single-family residences. An emphasis at this installation has been the measurement of CO_2 fluxes. The measurements show that the suburban area around the tower produces net CO_2 emissions to the atmosphere during winter but that during summer days, when CO_2 uptake by vegetation is high, the suburban surface becomes a net absorber of CO_2. This high uptake is undoubtedly owing to the high percentage of area, 67 percent, covered by vegetation around the tower. On average, including winter and nighttime, the area is a net CO_2 source, but the net annual CO_2 release to the atmosphere ($361 \text{ g C m}^{-2} \text{ y}^{-1}$) was an order of magnitude lower than had been reported for any other city in the world.

Study of the energy balance terms (box 10.3) was reported for a year of measurements at a tower in a low-density, low-rise residential suburban

---

BOX 10.3. URBAN ENERGY BUDGET MEASUREMENT

In both CAP and BES there has been a series of urban area vertical flux measurements from towers to estimate the total surface energy balance defined by

$$Q^* + Q_F = Q_H + Q_E + \Delta Q_S + \Delta Q_A,$$

where Q represents vertical energy transfer in Wm^{-2} and the subscripts or superscripts indicate the energy components. The net radiant energy into the urban surface, Q^*, is measured from the tower as the algebraic sum of downward solar radiation (+), reflected solar radiation (–), downward long-wave radiation (+), and upward emitted long-wave radiation (–). Eddy correlation instruments on the tower measure the sensible (Q_H) and latent (Q_E) heat fluxes. Energy terms that cannot be measured directly include the anthropogenic input (from heating and cooling systems, transportation, etc.) symbolized by Q_F; heat storage, mostly in soil or human-built surfaces, ΔQ_S; and advection, ΔQ_A, heat carried horizontally into the measurement area by wind. Quantification of the energy balance contributes to understanding of urban climate processes and makes possible analytical modeling of these processes. For example, sensible heat/latent heat flux, $Q_H/Q_E = \beta$, or Bowen ratio, defines the portion of incoming energy that goes into heating the air relative to energy that goes into evaporation of water or transpiration from plant leaves.

---

development in an arid area of Phoenix. The land cover was 48 percent impervious, 37 percent sandy, and 15 percent in grass and trees. Basically, the measurements agreed with studies in other urban areas that city built infrastructure induces higher proportions of stored heat in relation to net radiation as well as higher sensible heat flux in comparison to rural areas. However, the ratio β of sensible heat to latent heat flux differed greatly from less arid areas. In Phoenix under clear skies β averaged about 5, whereas at Cub Hill in Baltimore β averaged about 0.5 or less in summer. This indicates the possibility in Phoenix of decreasing sensible heat production, which warms the air, by increasing evaporation through irrigation and by fostering of vegetation.

Temperature Influences on Health

LTER urban climate research is important because human exposure to excessively warm weather is an increasingly important public health problem at a global scale. Heat stress can cause illness and even death. Heat waves have been considered the predominant cause of death from natural hazards both worldwide and in the United States. Heat-related deaths are a chronic problem in hot, arid climates such as that of Phoenix; but even more deaths are attributed to heat in temperate climates than in warm climates because people in temperate zones are less acclimated to high temperatures. A study that examined the relationship between types of warm air masses and increased mortality for forty-four major U.S. cities found that in Baltimore the most dangerous type of air mass caused an average of 2.7 excess deaths per day, whereas the most dangerous air mass in Phoenix caused an average of just 0.9 excess deaths per day. Thus, although the Phoenix general climate is warmer, extreme heat events are even more dangerous in Baltimore, where the population is less adapted to extreme heat. The UHI effect is a concern in both cities because it exacerbates the long-term effect of global warming and the shorter-term effect of large warm air masses.

A CAP study examined the potential for heat-related health inequalities within Phoenix populations. The study examined the relationships between the microclimates of urban neighborhoods, population characteristics, thermal environments that regulate microclimates, and the resources people possess to cope with adverse climatic conditions. It used a simulation model (box 10.4) to estimate an outdoor human heat stress index as a function of local climate variables collected in eight diverse neighborhoods during summer. There were statistically significant differences in temperatures and heat stress between the neighborhoods, and the differences increased during a heat wave. Lower socioeconomic and ethnic minority groups were more likely to live in warmer neighborhoods with greater exposure to heat stress. High settlement density, sparse vegetation, and lack of open space in the neighborhood were significantly correlated with higher temperatures and heat stress.

APPLICATIONS

Our urban climate research has been applied in university and general public education, public understanding of urban influences on climate, tools

---

BOX 10.4. HUMAN COMFORT MODELING

Several CAP studies, including the one by Harlan et al. (2006) used an outdoor human comfort model, OUTCOMES (Heisler and Wang 2002) to estimate heat stress based on a calculation of the total energy balance, EB, of a representative person:

$$EB = M + S + T_a + C - E - T_e$$

where M is internal metabolic heat, S is solar irradiance absorbed, T_a is thermal radiation absorbed, C is convective heat gain, E is evaporation, and T_e is thermal radiation emitted from the person. Each term has units of Watt m^{-2} of body area. If EB is near zero, most people will be comfortable; if it is significantly negative, most people will feel cold; and if it is strongly positive, people will feel too warm.

---

for administrative decision making, as well as in direct interaction with city officials. We are also convinced that we have provided inspiration and tools for further research on urban climate by other scientists.

Education

Climate research is an important educational tool because of the relevance of climate and climate change to many audiences. We contributed to education at the university level through collaborative studies of urban climate in Baltimore and Phoenix that were reported in several widely referenced publications aimed at interdisciplinary audiences. These works have been used as reading assignments in urban forestry and urban ecology classes at the State University of New York (SUNY-ESF), in urban planning and urban climate classes at Arizona State University, and in teaching of urban ecology as an undergraduate upper-division class at the University of California Davis. There was good student evaluation of a book chapter as an assignment for students in an urban ecology class at SUNY-ESF. Numerous graduate students benefited by working as assistants on urban LTER climate studies, including a master's thesis that made a major contribution to empirical modeling and imaging of the Baltimore UHI.

Contributions to primary and secondary education related to climatology included working with the BES education team in developing curriculum materials, visits to Baltimore elementary schools, and contribution of a weather station that we installed at a city school. The primary BES weather station is located at a school outside the city, and data are regularly made available to a science teacher there who uses them in his secondary classes. We have collaborated with the Maryland Science Center in teacher training and assisted them in engaging volunteers in reporting temperature measurements across the urban-rural gradient.

General audiences have been reached with a variety of climate research results. For example, modeling results that indicate the magnitude and areal extent of Baltimore park influences on temperature have been made available on the website of the National Recreation and Parks Association. BES research results are regularly presented to the public at meetings, such as the BES Annual meetings and community open house events. The BES contribution of a weather station sited in a forested Baltimore County park with the data display in the nearby nature center and an explanatory display are additional contributions to public understanding.

Decision Making

Our climate research has supported decision making in several ways. For example, our BES research on regimes of UV radiation in Baltimore, along with modeling of reductions in UV radiation below urban trees, inspired the development of a module to predict tree influences on UV radiation in the Version 6.0 of the popular and publicly available i-Tree model (https://www.itreetools.org/). The USDA Forest Service and cooperators developed i-Tree as a suite of programs to analyze urban and rural forestry structures, functions, and benefits. The i-Tree Tools have helped communities of all sizes around the world strengthen their forest management and advocacy efforts, which are highly relevant to urban climates. Models of spatial temperature patterns, such as we developed under BES, may be used to predict the effect of tree-planting programs on UHIs, and these modeling methods and information have been used to develop a prototype air temperature program for i-Tree.

We made available results of research on the UHI effect to the Office of Sustainability for the City of Baltimore to assist in their climate adaptation planning. The 2015 Annual Report of the Office of Sustainability illustrates the use of this information. A "Neighborhood Spotlight" in the report tells of the efforts of a volunteer community group, Cleaning Active Restoring

Efforts (CARE), to improve their neighborhood of about twenty city blocks. One of the CARE projects was to plant new trees on every street in the community. The report points out that the CARE area "is a noted heat island in the City of Baltimore, and tree planting efforts, as well as greening [vacant] lots will help the community become cooler and reduce the negative effects of high heat from asphalt, concrete, and tar roofs." Indeed, the CARE area appeared in our temperature pattern modeling to be among the warmest parts of Baltimore. It is just outside the influence of Patterson Park, which is a relatively cool zone in east Baltimore. The 2009 Sustainability Plan for the City as a whole includes the goal of doubling tree cover from 20 to 40 percent by 2037, with the specific goal of targeting areas with the greatest urban heat island impacts.

The BES and CAP studies of LST ("Upper Urban Surface Temperatures" section above) provide information on methods for focusing and tools for planning to prevent heat stress. Cities can implement strategies to minimize heat exposure and enhance the thermal comfort of their residents, but these intervention strategies are likely to be costly, so that focusing on the most vulnerable groups is most efficient. For example, in Baltimore the social characteristics of the residents living in census block groups with the highest LST had the most vulnerable resident populations defined as low income, high poverty, low educational attainment, a larger proportion of ethnic minorities, and elderly and at greater risk of crime. Using the average value of each of these social indicators of vulnerability as well as a threshold LST value identified block groups that should be prioritized for high-temperature mitigation strategies. This tool of prioritization can (1) be modified based on resources available by changing the thresholds or the variables considered to either increase or decrease the intervention efforts; and (2) be used to guide specific strategies of heat intervention that may be most beneficial for the specific social characteristics of the residents.

With support from NSF both through CAP LTER and outside of the LTER program (for example, the NSF project Decision Center for a Desert City), CAP climate scientists participated in research that was aimed at developing management decisions to alleviate extreme temperatures in Phoenix, to reduce the UHI, and to study heat vulnerability. A series of studies have been produced, sometimes in collaboration with City of Phoenix personnel, regarding local-scale climate outcomes related to water-use alterations, vegetation manipulation (irrigation and xeriscaping), and cool roof technology. The studies have potential to alleviate high-temperature risks and to help the city with decision making. They may be grouped into a number of categories: (a) land cover impacts on temperature, (b) mitigation assessment (water, vegetation, roof color), (c) spatial arrangement of land cover and impacts on temperature, and (d) heat vulnerability and socioecological patterns.

The story of CAP LTER collaboration with the city in xeriscaping is an excellent example of research applications. Xeriscaping, the practical use of low-water-demand plants in a yard (shrubs and trees) adapted to arid climates, is a unique neighborhood land cover phenomenon in the CAP LTER area. Several cities in the metro area have successful programs offering financial incentives to alter homeowner landscapes from water-guzzling turf or landscapes with virtually no trees and ground cover to xeriscaping, which results in considerable water savings. Improving arid surfaces to xeriscaping trees and low-water-use plants in a yard might add a water burden but have other benefits. Prior to the studies with CAP, city officials had no idea of temperature effects of these incentive programs, as the main goal was simply water savings. A major theme that became important to the city was increasing shade and increasing tree cover to upward of 25 percent from its current cover of about 13 percent. Depending on percent area treated in a currently dry and vegetation-devoid neighborhood, planting with xerophytic shade trees creates cooling of up to 2.5° C locally. This was arrived at using a microclimate model called ENVI met3. Conversely, replacing watered landscapes with xeriscaping in a moist or mesic neighborhood experiencing plenty of flood irrigation and high tree cover raised temperatures from 0.8° to over 1.0° C.

The issue of nonlinear or disproportional changes in temperature as a response over the full range of altering moist surfaces to dry ones remains for future research. For example, an initial study using the LUMPS energy budget model included authors from the Phoenix water department. The study found that as surface cover was converted from dryness to moistness by increasing fractions of well-watered vegetation, the rate of temperature change was nonlinear; cooling quickly at first as vegetative cover increased up to 20 percent. However, there was very little further cooling with higher percent vegetative cover beyond this threshold. The thresholds of these kinds of changes were investigated in more detail at a finer resolution using LUMPS along with an index of daytime cooling efficiency; and again there was a threshold of about 20 percent cover beyond which cooling efficiency dropped off. In other words, there was very little realized cooling with further increases in percent watered acreages. However, the rates of mitigative cooling day and night are conditioned by configuration as well as fraction of the various land cover components as determined by remotely sensed surface temperature data and satellite imagery. These findings are in concert with larger-scale findings from BES. An overall useful conclusion that remains paramount for city officials was a takeaway statement from a CAP report: "Any one [mitigation] strategy [using water] will have inconsistent results if applied [equally] across all urban landscape features and may lead to an inefficient allocation of scarce water resources."

Overall, a series of studies in both BES and CAP add a social dimension to the above findings, namely, that people in warmer neighborhoods are more vulnerable to heat exposure because they have fewer social and material resources to cope with extreme heat. This suggests that UHI reduction policies should specifically target more heat-prone and vulnerable residential areas. Especially in Phoenix it is important to heed the message to consider the various landscape features in utilizing valuable water resources in any mitigative strategy for the city.

FUTURE NEEDS

One frequent result of research activity is that more questions are discovered than answers, which is certainly the case in our LTER urban climate research. A sample of needs for future research includes the following:

1. An important future goal should be to examine in more detail the long-term trends in air temperature means and especially extremes in Baltimore and Phoenix to include the goal of separating out any global climate-change influences from the urbanization influences. This will be facilitated by the more than twenty years of additional observations since the data used in the publication of trends in Baltimore and Phoenix over the course of the twentieth century. New data include those from the BES primary weather station that was established in April 2000. A challenge will be to account for the closing and moving of some stations and the encroachment by new development toward rural stations.

2. Better ways are needed to characterize urban structure related to

effects on climate. This is true for any method of computer modeling of temperature patterns as a function of structure. For example, for semi-empirical modeling of Baltimore temperature influences, we based predicted temperature differences across the city and surrounding suburbs on land cover categorized as percentage of impervious, tree, and water. This made possible description at a relatively fine scale of a large area at low-cost, but with the assumption of a two-dimensional surface. Additional relatively simple descriptors of urban structure could include average heights of buildings and vegetation. Especially to focus on portions of the total area, more refined descriptive characterization might be based on the HERCULES land cover patch system that has been used to resolve the landscape of Baltimore's Gwynns Falls watershed into buildings, pavement, woody vegetation, herbaceous vegetation, and bare soil. Another patch characterization is the Local Climate Zones (LCZ) system, as used in a CAP study. Consideration might be given to overlaying a patch characterization, either HERCULES or LCZs, on high-resolution impervious and tree cover in order to develop predictor variables for empirical modeling of temperature differences.

3. In order to assess net annual impacts of UHIs, analysis needs to be extended for the winter season. Much previous research, including our own, considered UHI temporal patterns only in months when large UHI effects were likely; future study should compare summer and winter.

4. Methods are needed to enhance application of LTER UHI results to other cities. This is important not only to justify the substantial research effort expended in CAP and BES but also because few cities will have the resources to carry out similar investigations for their particular community. One possibility would be to associate urban climate results with one of the patch classification systems, which are more intuitively visualized.

CONCLUSIONS

Since the urban LTER research programs began in Baltimore and Phoenix, their priorities have been shaped in large measure by the background climates of the two regions. Although their history of development differs significantly, these two urban areas are similar today in being of the same order of magnitude in the features that research elsewhere has shown to be related to urban climate modification—size and population. The measure of the urban heat island that has been applied in our research and in other cities around the world, urban minus rural temperatures, differs between the two cities in large part because background climate has determined the natural vegetation of the rural areas: temperate forest around Baltimore and desert around Phoenix. However, despite the tendency of homogenization of the built and vegetative structure of urban areas across a wide range of climate, the higher temperatures and drier atmosphere of Phoenix create somewhat more concern for human health, somewhat greater restrictions on the use of vegetation to control high temperatures, and greater concern for water quantity than in Baltimore. In both cities, the combined pattern of built structure and tree structure has led to cooler temperatures in higher-income neighborhoods.

Research methods in the urban climate studies benefited from a range of subdisciplines of atmospheric science among researchers in BES and CAP LTER. Expertise included classical physical geography, general climatology, forest influences, microclimatology, and micrometeorology. However, particularly in problem selection and application of results, expertise from the wide-range Urban LTER scientific disciplines aided urban climate research; among these disciplines are ecological and social theory, urban design, computer science, demographics, soil science, watershed management, education, traditional forestry, and urban forestry. For example, in BES, because of educators in the program, we collaborated in public school programs related to weather and climate. Both BES and CAP research has involved social scientists in developing methods to prioritize locations for urban heat island remediation for especially vulnerable populations. We reaffirm a conclusion from one of our 2009 publications: "The BES and CAP LTER urban sites should remain excellent laboratories to refine concepts in urban ecology and climatology and links to social, political, economic, and ecological processes over a long time period."

Acknowledgments   Too many individuals to name have contributed to BES and CAP LTER climate research, but we especially wish to acknowledge the participation and assistance of Kenneth Belt, Dan Dillon, Alexis Ellis, Richard Grant, Sue Grimmond, Peter Groffman, John Hom, Karla Hyde, Hang Ryeol Na, Jarlath O'Neil-Dunne, Emma Powell, John Stanovick,

Yingjie Wang, and Ian Yesilonis, Nancy Grimm, Steven Earl, Marcia Nation, Chris Martin, Sharon Harlan, Darren Ruddell, Winston Chow, Chao Fan, Donna Hartz, Brent Hedquist, Darryl Jenerette, Larry Baker, Susanne Grossman-Clarke, Nancy Selover, Juan Declet-Baretto, David Greenland, William Stefanov, Ariane Middel, Soe Myint, Alex Buyantuyev, Andrew Ellis, Baojuan Zheng, Benjamin Ruddell, Shai Kaplan, and Xiaoxiao Li.

CHAPTER ELEVEN

Responses of Biodiversity to Fragmentation and Management

Christopher M. Swan, Anna L. Johnson, and Steward T. A. Pickett

IN BRIEF

- Local environmental constraints and regional dispersal patterns inform our understanding of how biodiversity unfolds at multiple scales.
- Spatial disaggregation of the urban landscape generates heterogeneity in habitat quality that varies as a function of both direct and indirect effects of human behavior.
- People indirectly alter habitat quality to either eliminate or promote species coexistence, and also sidestep natural dispersal pathways to maintain desirable species associations.
- A consequence of spatial variation in human behavior is high turnover in species composition, generating substantial biodiversity in the urban environment.
- Complementing studies of turnover with those of local and regional diversity yields insights into the mechanisms that generate biodiversity at multiple scales.

INTRODUCTION

Community ecology is the study of the patterns of diversity and the processes that maintain species coexistence in space and time. In recent years there has been an increasing focus on how global climate change and habitat loss will alter global levels of biodiversity and linked ecosystem functions. Additionally, there has been an increasing interest in, and controversy over, "novel ecosystems," or locally nonhistorical combinations of abiotic environments and altered species assemblages, primarily as a result of human changes to the environment. Determining whether or not these increasingly recognized human inputs are actually novel and what the magnitude of the change in many ecosystems is as well as the recognition of the ubiquity of human effects on ecosystem structure and function are all valuable contributions to many areas of biodiversity research. Cities are a study system where the human influence is clear and where the consequences of findings often have direct relevance to most people's lives: more than half of the global human population now lives in urban areas. Therefore, we contend that urban systems are excellent places to both develop and test new theory in community ecology and to do policy-relevant science (box 11.1).

Humans both modify and create habitats in urban ecosystems by altering the abiotic environment. Furthermore, they alter the biotic community of urban environments by changing species pools at the scale of the entire city and also by intervening in community structure locally. The complex human components of the creation of novel ecological communities include management activities such as restoration and park management, street tree plantings, mowing regimes, and gardening and cultivation. Management activities can lead to the relaxation of interspecific interactions through such interventions as weeding and pest management and intervening in species life-history processes (for example, human-aided dispersal via propagation and seeding or dead-heading flowers to limit natural reproduction). Across cities, large-scale plantings for erosion control or aesthetics alter regional propagule supply and change the structure of local communities by eliminating undesirable species and promoting the presence of desirable, ornamental species, some of which are or may become invasive (for example, the Callery or Bradford pear [*Pyrus calleriana*] in the United States; black cherry [*Prunus serotina*] in Europe; English ivy [*Hedera helix*] in the United States). All of these actions occur in a landscape that is spatially disaggregated and often exhibits extreme heterogeneity in environmental conditions and habitat quality as a result of the complex history of human interventions. As such, not only does diversity respond locally to human influence, but so does species compositional turnover in space. To-

BOX 11.1. KEY TERMS

Diversity: In general, the variety of form of life on the planet. Can encompass taxonomic, trait, and genetic variability. Diversity is a multiscale concept, with local (or α) diversity describing an ecological
assemblage at the local scale, and regional (or γ) scale, representing
the collective diversity at the largest scale of inquiry. Turnover, or
β-diversity, is the compositional change in the elements of an ecological community from one location to the next.

Landscape: A mosaic of interacting ecosystems, at any scale. Such an
area must be spatially heterogeneous in at least one factor of interest.

Life history strategy: Traits that affect the pattern of birth and death of
an organism. Strategies can be imagined as various investments in
growth, reproduction, and survivorship.

Metacommunity: A set of local communities that are linked by dispersal of multiple interacting species.

Patch: A discrete area of habitat. Patches have variously been defined
as microsites or localities capable of holding populations or communities.

Productivity: The fixation of solar energy by plants and the subsequent
use of that fixed energy by herbivores, carnivores, and the detritivores
that feed on dead biomass.

Species pool: All species available to colonize a focal site.

Vagile: Organisms are said to be vagile if they have the ability to move
about freely and migrate.

gether, these two components of diversity make up the urban regional species pool. Sustainability action plans often include efforts to maintain native
biodiversity in cities, as urban society increasingly recognizes the importance of biodiversity for providing urban ecosystem services. Therefore,
understanding the mechanisms that maintain ecological community structure in space and time can increase the effectiveness of urban environmental management plans as people both continue to urbanize the landscape
and respond to the ecological implications of such development.

A CONCEPTUAL MODEL OF COMMUNITY ASSEMBLY
IN THE URBAN ENVIRONMENT

Urban biodiversity is largely described at either local or regional scales, with less focus on the mechanisms that generate the patterns across scales that lead to compositional turnover between locations. Three drivers have been invoked from general ecological theory to explain local diversity patterns in urban ecosystems, as represented by human influence. First, as productivity increases owing to human enrichment of the environment, species richness is expected to increase. This is thought to be because more productive habitats attract both humans and other species. Second, the intermediate disturbance hypothesis has been suggested to explain richness patterns. Systems under intermediate levels of human impact or disturbance have higher habitat/resource diversity compared with habitats that are either pristine or human-dominated. Therefore, these intermediately impacted habitats should support higher diversity and have been observed along rural-to-urban gradients. Last, ecosystem stress is hypothesized to result in a decline in richness as humans impose a harsh geophysical environment. Humans remove habitat and resources, therefore predicting a negative relationship between species diversity and human influence. These hypotheses can bear out on the landscape but focus almost entirely on local-scale effects.

One of the most striking features of the urban landscape is the substantial disaggregation of habitat and subsequent increase in spatial heterogeneity. Even over very short distances patterns in natural and remnant habitat structure can change dramatically as human-built infrastructure interrupts once well-connected environments. Therefore, urban places exhibit a juxtaposition of patch types that support a diverse set of ecological communities governed by a variety of social-ecological processes. Such patch dynamics were identified early on in the development of urban ecological research. This includes not only the remnant, often interpreted as "disturbed," environments such as roadside plant communities but also designed landscapes and created habitat, such as urban stormwater ponds (figure 11.1). The resulting biodiversity in each location integrates both the direct and indirect effects of human activity on the landscape.

In the 1960s ecologists first conceptualized biodiversity as a multiscale phenomenon. The regional species pool, or gamma (γ) diversity, is composed of local, or alpha (α), diversity within specific habitats and compositional turnover, or beta (β) diversity, between different habitats. For urban

FIGURE 11.1 Local ecological communities and representative urban taxa exhibiting a range in human facilitation of coexistence patterns. (Photographs, C. M. Swan)

ecosystems the focus has been in general on either α- or γ-diversity patterns. The metacommunity concept was introduced in the early 1990s and codified a large body of work generated over decades to understand the mechanisms that explain patterns α-, β, and γ-diversity. This framework recognizes that ecological communities are not isolated entities structured completely by local dynamics, but that dispersal from neighboring patches can substantially influence local assemblage structure. An explicit incorporation of dispersal aids in our understanding of why species ill-matched to local habitats can nonetheless occur. One can envisage the simple case where emigrants from a productive habitat disperse to an unproductive habitat. The colonizing species may establish in the less productive community but not be sustained there—internal dynamics cannot explain its presence. This framework is powerful in that it addresses patterns and processes that arise across different spatial scales.

Species vary in their dispersal ability. In urban ecosystems people not only actively move species around but also facilitate the establishment of very vagile, nonnative species. Dispersal is hypothesized, and has been shown empirically, to alter the relative contribution of local diversity and compositional turnover to regional diversity patterns (figure 11.2.). In general, as dispersal increases, local assemblages become saturated with all species from the regional species pool, decreasing turnover but increasing local diversity. Eventually, competitive dominants eliminate less competitive

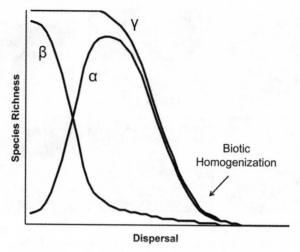

FIGURE 11.2 The hypothesized relationship between dispersal and α-, β-, and γ-diversity. (Developed based on empirical findings of Mouquet et al. [2003])

species, resulting in a decline in regional diversity. In urban ecosystems the replacement of native species with generalist, competitively superior non-natives in part explains declines in urban regional biodiversity. However, we know diversity does not decline equally in all locations in urban environments. Compositional turnover between habitats, or β diversity, and interspecific variation in dispersal help us understand why this is the case.

To illustrate this pattern, consider three plant communities under different property management regimes: brownfields that were previously building sites, designed and maintained lawns, and unmanaged gardens or yards after housing is abandoned (figure 11.3). The predictions for β diversity between these types of plant communities are quite high, as divergent histories and contemporary local controls on composition cause species compositional trajectories to diverge. This hypothesized result is associated with human management of both the local environment and direct facilitation of species coexistence, for example, via soil amendments and planting and weeding. However, assemblages at other trophic levels co-occur with plants. Soil biotic communities are under less human "active" management, are very vagile, and are quite adapted to the urban environment. In this case, turnover is predicted to be quite low, particularly in the most abundant groups of taxa. Common invertebrate pests are also vagile, but, depend-

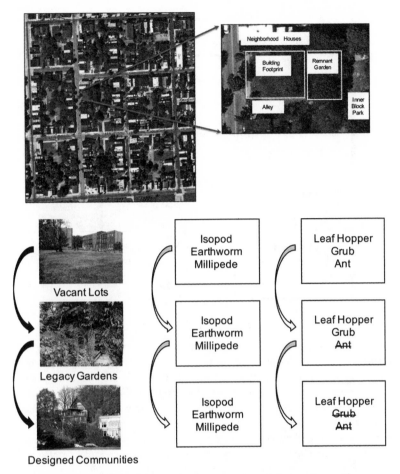

FIGURE 11.3 Google Earth image showing extraction of photos of vacant lots, designed lawns, and legacy gardens (top); ecological communities with different levels of dispersal (bottom). (Imagery © 2015 Google Map data © 2015 Google; photographs, C. M. Swan)

ing on the stage of land abandonment and/or management regime, the magnitude of turnover in these pest communities is likely to vary.

We distinguish between two extremes in assembly mechanisms (figure 11.4). Self-assembled communities are those responding to the urban geophysical template. Here, community composition is largely governed by environmental filtering and niche effects. These communities are generally low in richness and, within the same taxonomic group, similar also in com-

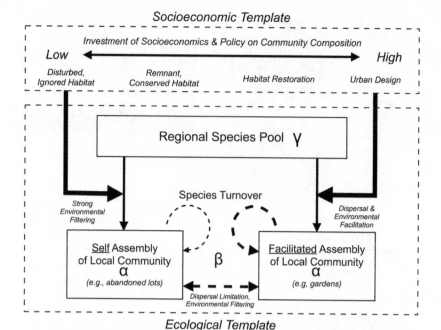

FIGURE 11.4 A conceptual model of urban community assembly. Self-assembly and facilitated assembly are the processes generating local diversity patterns (α) from the regional species pool (γ), with assembly governed either by the geophysical template or by humans actively promoting coexistence. The magnitude of species turnover (β) is predicted to be high among facilitated communities and between self-assembled and facilitated communities. However, strong environmental filtering due to the harsh urban geophysical template should constrain species composition. We suggest that low social investment in community composition is associated with the disregard for ecological integrity and results from historical development. As such, it is a unidirectional constraint on local species composition (*left*). However, sustainability has the potential to create a feedback between desirable ecological assemblages and human behavior (*right*). (Redrawn from Swan et al. 2017 and used with permission of John Wiley and Sons)

position. In contrast are communities whose assembly is facilitated by humans. Communities experiencing facilitated assembly are in many ways designed. As such, they are locally structured as a function of human preferences, and, therefore, management for species identity is hypothesized to be favored over species richness. The most straightforward examples are gardens; horticultural activity has historically been a rich source of biodiversity in the urban environment. But this concept can be extended to other types of management activities. For example, restoration of streams and rivers is designed to attract desirable fish, such as trout, and certain inverte-brate communities, and public tree plantings are a function of city or neigh-borhood preferences and species availability at local nurseries, in addition to bioclimatic variables.

Community assembly is generally conceived as the process by which species come to co-occur locally as a subset of the regional species pool, with site-to-site variation in species composition describing species turnover. In urban environments there is a strong traditional, although not exclusive, ecological focus on remnant or highly disturbed habitats such as abandoned lots. We conceive these locations as being self-assembled, in that communi-ties develop similarly to nonurban ecosystems but in response to the harsh urban geophysical template. Yet many assemblages occur in the urban land-scape as a function of a multitude of social-ecological processes that may collide. Therefore, imposing a hypothetical socioeconomic template we de-scribe as a gradient in human investment in community composition aids in our understanding of the human role in the community assembly pro-cess. Where investment is low, composition and/or diversity is not a priority or is ignored, the harsh geophysical template imposes strong environmental filtering, and habitats are similar in condition, leading to low compositional turnover. Where we see the potential for this hypothesized gradient to inte-grate relevant socioeconomic processes is in how they translate into envi-ronmental decisions and management activity at the local scale. For exam-ple, as investment increases, focus may turn to conservation of habitat, restoration of habitat, and ultimately the design of novel ecological commu-nities. In this case, humans impose substantial constraints on the assembly process: habitat is altered to promote the desired species composition, spe-cies are directly assembled and/or maintained, and/or habitat is conditioned to attract desirable species from the regional species pool. Among such hab-itats turnover is predicted to be quite high owing to differential human de-cision making and preferences.

We acknowledge that these "self-assembled" and "human-facilitated" communities are likely endpoints along a gradient. Different trophic levels could react differently to this socioeconomic template. For example, soil organisms are quite vagile, not highly managed, and likely to occur across the urban landscape. Thus the relative strength of the difference in turn-over would not be expected to be so high as it is for plants, which are actively manipulated by human management actions. This "leakage" of species between self-assembled and human-facilitated communities is central to our understanding of the multitrophic complexity in the urban environment.

The perspective described above is a shift in the traditional focus, from a study of biodiversity at the local scale to compositional turnover, because we recognize that the urban environment is composed of a juxtaposition of patch types. Furthermore, patch to patch variation in the biophysical environment is known to be strongly connected to relevant social processes. Identifying differential compositional turnover among patch types suggests different assembly mechanisms, such as those that depend more on dispersal versus local constraints. The discoveries outlined above provide important information about how biodiversity is generated and maintained in the urban environment and therefore how sections of city sustainability action plans with a focus on urban biodiversity can be developed, modified, and implemented.

CASE STUDY: URBAN PLANT METACOMMUNITY STRUCTURE

As an empirical test of the concept of self- versus facilitated-assembly, we examined patterns in woody plant biodiversity across a gradient in human land use. We focused primarily on β diversity at the taxonomic level. We predicted that turnover should be highest in areas where humans exercise relatively greater control over composition. We analyzed turnover among plots within six land use categories, arranged along our hypothesized gradient of human investment in species composition (table 11.1). Vacant lots and open space are amenable to self-assembly from the regional species pool, while parks and plots with significant human management are prone to strong manipulation of the environment and placement of species. We made use of a large data set created as a part of a tree inventory effort in the Baltimore metropolitan region. Forest composition of planted and naturally recruited individuals were identified to species using plot-based random sampling. Dates of inventories ranged from 1999 to 2009. The data set we analyzed comprised 209 0.1 ha plots randomly located across Baltimore City and County (box 11.2).

We found 116 woody plant species across all plots. Local diversity was variable across land use types but generally higher where humans were considered to exhibit less active control over coexistence patterns (for example, in vacant lots, open space, and parks) compared with designed spaces such as institutional, commercial, and residential parcels (figure 11.5). However,

Table 11.1

Land use classes as defined by Cadastral Geodatabase LANDUSE.TaxParcel

LAND USE	DESCRIPTION
Vacant Lots	Parcel that does not have a principal building, at least ten years since created through the subdivision process, and is not predominantly covered with accessory uses such as garages and swimming pools. Also includes unbuildable land.
Open Space	Unimproved space transferred to the local government through the development process or acquired by other means; space that contains surface or underground water storage and also provides open space/recreational amenity.
Parks	Maintained space containing a permanent recreational improvement
Institutional	Places of worship, colleges/schools, medical facilities, police/fire, cemeteries, libraries, other government facilities.
Commercial	Structure containing retail/service or office uses. Includes parking lots.
Residential	Single-family detached, attached or multifamily (3+) homes.

Data Dictionary (Swan et al. 2017)

BOX 11.2. METHODS FOR METACOMMUNITY ANALYSIS

Woody plant communities were sampled following protocols outlined in http://www.itreetools.org/. Abundance data for trees were used for all analyses. The Shannon Index was used to describe local taxonomic diversity. In this study we considered beta diversity as variation within groups from one plot to the next. Using the betadisper function in the R programming environment, we tested for the multivariate homogeneity of group dispersion on a Bray-Curtis dissimilarity metric. This method calculates the average distance of each plot to the group median in multivariate space and assessed the differences among land use groups using a permutation test. Pairwise comparisons between land use categories were carried out using Tukey's HSD test and significant evaluated at $\alpha = 0.05$.

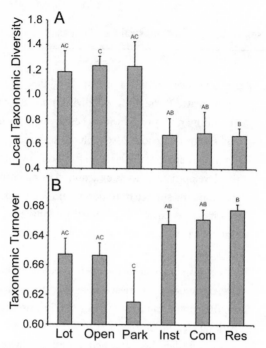

FIGURE 11.5 Diversity of woody plant communities across land use categories of vacant lots (Lot), open space (Open), parklands (Park), institutional grounds (Inst), commercial (Com), and residential (Res). Panel A is alpha, or local, taxonomic diversity, while Panel B is taxonomic turnover. Bars with the same letters are not significantly different. Land use categories are arranged from left to right based on whether they are hypothesized to exhibit self-assembly (e.g., vacant lots) or facilitated assembly (e.g., residential parcels). (From Swan et al. 2017 and used with permission of John Wiley and Sons)

consistent with our concept of urban metacommunity structure, β-diversity was higher in designed spaces. We interpret this as variation in human choices at the parcel level leading to significant taxonomic shifts in space.

MOVING FORWARD: PHYLOGENETIC AND FUNCTIONAL ORGANIZATION OF URBAN METACOMMUNITIES

General ecological theory organizes studies of broad biogeographical patterns in natural communities using such tools as species richness-latitude relationships and body-size temperature relationships. Such underlying theoretical models are needed to structure studies of urban biodiversity as well and to allow results to be comparable across cities. To date, the study of species coexistence in urban areas has focused primarily on patterns of biodiversity but has lacked broad consideration for mechanisms, resulting in a lack of generalizability across studies. This has also led to a mismatch between available urban ecological research and information needed to address problems of urban sustainability and environmental management.

In an attempt to address this problem, modern coexistence theory, which focuses on mechanisms of community assembly, is increasingly being incorporated into urban studies to combine our understanding of natural assembly processes with predictions about the way humans alter connectivity, species pools, the local abiotic environment, and species interactions. This includes a focus not just on taxonomic but also on functional and phylogenetic dimensions of biodiversity. Functional and phylogenetic diversity metrics are a more information-rich way to describe the diversity of biological communities than species richness and allow comparisons to be made between cities differing in species identities but potentially sharing functional and phylogenetic groups. In addition to giving a more nuanced description of biodiversity patterns, functional and phylogenetic metrics allow compositional changes to be correlated to underlying environmental or human impact gradients in biologically meaningful ways.

This approach leads to an ability to make predictions about how shifts in species pools, for example, nonrandom species introduction patterns, might affect species interactions within communities. Such predictions would facilitate the incorporation of human impacts into conceptual models at multiple scales. For example, human preferences for ornamental plant species might lead to selection for increased floral display, thus increasing the fecundity and likelihood that introduced species will naturalize in novel habitat ranges. Human neighborhood-level norms can lead to spatial autocorrelation in horticultural plant species selection and garden management. This in part explains levels of observed species diversity as well as locational turnover within cities, in addition to illuminating appropriate spatial scales for management action. Depending on the type of biodiversity considered, historical neighborhood characteristics can be more important than contemporary conditions for explaining variation.

One example of a well-documented pattern that is just beginning to be explored mechanistically is that of biotic homogenization. Biotic homogenization is defined as a reduction in biodiversity due to an increase in the similarity of species composition that results from both the introduction of nonnative species and the extinction of endemic species. Species with smaller ranges or of lower abundance that serve to distinguish one location from another become a smaller proportion of the individuals present in a location. One of the primary drivers of this pattern is urbanization. Another way of measuring homogenization, apart from a decrease in native or rare species, is by studying patterns of functional or phylogenetic diversity in species pools and describing homogenization as a reduction in represented evolutionary lineages or functional roles in a community. This pattern has been shown in many taxa, including plants, fish, and birds, along urban-rural gradients. Other taxa, such as herbivorous arthropods, have much more variable relationships with urbanization and tend to respond to within-city environmental heterogeneity more strongly than they respond to urban-rural gradients.

Incorporating functional and phylogenetic metrics begins to help us move beyond these patterns to begin to understand mechanisms of biotic homogenization. For example, researchers in Germany, by demonstrating shifts in functional trait distributions between urban and rural plant communities, were able to hypothesize which plant functions were most strongly selected for versus against by urbanization. Others used a combination of climatic, demographic, and economic variables to explain urban tree composition. Another study compared the relative importance of local exotic tree dominance to levels of surrounding urbanization and habitat connectivity on the alpha and beta diversity of understory plant communities. A study that contrasted the relative importance of local environmental conditions, urban land matrix composition, and habitat connectivity found that the structuring influence of these environmental variables on urban arthropod communities varied depending on the dispersal ability of the species. The contributions of the Baltimore Ecosystem Study toward understanding community structuring include studies of variation in human legacy effects in vacant lots on herbaceous plant community structure and investigations of turnover in woody plant composition across land use gradients. Both studies demonstrated that variation in beta diversity, as a result of either contemporary or legacy land use differences, contributes more to regional biodiversity than variation in alpha diversity; that is, heterogeneity in human impacts creates strong patterns of heterogeneity in biodiversity.

CONCLUSION

We offer here a conceptual model for understanding the role people play in shaping urban biodiversity patterns at multiple spatial scales. Assemblage structure is distinguished between two extremes of a continuum of human intervention. Where humans intervene little in the assembly process, species coexist as a function of local environmental constraints. However, when humans are quite involved in assemblage structure, high variability in human behavior drives compositional turnover up. This is one explanation for high biodiversity in cities. Complementing studies of turnover with those of local and regional diversity yields insights into the mechanisms that generate biodiversity at multiple scales.

CHAPTER TWELVE

Lawns as Common Ground for Society and the Flux of Water and Nutrients

Peter M. Groffman, J. Morgan Grove, Dexter Locke, Austin Troy, Jarlath O'Neil-Dunne, and Weiqi Zhou

IN BRIEF

- Lawns and residential lands are an important and increasing land cover and land use type, and there is great concern about their ecological value and environmental performance.
- Long-term integrated social-ecological studies show that the socio-ecology of lawns and residential land use is much more complex than originally thought.
- Lawns are less intensively managed and have produced less nitrogen pollution of water and air than originally expected by Baltimore Ecosystem Study (BES) investigators.
- Long-term studies of lawns and residential lands have facilitated integration of biophysical and social science research within BES.
- Long-term studies of lawns and residential lands have produced results that can improve the environmental performance of urban ecosystems and landscapes.
- Novel methods have been required to examine social-ecological dynamics of residential land uses because of the fine-scale, spatial heterogeneity of urban areas. These methods include high-resolution characterization of land cover; integration with demographic, social, economic, and built data; and characterizing individual property parcels with these physical, biological, social, and built data.

INTRODUCTION

Lawns and residential lands are two of the most obvious components of urban ecosystems. While urban ecosystems are a heterogeneous and variable mix of paved surfaces, trees, shrubs, and grass, grass itself is a dominant land cover, representing 20–30 percent of typical residential parcels. There are over 150,000 km^2 of lawns in the United States. This is larger than the area of any irrigated crop in the country and is roughly equal to the area of the northern forest of Maine, New Hampshire, Vermont, and New York.

Early ideas about lawns in BES were rooted in theory and data relevant to highly disturbed, heavily managed ecosystems with significant inputs and outputs to surrounding environments, such as agricultural ecosystems. Early ecological analyses of lawns focused on concerns about their environmental performance, especially outputs of nutrients and pesticides and intensive use of energy and water. These analyses also addressed some of the social-ecological aspects of lawns and residential lands with ideas about the benefits or ecosystem services that people derive from lawns and some of the philosophical, emotional, social, and political motivations behind the establishment and maintenance of lawns.

More recently, we have attempted to use theories and concepts from grasslands and rangelands in our studies of lawns, developing the term "urban grasslands," which we define as ecosystems dominated by turf-forming species created and maintained by humans for aesthetic and recreational (not grazing) purposes. We use this term to indicate that urban grassland ecosystems cover significant areas and have coherent patterns of ecosystem processes that can be evaluated using the same approaches used to study other ecosystem types like forests, rangelands, and prairies.

While ecologists have paid some attention to lawns over the past fifty years, other disciplines have paid much greater academic and commercial attention. Turfgrass science research and management programs are active at most land grant universities in the United States, and there is a multibillion dollar industry associated with the production, establishment, and management of lawns. The focus of these efforts has primarily been on aesthetics and practical uses of lawns, but there has also been extensive analysis of environmental performance, driven by concerns identified by ecologists and others who have noted the potential for high inputs of fertilizer, pesticides, and water on lawns. Many of these analyses have suggested that the environmental performance of lawns is better than expected, with less runoff of water and fewer contaminants and less leaching of contaminants to groundwater than expected given the amount of fertilizer applied. Thus two very different threads of theory and practice, one from basic ecology and one from applied turfgrass science, have provided a foundation for lawn research in BES.

We have taken multiple approaches to study lawns in BES. Our first effort was to establish a network of long-term biogeochemical study plots (box 12.1) so that we could compare the soil carbon and nitrogen cycles in lawns with forests, the dominant natural ecosystem type in the region, and with agriculture, the other dominant human land use in the region over the past three hundred years. These plots have provided long-term data on hydrologic and gaseous outputs of nitrogen, a great environmental concern in the region. We also established an early focus on lawn management practices and how they varied within our study watersheds. These early efforts were driven initially by the need for detailed input data for watershed nitrogen budget analyses (chapter 9) but rapidly developed into a key platform for integrated social-ecological research. These early studies have led to a series of efforts to conduct integrated research with a focus on practices on actual lawns across socio-demographic gradients within Baltimore and across the United States. In the sections below we trace the evolution of this research and develop the idea that lawn research is an ideal platform—or a common ground—for social-ecological research addressing society and the flux of water and nutrients.

LONG-TERM BIOGEOCHEMICAL RESEARCH PLOTS

One of the earliest efforts in BES was the establishment of long-term plots for comparative analysis of soil biogeochemical variables in lawns and forests. We established eight plots in urban (Leakin and Hillsdale Parks) and rural (Oregon Ridge Park) forested parks and four grassland plots on two academic campuses (McDonogh School and University of Maryland Baltimore County [UMBC]). The forest plots provided two contrasts: urban versus rural atmospheric conditions and soils, as they encompassed the two most common soil types in the region. The lawn plots provided a gradient of management intensity, with the McDonogh plots receiving no fertilizer or pesticides, one of the UMBC plots receiving moderate fertilizer (~100 kg N ha^{-1} y^{-1}) and occasional herbicide applications and the other UMBC plot

BOX 12.1. KEY TERMS

Biogeochemistry: The study of how biological, chemical, and geological factors interact to control the fluxes of energy, water, and matter across the surface of the Earth.

Carbon dioxide (CO_2): A gas present in the atmosphere, presently at ~400 ppm, produced by biological respiration and consumed by photosynthesis. It plays an important role in absorbing infrared radiation in the atmosphere, i.e., the greenhouse effect.

Lysimeters: Devices used to sample water that percolates through the soil profile. These can involve suction that pulls water out of the soil (tension lysimeters) or passive devices that sample water percolating by gravity (zero tension lysimeters),

Methane (CH_4): A gas present in the atmosphere at ~1.7 ppm produced by anaerobic respiration and consumed by a specialized group of aerobic bacteria in soil. It plays an important role in absorbing infrared radiation in the atmosphere, i.e., the greenhouse effect.

Nitrogen retention: The ability of ecosystems to absorb nitrogen added from the atmosphere, fertilizer, or hydrologic sources.

Nitrous oxide (N_2O): A gas present in the atmosphere at ~300 ppb produced and consumed by several nitrogen cycling processes in soil. It plays an important role in absorbing infrared radiation in the atmosphere, i.e., the greenhouse effect.

Soil to atmosphere gas flux: The ability of soil to produce and consume greenhouse gases (CO_2, CH_4, N_2O) is an important ecosystem function and service.

Urban grasslands: Ecosystems dominated by turf-forming species created and maintained by humans for aesthetic and recreational (not grazing) purposes.

receiving more intensive management with higher fertilizer (~200 kg N ha^{-1} y^{-1}) and more regular pesticide application. This network of plots is clearly not an ideal experimental design providing controlled contrasts of multiple factors in representative components of the urban environment. Rather, these were plots we considered to be representative of the major ecosystem types in our study region where we could have access and control for long-term studies. The limitations of these plots have driven us to constantly compare them with a wider range of lawns in the region to ensure that the information they produce is relevant.

The long-term study plots were instrumented with lysimeters (tension and zero tension) to sample water that percolates through the soil profile, with chambers that allow for quantification of fluxes of gases (carbon dioxide—CO_2, nitrous oxide—N_2O, methane—CH_4) from soil to the atmosphere and probes for measurements of soil temperature and moisture. This instrumentation has produced some interesting and surprising results. First, leaching of nitrate (NO_3^-), the most mobile form of nitrogen, was not as high as expected. Lawns had higher leaching than forests (figure 12.1), but the differences were not as large as expected given the differences in input. Even more interesting, there were no systematic relationships between inputs and outputs. While the McDonogh grass plots received less fertilizer than the UMBC plots, they had higher NO_3^- leaching. And although the two UMBC plots differed markedly in fertilizer input, they had similar leaching. These data as well as results from studies conducted by other investigators signaled to us that the nitrogen cycle in these urban grasslands was more complex and retentive than we had originally thought.

The lawns nevertheless had significant hydrologic losses of NO_3^-. If the concentration data are combined with estimates of water flow, the forest plots consistently yielded less than 3 kg N ha^{-1} y^{-1} of leaching, considerably less than estimates of atmospheric deposition in the region (8–12 kg N ha^{-1} y^{-1}). The lawns produced from 1.4 (in a very dry year) to 25 kg N ha^{-1} y^{-1}. So even though the differences between lawns and forests were not as big as expected and there was evidence of significant retention of nitrogen in the lawns, they are still important sources of reactive nitrogen to the environment.

Surprising results were also evident in our soil to atmosphere gas flux data. We had expected to find high fluxes of N_2O, a potent greenhouse gas, from the grass plots, as fertilizer input is a strong driver of such fluxes. Studies in other areas, especially irrigated lawns in the arid U.S. West, have found very high N_2O fluxes from lawns. Instead, we found that N_2O fluxes were generally lower in lawns than in forests (figure 12.2). And, as with the leaching data, there was no systematic response to fertilizer input; that is, the McDonogh plots which received less fertilizer than the UMBC plots had

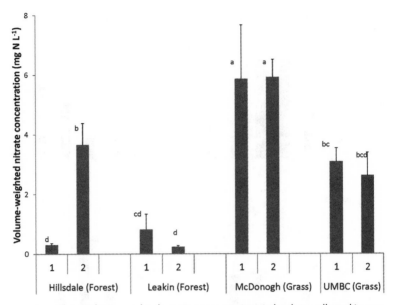

FIGURE 12.1 Volume-weighted nitrate concentrations in leachate collected in zero tension lysimeters in four forest and four grass plots in the Baltimore metropolitan area. Values are mean (standard error) of three water years (2002–4). Bars with different superscripts are significantly different at $p < 0.05$. (Reprinted with permission of the American Society of Agronomy, from Groffman et al., *Journal of Environmental Quality* 38, © 2009, conveyed through Copyright Clearance Center)

higher N_2O fluxes, and the two UMBC plots that had very different fertilizer input had very similar (and low) N_2O fluxes.

The long-term data series of N_2O flux was also interesting and surprising. We observed an increase in flux in 2003 and 2004 (figure 12.3), which were wet years with very dynamic nitrogen cycling and loss in our study watersheds. However, the increase was much more marked in the forest plots than in the grass plots, suggesting the nitrogen cycle in lawns may be less susceptible to hydroclimatic disruption than that in forests.

A major factor underlying the complexity of nitrogen cycling and retention in the lawn plots is likely high flux of carbon from the atmosphere into plants and then into soil microbial populations. High carbon flux in lawns is evident from the long-term patterns of soil to atmosphere CO_2 flux (figure 12.2b), that is, this flux was consistently higher from lawns than from forests, suggesting that there are high rates of both carbon and nitro-

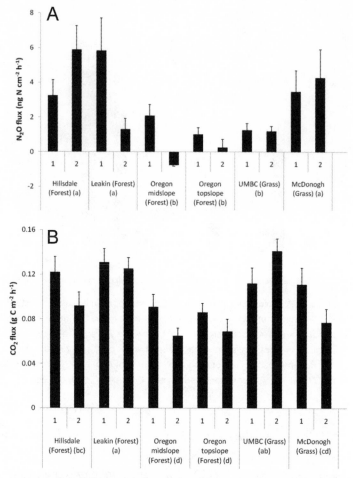

FIGURE 12.2 Soil to atmosphere fluxes of N_2O (A) and CO_2 (B) from forest and grass plots in the Baltimore metropolitan area. Values are means of all fluxes measured in three chambers per plot from June 2001 through May 2004. Sites followed by different letters parenthetically in the site labels are significantly different at $p < 0.05$. (Reprinted with permission of the American Society of Agronomy, from Groffman et al., *Journal of Environmental Quality* 38, © 2009, conveyed through Copyright Clearance Center; for other forest versus lawn comparison, see Raciti et al. [2008], Raciti et al. [2011a, 2011b, 2011c], Martinez et al. [2014], and Waters et al. [2014])

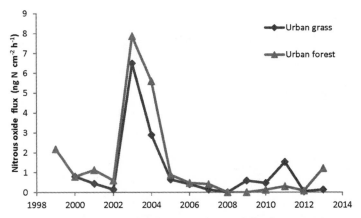

FIGURE 12.3 Mean annual soil to atmosphere N_2O flux from two urban forests (1998–2014) and two urban lawns (2001–14) in the Baltimore metropolitan area. (Data from Groffman et al. [2009]; also see Martinez et al. [2014])

gen cycling in lawn soils. High carbon cycling is also likely driven by the higher temperatures in lawns than in forests, that is, mean annual average temperature at 10 cm depth ranged from 13.5° to 15.0° C in lawns and 12.2° to 12.6° C in forests over an eight-year period.

The major data stream that behaved as expected in our long-term study plots was soil to atmosphere CH_4 flux. Upland soils are known to have the capacity to remove CH_4, a potent greenhouse gas, from the atmosphere. However, this capacity is known to be susceptible to inhibition by soil disturbance and especially by nitrogen inputs. We found that CH_4 uptake was reduced by approximately 50 percent in the urban forest plots compared to the rural forest plots and was completely eliminated in the lawn plots (figure 12.4). Subsequent studies determined that this inhibition was linked to long-term increases in nitrogen enrichment and cycling in the urban forest and lawn soils.

PROCESS STUDIES

As is common in long-term ecological research projects, the long-term study plots were an effective platform for more detailed process-level research. The surprisingly high nitrogen retention that we observed motivated an isotope tracer study, where small amounts of fertilizer enriched with the stable isotope ^{15}N were added to a series of forest and lawn plots, and the

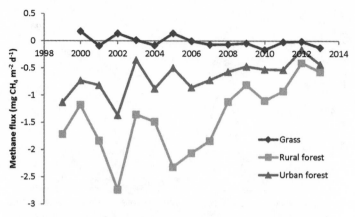

FIGURE 12.4 Mean annual soil to atmosphere CH$_4$ flux from two urban and two rural forests (1998–2010) and two urban lawns (2001–10) in the Baltimore metropolitan area. (Data from Groffman et al. [2006], Groffman and Pouyat [2009], and Costa and Groffman [2013])

movement of this tracer was followed into aboveground and belowground plant material and detritus, soil microbial biomass, and soil organic matter. These studies confirmed that lawns have a high potential for nitrogen retention. After one year we were able to recover more of the tracer in the grass plots than in the forest plots, suggesting that more of the N that we added was lost to the environment from the forests than from the lawns.

We were also keen to make measurements in actual residential parcels to determine whether our long-term study plots were relevant to "real-world" conditions. A first step in this effort was an assessment of home-owner practices through a detailed, door-to-door survey in two of the main BES long-term study watersheds; one exurban area with large lots and expensive homes and one suburban area with older, smaller, less expensive homes. This survey produced surprising results in that fertilization was less common than we expected, ranging from approximately 50–75 percent and was higher in the older, denser, less wealthy neighborhood (Law et al. 2004).

The survey results motivated a more comprehensive analysis of variation in lawn carbon and nitrogen cycling. The High Ecological Resolution Classification for Urban Landscapes and Environmental Systems (HERCULES) system was used to produce an experimental design comparing 32 actual residential parcels with different tree density (driven largely by previous land use, that is, forest versus agriculture) and structure density (larger versus smaller lawns). These studies, which included comparison with the 8 forested long-term study plots, confirmed that lawns in the Baltimore region have high capacity for nitrogen retention, driven by active carbon cycling, and that our long-term study plots are generally representative of lawns in the region.

These studies on actual residential parcels also produced some surprising mechanistic insights into lawn biogeochemistry. First, we were surprised that residential soil profiles were largely intact, as we had expected lawn soil profiles to be highly compacted and to show evidence of imported or altered soil materials. Second, residential soils had higher carbon and nitrogen content than soils from the forested reference plots (6.95 versus 5.44 kg C/m^2 and 552 versus 403 g N/m^2), which was unanticipated given the much higher aboveground biomass and lack of disturbance in the forested plots. It was particularly notable that much of the carbon and nitrogen accumulation in the residential soils occurred at depth (30–100 cm) in the soil profile. We also observed strong relationships among carbon and nitrogen content and lawn age, but only at sites that were previously in agriculture. Rates of N accumulation at these sites were roughly equal to estimated fertilizer N inputs at the sites, confirming a high capacity for N retention.

Detailed nitrogen cycle process results from the actual residential parcels also confirmed results from the long-term study plots and increased our understanding of lawn biogeochemistry. Consistent with the surprisingly low NO_3^- leaching we observed in the long-term study plots, soil NO_3^- pools and internal NO_3^- production by nitrification were higher in residential parcels than in forested reference plots, but they were not as high as expected, that is, they were comparable to deciduous forest stands in other studies. Also consistent with results from the long-term study plots was the observation that homeowner management practices—fertilization and irrigation—were not predictive of NO_3^- availability or production, suggesting that active carbon and nitrogen cycling are key drivers of the environmental performance of lawns.

THE SOCIAL ECOLOGY OF LAWNS AND RESIDENTIAL LANDS

Novel methods have been required in BES to examine the social-ecological dynamics of lawns and residential lands because of the fine-scale, spatial

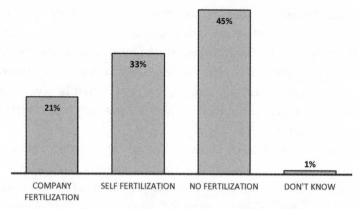

FIGURE 12.5 Fertilization practices of 496 homeowners in the Baltimore metropolitan region surveyed in summer 2008. (Data from Fraser et al. [2012])

heterogeneity of urban areas. These methods have included a combination of intensive surveys of household perceptions, attitudes, and behaviors; improving the characterization of urban heterogeneity patterns using remote sensing and geographic information systems; and integration with existing administrative and marketing data.

Our survey of approximately 500 households across Baltimore found that only 50 percent of households applied fertilizer (figure 12.5) and that the amount of fertilizer varied widely from 10 kg to 679 kg N ha^{-1} y^{-1}, with a mean of 116 kg N ha^{-1} y^{-1}. The high variation in the amount of fertilizer applied was likely driven by the large percentage of households that carried out their own fertilization. We assume that individual practices vary much more than commercial practices, and of the households who applied fertilizer, more than 60 percent of those applications were performed by the homeowner.

The most challenging result from the lawn care practice survey was the lack of relationships between fertilization rate and a variety of potential explanations, including lawn area, lot size, house size, house age, house value, and lawn greenness as indicated by normalized difference vegetation index (figure 12.6). None of these factors was significantly related to the amount of fertilizer applied, raising fundamental questions about the social-ecological factors influencing fertilizer use.

The uncertainty in the factors influencing fertilizer use led us to con-

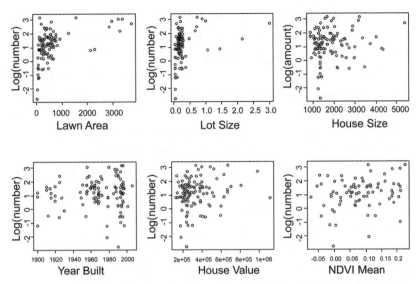

FIGURE 12.6 Relationships between the amount of fertilizer applied and lawn area, lot size, house size, year built, house value, and normalized difference vegetation index (NDVI). (Data from Fraser et al. [2012]; for other social science and related analyses of lawns, see Law et al. [2004], Groffman et al. [2014], Polsky et al. [2014], Groffman et al. [2016], Larson et al. [2016], Groffman et al. [2017], and Locke et al. [2017 and 2018])

sider alternative explanations and formulate new theories of lawn care behaviors in residential lands. We developed the idea that a household's land management decisions are influenced by its desire to uphold the prestige of its community and outwardly express its membership in a given lifestyle group. From this perspective, housing and yard styles, green grass, and tree and shrub plantings are status symbols, reflecting not only the different types of neighborhoods to which people belong but also people's propensity to use their yard to express their belonging in a certain social group or class. We called this explanation the Ecology of Prestige theory, drawing from classic reference group behavior theory from sociology.

To test our ecology of prestige theory we developed detailed studies that focused on specific household and neighborhood socioeconomic characteristics as predictors of residential lawn care expenditures and lawn greenness. These studies examined relationships between population, social stratification (income, education, and race), lifestyle behavior, and housing age as predictors of lawn care expenditures and lawn greenness. Lifestyle behavior was indexed using Claritas, Inc.'s PRIZM™ (Potential Rating Index for Zipcode Markets) marketing categorization system. Despite the name, PRIZM is also a block group-level product, which is a much smaller spatial unit, more socially homogeneous than zipcodes, and allows for linkages to other U.S. Census data. This system segments the American population and their urban, suburban, and rural neighborhoods into clusters using census data related to household education, income, occupation, race/ancestry, family composition, and housing and classifies neighborhoods by social rank (income, education), household (life stage, size, etc.), mobility (length of residence), ethnicity (race, foreign versus U.S.-born, etc.), urbanization (population, housing density, etc.), and housing (owner and renter status, home values, etc.). The second objective of the PRIZM classification system is to associate these clusters with consumer spending patterns and household tastes and attitudes using additional data such as market research surveys, public opinion polls, and point-of-purchase receipts.

In our analysis the PRIZM lifestyle segmentation was a useful predictor of both lawn extent, lawn care expenditures, and lawn greenness. The ecology of prestige theory has provided a basis for understanding the variation in lawn management within cities and may be key to understanding how this management can be altered to achieve social and environmental objectives.

While we have found the ecology of prestige to be important for understanding lawn care behaviors, it is also important to recognize that management of residential lands is a multiscale process. For instance, neighborhood-level factors such as membership in homeowners' associations (HOAs) may influence yard care decisions such as fertilization. For example, we found in Baltimore that households in neighborhoods with HOAs had a greater probability of higher rates of fertilizer application than households in neighborhoods with neighborhood associations, which did not have legally binding rules. Further, the probability of applying lawn fertilizers was greater in neighborhoods with higher neighborhood social cohesion, suggesting that social norms and expectations affect lawn care practices.

Our production and use of high-resolution land cover data (<1 m) and integration with high-resolution, parcel-based data on landownership have enabled us to analyze how land cover varies among land use types and, within residential lands, among different neighborhood types. Using these methods, we discovered that most of the existing canopy cover and opportunities for increasing canopy cover are found in residential lands, a finding in BES that has since been supported in many other cities. Further, using our ecology of prestige theory and the PRIZM marketing data, we found that lifestyle factors, such as family size and life stage and ethnicity appear to be stronger predictors of variations in residential canopy cover than population density or socioeconomic status. Second, different social groups' need for status and group identity produces neighborhood-based and geographically coherent differences in ecological structures and functions. Finally, there are important temporal lags and legacies associated with the distribution of residential canopy cover because trees can live for long periods of time, while the neighborhood characteristics of residential land uses may change over shorter periods. For example, neighborhood lifestyle characteristics in the 1960s were the best predictors of canopy cover existing in the 2000s. These results suggest the need not only to understand the heterogeneous legacies of past housing markets and land management and their effects on the current spatial heterogeneity of ecological structures and functions, but also to make predictions about changes in the future. For instance, because the data on the 1960s improve our understanding of the present distribution of trees in Baltimore, the 2000s will matter for city managers as they attempt to green the city over the next forty years.

The distribution of vegetation structure and land management practices on residential lands has important implications for social processes like social cohesion, neighborhood satisfaction, and social order (crime) and for ecological processes such as the urban heat island. By combining our land cover data, parcel data, and household telephone survey data, we found that differences in neighborhood desirability, environmental satisfaction, quality of life, and social cohesion were positively associated with variations in canopy cover. Our research relating residential vegetation and social order found that while a ~10 percent increase in canopy cover reduced crime by ~11 percent (controlling for other factors such as income and race), residential land management had significant effects, supporting Joan Iverson Nassauer's theory of "cues to care" (chapter 6). Factors that were negatively associated with crime included the presence of yard trees, garden hoses/sprinklers, and lawns in addition to the percentage of pervious area in a yard, while factors positively associated with crime included presence of litter, desiccation of the lawn, and lack of lawn cutting. The presence of trees also has a significant effect on the urban heat island and household exposure to heat stress, with a variation of ~30° F from the coolest neighbor-hoods (76° F) and the hottest (106° F) on a summer day.

THE ONGOING CHALLENGE OF LAWNS AND RESIDENTIAL LANDS

Lawns and residential lands will continue to be an important topic of re-search in Baltimore for at least the next ten to twenty years. A major con-cern is whether our past and current studies, which have painted a surpris-ingly optimistic picture of the environmental performance and social value of lawns, are complete and accurate. We are initiating studies to determine whether we have missed lawns with poor environmental performance, es-pecially those developed on disturbed and compacted subsoils that might have much less active carbon cycling and nitrogen retention than the lawns we have studied so far, which overwhelmingly have relatively intact soil profiles.

We are also keen to resolve ongoing biogeochemical and social-ecolog-ical mysteries that have emerged from our past and current studies. There is great uncertainty about why carbon is accumulating at depth in the soils beneath lawns and whether this is driven by root production, decomposi-tion, or translocation of dissolved organic carbon. There is great interest in determining the limits of nitrogen retention. As lawns age and/or fertiliza-tion rates are sustained or increased, will lawn soils become nitrogen satu-rated, leading to increases in hydrologic and/or gaseous losses? Gaseous losses are a particular uncertainty, and there is a great need to reconcile our results showing low N_2O fluxes in lawns relative to forests with studies from other (mostly arid) regions that have shown higher fluxes in lawns relative to the native ecosystems that they replaced.

There is a great need to develop landscape-, regional-, and continental-scale contexts for studies of lawns and residential lands. While most of our research has focused on the parcel scale, each parcel is hydrologically con-nected to other parcels and to natural and/or anthropogenic drainage sys-tems. In future studies we hope to understand the nature and extent of these connections and determine how they influence the environmental performance of neighborhoods and watersheds. A key question is whether interventions at the parcel (for example, apply less fertilizer) or neighbor-hood (for example, engineer nitrogen sinks in drainage swales) scale are more or less efficient for achieving environmental objectives.

At coarser scales there is interest in determining whether the biogeochemical and social patterns we observe in Baltimore occur in other cities in the region, in the United States, and across the globe. A current project is comparing biogeochemical and social variables in six cities, Boston, Baltimore, Miami, Minneapolis-St. Paul, Phoenix, and Los Angeles, to determine whether residential land use change is homogenizing the continent by creating ecosystems that are more similar to each other than to the native ecosystems they replaced. A major component of this research is to test whether our ideas about lifestyle as a driver of lawn management are robust at larger scales by determining whether practices in similar neighborhoods in different cities are more similar than practices in different neighborhoods in the same city. This research also includes detailed interviews with a set of homeowners in each city, allowing us to explore their values and motivations related to lawn management more intensively. Lawns and residential lands are likely to continue to be an important platform for theoretical and practical advances in social-ecological science.

CHAPTER THIRTEEN

Long-Term Trends in Urban Forest Succession

Grace S. Brush and Daniel J. Bain

IN BRIEF

- The forested watershed that existed at the time of colonization has been reduced over time to many small patches of forest cover.
- The species composition of the patches has remained closely related to the geology and soils.
- Afforested patches of early cleared land consist primarily of species with late successional life characteristics, whereas afforested patches on more recently cleared land consist of species with early successional characteristics.
- Wet-adapted floodplain species are being replaced by dry-adapted upland species in streamside, riparian areas.
- Exotic species make up approximately half of the non-woody vegetation but only a small percentage of the woody vegetation.

THE PRECOLONIAL LANDSCAPE

The forest communities that occupied the area that is Baltimore in precolonial time developed some twelve thousand years ago as plants migrated to their present locations from areas south and east of the glacial border during Holocene time (box 13.1). This chapter follows the changes in plant cover from before European colonization through the period of deforesta-

BOX 13.1. KEY TERMS

Afforestation: In modern American usage, the growth of new forest on lands that had been cleared or subjected to large natural disturbance. The opposite of deforestation, or the cutting or other removal of forest. Note that Europeans classically use this word to refer to a forest land use rather than to an actual stand of trees.

Amphibolite: A coarse-grained metamorphic rock type.

Early successional species: Plant species adapted to colonize or flourish in areas newly opened by intense human or natural disturbances. These species are usually characterized by high resource demand, adaptation to high light levels, rapid growth, and abundant reproduction by small seeds. In contrast, late successional species are adapted to lower resource availabilities, grow more slowly, and reproduce via fewer, highly provisioned seeds.

Endemic: An organism restricted to a locality or region.

Exotic species: Species accidentally or purposefully introduced from regions or continents beyond a focal area. Some may become invasive, and some may become naturalized in the target environment.

Flora: The collection of plants characteristic of a region.

Fragmentation: Splitting a continuous forest or other vegetation type into discontinuous islands or other spatially separated units.

Geomorphology: The physical features of the Earth's surface, relating to the underlying geology and the surface forms of the land.

Gneiss: An intensely metamorphosed rock, showing banding of different mineral layers. Often high in quartz and mica.

Holocene: Referring to the Earth's geological time horizons, the Holocene began with the retreat of the glaciers at the end of the last, or Pleistocene, ice age, approximately twelve thousand years ago. Although the term translates as "whole recent," many experts consider that the intense global alterations of the earth's environment and surface by human presence and activities have established a new geological era, the Anthropocene.

Hydrological drought: Dry conditions created not by natural topographic features and drainage or by climatic patterns of rain or snow fall but by human modification of hydrological flows on the surface and below ground. Water and drainage infrastructure, spatial and temporal patterns of irrigation, and impervious surfaces are elements contributing to hydrological drought.

Monospecific: Dominated by one plant species.

Riparian: Streamside habitats, often naturally interacting with the water table or with stream flooding.

Saprolite: Partially decomposed bedrock at the base of the soil profile. Saprolite is typically porous and likely to have high clay content.

Schist: A moderately metamorphosed rock, having a conspicuously grained texture and layering, with quartz an important layering mineral.

Serpentine: A group of minerals especially rich in magnesium and silicates and sometimes containing heavy metals or forming asbestos. They generally yield soils that are not valued for agriculture and, in the Baltimore region, support sparse woodlands or grasslands.

Succession: The process of change in plant communities through time. Succession is based on the fact that different species arrive at a disturbed site at different rates and are differently adapted to use highly available versus biologically conserved nutrients. Light availability at ground level also tends to decline in uninterrupted successions.

UFORE: Urban Forest Effects Model, replaced by iTree, developed by the USDA Forest Service to characterize urban forests and estimate various potentially useful functions and economic values of trees in cities, towns, and suburbs (https://www.itreetools.org/eco/).

Ultramafic: Igneous rocks with low silica content, predominated by minerals containing iron and magnesium and sometimes also heavy metals. Serpentine is a type of ultramafic rock.

tion and agriculture in postcolonial time and the eventual shift to an urban landscape.

Evidence of precolonial forests based on pollen and seeds preserved in sediment cores collected from various tributaries of the Chesapeake Bay indicate that climate and geomorphology controlled the distribution of plant species on the precolonial landscape. Plant cover progressed from components of an Arctic flora, including spruce and fir twelve thousand years ago, to a warmer- and wet-adapted flora about nine thousand years ago, to the modern flora characterized by drier-adapted taxa, including members of the blueberry family and an increase in oak numbers around six thousand years ago. Human populations in the Chesapeake region were small and had little detectable effect on plant distributions. Following European colonization some four hundred years ago, the sediment cores show widespread changes in the species composition of forests, unrelated to climate and geomorphology. What were the changes and when and why did they happen?

THE POSTCOLONIAL LANDSCAPE

Postcolonial human activities both threatened and enhanced long-term forest resources. The early colonists cut forests to grow food crops, primarily corn and wheat as well as tobacco, to erect buildings for shelter, and to build roads for transport. In order to farm some of the richest land, colonists first had to drain portions of it. Draining the land resulted in rapid changes in the herbaceous plant cover from a predominantly wet to a dry flora, as represented in the pollen profiles of the Chesapeake region (figure 13.1). Human disturbances—cutting of forests and draining of land for agricultural crops—are superimposed on a diverse geologic and hydrologic substrate that controls the distribution of plant species. The Gwynns Falls watershed straddles a great deal of this substrate, stretching from fine-grained schists in the upper part of the watershed to the unconsolidated sediments

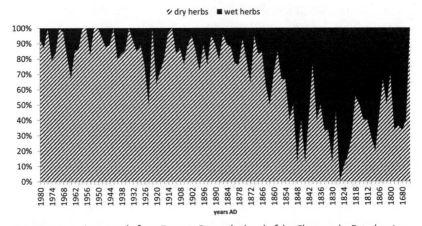

FIGURE 13.1 An example from Furnace Bay at the head of the Chesapeake Bay showing the change from wet, nonarboreal pollen (NAP) to predominately dry NAP approximately two hundred years ago.

of the Coastal Plain at its mouth. Because of the spatial heterogeneity of that substrate, the response of vegetation to disturbance varies widely across the Baltimore region.

Postcolonial deforestation that accompanied agriculture resulted in unmanaged forest patches with greatly reduced timber production. In order to conserve and manage the forest economy, Frederick Besley was appointed the state forester for Maryland in 1906. During his tenure Besley mapped the forests of Maryland, which showed occurrences and sizes of predominantly timber species throughout the state. However, some of these efforts had unintended consequences. For example, Besley was instrumental in having laws passed to control fires, which he felt were a prime contributor to the poor state of the forests. One result of this legislation was the replacement by *Pinus virginiana* (Virginia pine) of the native diverse grassy barrens growing on serpentine rock at Soldiers Delight. In the early 1990s prescribed burns were allowed in Soldiers Delight in the upper Gwynns Falls watershed to restore the grasslands, which contained numerous endemic species.

The history of fragmentation that accompanied deforestation has been compiled for the past ninety years by mapping tree cover for 1915, 1938, 1971, and 1999 and calculating the changes in cover over time. Based on tree cover and the characteristics of different species, the leaf area index has been reconstructed and synthesized in land cover maps for the years sampled. Using Besley's 1916 forest map and aerial photographs from 1938 to 2004, Zhou et al. reconstructed changes in the Gwynns Falls watershed as it became increasingly urbanized. While the total area of forest cover did not change during the period from 1938 to 2004, the mean size of forest patches decreased continuously while the number of patches increased. The configuration and number of patches resulted from forest regrowth on abandoned agricultural fields and land that was initially cleared for urban development. The latter resulted in small regenerated patches ending in greater fragmentation, as the urban areas expanded out from Baltimore City into Baltimore County. Hence the landscape overall has been characterized by an increase in forest edges and a decrease in interior habitat.

Documentation of plant species following widespread deforestation was compiled in two studies: one by Shreve et al. published in 1910, which described all species on each soil type throughout Maryland; and the second by Brush et al. in 1980, which documented relations between woody species and geology/soils also throughout Maryland. Brush et al. showed that species composition is similar both in small and large patches on similar geologic substrates, indicating that for the most part species composition did not shift with change in cover. The study by Shreve et al., carried out in the late 1800s, documented distributions of species after one hundred to two hundred years following colonization, when 70 to 80 percent of the land was deforested. The relationships described in the late 1970s by Brush et al. are, with few exceptions, similar to those reported by Shreve et al. in 1910 for the late 1800s. Disease, changes in fire control, and some unknown causes have resulted in altered distributions of a few species since the time of Shreve. For example, *Castanea dentata* (American chestnut) was the most abundant species in many nineteenth-century forest stands, constituting up to 30 percent of forests on the mesic-dry and dry or well-drained substrates in the Coastal Plain and Piedmont provinces and the Appalachian highlands. However, *C. dentata* was eliminated as a canopy tree in the 1930s by an infection caused by the fungus *Endothia parasitica*. Today, small trees up to about twelve feet in height originating from root sprouts are present, though sparse throughout the forests. Most are infected with the disease and die before they are of sufficient age to flower. More recently, within the last twenty to thirty years, *Tsuga canadensis* (hemlock) stands described by both Shreve and Brush are being decimated by the introduced *Adelges tsugae* (wooly adelgid) infestation. *Tsuga canadensis* occurs generally in monospecific stands; its demise leaves large gaps in the forest. Its replacement is confounded by intensive deer grazing and a huge infestation of the exotic *Microstegium vimenium* (stilt grass).

GWYNNS FALLS VEGETATION TODAY

The Uplands
The spatial patterns of forest fragmentation reflect both human decisions and underlying soil and topographic gradients. Following the arrival of European settlers, the Gwynns Falls watershed was deforested over relatively short periods of time. This clearance was spatially variable. In the lower basin, charcoaling for the Baltimore Company resulted in almost complete deforestation in areas owned by the company. In other parts of the basin, deforestation was likely less complete, as it focused on more fertile portions of the landscape that were preferentially claimed. As agricultural activity

continued, these more fertile areas would have remained open and in use. The differentiation between cleared areas of poor and fertile agricultural soils in all likelihood grew stronger. So, as suburbanization was poised to begin, the landscape was covered largely with forest patches that had been growing since early clearance, areas that had begun to regenerate as they were abandoned through time, and areas that had remained in agricultural production (figure 13.2).

As Baltimore expanded upstream into the watershed, urban areas tended to preferentially occupy flat areas that had been prime agricultural land. Thus while there was a period of afforestation following peak agricultural clearance, patches in early successional stages were more likely to be developed, given that level, well-drained sites were preferred for residential development as well as for agriculture. Therefore, over the period of suburbanization development occurred in areas where younger forests growing on abandoned farmland were removed or on recently abandoned agricultural fields.

However, if we examine the composition of forest communities remaining in these forested patches, the continued influence of the geologic/soils substrate remains. If tree species are divided based on their life cycle characteristics into early successional, for example, most maples, and late successional, for example, most oaks, we can evaluate the status of forest patches. We compared 1938 forest cover in upland patches (our best, oldest available data; Besley does not map patches <5 acres) with early successional and late successional biomass measured in our plots. Plot data show that biomass arising from late and early successional species is not significantly different in the open or forested plots in 1938. However, areas with low available water have more late successional biomass. Thus soil moisture conditions arising from the geologic substrate are influencing the speed and nature of succession.

In order to determine the effect of land use on species composition, we studied the occurrences of fifty tree species in the Gwynns Falls uplands in 2004. Among these fifty species, eight are exotic: *Prunus avium* (sweet cherry), *Pyrus baccata* (Siberian crab apple), *Acer palmatum* (Japanese maple), *Ailanthus altissima* (tree of heaven), *Acer platanoides* (Norway maple), princess tree (*Paulownia tomentosa*), and paper mulberry (*Broussonetia papyrifera*) and *Morus alba* (white mulberry). None have spread widely. The greatest species richness is on amphibolite and schist bedrock types. *Liriodendeon*

Pre European Settlement

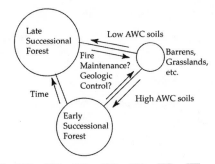

European Agricultural Activity

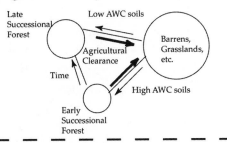

Agricultural Abandonment

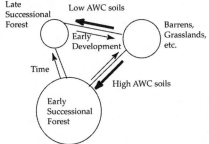

Urbanization

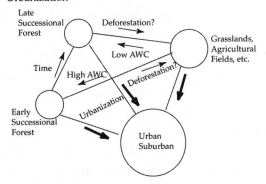

FIGURE 13.2 Conceptual model of forest dynamics in the Gwynns Falls watershed. (AWC is available water content)

Table 13.1

Species restricted to one or two substrates

Those restricted to one substrate are in bold. There are no species that exist only on gneiss.

MARBLE	GNEISS	AMPHIBOLITE	SCHIST	ULTRAMAFICS	QUARTZITE, ETC.
shagbark hickory		**sugar maple**	**silver maple**	**blackjack oak**	shagbark hickory
honey locust		witch hazel	**fire cherry**	**post oak**	
		bitternut hickory	**black birch**	persimmon	
		slippery elm	**pawpaw**	witch hazel	
		beech	**red mulberry**	slippery elm	
		chestnut oak	persimmon	choke cherry	
			bitternut hickory	beech	
			choke cherry	Virginia pine	
			chestnut oak		
			Virginia pine		

tulipifera (tulip poplar) is the dominant tree in the study area, occurring on all geologic substrates and reaching a maximum size for all trees sampled of 104 cm dbh. *Fagus grandifolia* (beech) is the dominant species on soils derived from amphibolite rock as well as on soils formed from ultramafics; there, it occurs as a few small trees. The greatest density of stems per area of all species is on the ultramafics, followed by schist and gneiss (table 13.1).

The lowest biomass (m²/hectare) is on the dry ultramafic substrates, and the highest biomass is on the mesic gneiss and schist. Schist and ultramafics support mostly drought-tolerant species, marble and gneiss a majority of mesic species, and amphibolite many mesic and wet species. Forests in the upland regions where the subsurface consists primarily of thick saprolite weathered from metamorphic rocks with a large water holding capacity, follow a predictable pattern of succession from opportunistic species such as red maple to the more competitive beech and oak species (table 13.2).

Table 13.2

Occurrence of different successional groups in plots arranged by erodibility and forest cover

Species are assigned to different successional stages depending on their life history characteristics.

SUBSTRATE (% PLOTS)	ERODIBILITY INDEX	PERCENT FOREST COVER	EARLY (6 SPECIES)	EARLY TO INTERMEDIATE (5 SPECIES)	INTERMEDIATE (6 SPECIES)	LATE (18 SPECIES)
Marble (12%)	1.5	47	0	4 (80%)	3 (50%)	7 (39%)
Gneiss (8%)	1.2	24	0	2 (40%)	2 (33%)	7 (39%)
Schist (34%)	1.5	43	5 (83%)	4 (80%)	4 (66%)	11 (61%)
Amphibolite (31%)	2.4	51	2 (33%)	3 (60%)	4 (66%)	14 (77%)
Ultramafic (8%)	1.2	24	0	2 (40%)	4 (66%)	14 (77%)
Quartzite, sand, clay, fill (7%)	1.1	32	2 (33%)	5 (100%)	4 (66%)	7 (39%)

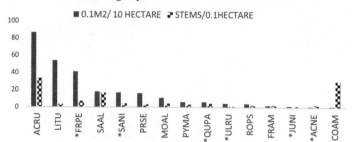

Tree demographics at 0 - 1 m elevation

■ 0.1M2/ 10 HECTARE ⚐ STEMS/0.1HECTARE

ACRU, LITU, *FRPE, SAAL, *SANI, PRSE, MOAL, PYMA, *QUPA, *ULRU, ROPS, FRAM, *JUNI, *ACNE, COAM

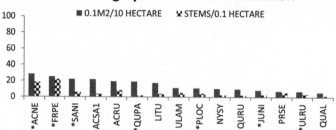

Tree demographics at 1 - 2 m elevation

■ 0.1M2/10 HECTARE ✶ STEMS/0.1 HECTARE

*ACNE, *FRPE, *SANI, *ACSA1, ACRU, *QUPA, LITU, ULAM, *PLOC, NYSY, QURU, *JUNI, PRSE, *ULRU, QUAL

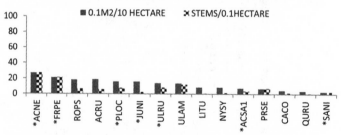

Tree demographics at > 2 m elevation

■ 0.1M2/10 HECTARE ✕ STEMS/0.1HECTARE

*ACNE, *FRPE, ROPS, ACRU, *PLOC, *JUNI, *ULRU, ULAM, LITU, NYSY, *ACSA1, PRSE, CACO, QURU, *SANI

FIGURE 13.3 Demographics of the fifteen most common species at the different floodplain elevations. Abbreviated names of floodplain species on the charts are preceded by an asterisk and are listed in bold in the legend. Species in alphabetical order are **ACNE** *Acer negundo* (box elder), ACPL *Acer platanoides* (Norway maple), ACRU *Acer rubrum* (red maple), **ACSA1** *Acer saccharinum* (silver maple), AIAL *Ailanthus altissima* (tree-of-heaven), CACO *Carya cordiformis* (bitternut hickory), CATO *Carya tomentosa* (mockernut hickory), CORA *Cornus racemosa* (red-panicle dogwood), FRAM *Fraxinus americana* (white ash), **FRPE** *Fraxinus pennsylvanica* (green ash), **JUNI** *Juglans nigra* (black walnut), LITU *Liriodendron tulipifera* (tulip poplar), MOAL *Morus alba* (white mulberry), NYSY *Nyssa sylvatica* (black gum), **PLOC** *Platanus occidentalis* (sycamore), PRPE *Prunus pensylvanica* (pin cherry), PRSE *Prunus serotina* (black cherry), PYMA *Pyrus malus* (domestic apple), QUAL *Quercus alba* (white oak), **QUBI** *Quercus bicolor*

The erodibility index for each plot was compared with percent forest cover. The erodibility index is a measure of soil fertility obtained from the Universal Soil Loss Equation and the Wind Erosion Equation made available for all soils by the National Resources Inventory. The higher the erodibility index number, the more erodible and therefore less fertile are the soils. In the Gwynns Falls, substrates with the highest erodibility index have the greatest percent forest cover. For example, soils derived from the rock amphibolite have an erodibility index of 2.39 and are 51 percent forested, while soils derived from ultramafic rocks have a 1.19 erodibility index and are 24 percent forested. Quartzite, sand/clay, of which there are only five plots, has the lowest erodibility index and is 32 percent forested. Our interpretation is that soils with the lowest erodibility index and hence most fertile would have remained in agriculture for a longer period of time and therefore would have less forest than the more erodible soils, which would have been abandoned as farmland earlier and would have more forest cover today. Furthermore, we hypothesized that the less fertile plots, abandoned earlier, would support late successional species today, whereas early successional species would occur on the more fertile plots with the lower erodibility index.

Riparian Areas

The hydrology of the Gwynns Falls riparian areas has been greatly altered by human occupation. New flooding regimes resulting from impervious surfaces have resulted in thick floodplain deposits separating the water table from the surface. The majority of species on point bars and floodplains, where the bank elevations are < 1 m, are predominantly dry-adapted species. We have referred to this phenomenon as a "hydrological drought" unrelated to climate but instead related to human influence. Riparian areas > 2 m elevation are populated primarily by wet-adapted floodplain species.

A total of forty-eight species were identified from riparian plots, seventeen at 0–1 m elevation above the streambed, thirty-five at 1–2 m elevation, and thirty-seven at > 2 m elevation (figure 13.3). There are a total of six exotic

---

(swamp white oak), **QUPA** *Quercus palustris* (pin oak), QURU *Quercus rubra* (red oak), ROPS *Robinia pseudo-acacia* (black locust), SAAL *Sassafras albidum* (sassafras), **SANI** *Salix nigra* (black willow), ULAM *Ulmus americana* (American elm), **ULRA** *Ulmus rubra* (slippery elm).

tree species and three species found in the eastern United States but not native to the area. The exotic species are *Acer platanoides* (Norway maple), *Ailanthus altissima* (tree of heaven), *Albrizzia julibrissin* (mimosa), *Brousson-etia papyrifera* (paper mulberry), *Pawlonia tomentosa* (princess tree), and *Salix babylonica* (weeping willow). The North American species not native to the area include *Maclura pomifera* (osage orange), *Populus deltoides* (common cottonwood), and *Populus heterophylla* (swamp cottonwood). None of the exotic, nonnative species were found on the lowest elevation. *Acer rubrum* and *Liriodendron tulipifera,* mesic species that occur predominantly in the uplands, are the dominant and co-dominant species at elevations > 1 m. *Acer negundo* (box elder) and *Fraxinus pennsylvanica var. subintegerrima* (green ash), typical floodplain species, are the dominant and co-dominant species on the higher elevations. The largest trees at all three elevations are *L. tulipifera, Quercus rubra* (red oak), and *Robinia pseudo-acacia* (black locust), all of which are considered upland mesic species.

PARKS IN THE GWYNNS FALLS

The history of forests in the Gwynns Falls diverges between parks promulgated by Besley that were established primarily to entice the public into supporting his forestry work compared to stands where native, primarily timber trees were planted and maintained. Large tracts of land were also being donated to or acquired by the city for the purpose of establishing parks where trees were planted and maintained for beautification and to improve city living. The result is that over time Baltimore has become a city of large and small parks, where both native and nonnative trees were planted. Trees were also planted along streets and highways. Decisions on what species to plant were made by various groups, including private organizations, schools, churches, and individual homeowners along with city agencies. Areas of the city developed by the Olmsted Brothers landscape architecture firm in 1904, for example, are characterized by a majority of native species, whereas in other neighborhoods fruit trees were planted in backyards, and in still others exotic cedars, conifers, and species imported for their form and color from all over the world occupy some estates and church properties. In addition to imported trees, vine and shrub species were introduced, including *Hedera helix* (English ivy) and *Ligustrum vulgare* (privet). Hence exotic species have become an important component of the urban plant cover. Unlike in nonurban areas, trees and shrubs planted along streets and in city parks require maintenance, which uses significant resources, and this affects what grows in the city. For a while during the 1950s and 1960s trees were removed and new ones planted, some years more removed than planted and other years vice versa. Since that time city tree cover has been declining. Recent studies using the Urban Forest Effects Model, based on permanent field plot measurements and analysis of historical imagery, show that between 1999 and 2001 the number of trees declined by 4 percent, while between 2001 and 2005 tree cover in Baltimore dropped from 30.4 percent to 28.5 percent. Even though the tree population has been declining in recent years, new trees are coming into the population at an average rate of about four trees per hectare per year, most of this new influx coming through natural regeneration dominated by the native *Fagus grandifolia* (American beech) and the exotic *Ailanthus altissima* (tree of heaven). The current goal of the city is to double its tree canopy within thirty years. This goal is to be accomplished by a number of projects such as TreeBaltimore, an umbrella organization for all city agencies that oversees planting of trees in the city. Decisions of what to plant are made by the people and institutions where the trees are being planted, whether a schoolyard or some other property. The priority is to plant a majority of shade trees, which include mostly native species but also some exotics that were previously planted and are desired by property owners.

EXOTIC SPECIES

Analysis of data from vegetation plots collected throughout the Gwynns Falls watershed in 2004 and 2009 (which do not include parks and trees planted along streets) shows that exotic species are abundant among the herbaceous flora and less so among the woody species. For example, of a total of 153 species recorded over the two years of sampling, 49 are trees, 37 are shrubs, and 67 are herbs. Only 8 of the trees and 9 of the shrubs are nonnative. Of the 67 nonwoody plants, 44 are native and 23 are exotic. The distributions of exotic species show that only a few are present in a majority of plots (figure 13.4). The remainder are restricted to one to three locations. Those that are present throughout, that is, occur in the greatest number of plots or with the greatest frequency, include *Glechoma hederacea* (ground ivy), *Lonicera japonica* (Japanese honeysuckle), and *Ligustrum vulgare* (privet).

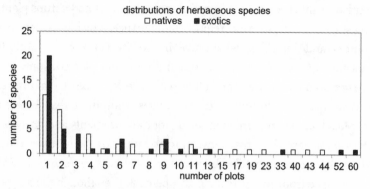

FIGURE 13.4 A few herbaceous species occur in the majority of plots, while the majority of species occur in one or two plots.

CONCLUSIONS

The available forest data in historical records as well as plot studies in this urban-suburban watershed show that the connection between plant species and the underlying substrate from which they receive water and nutrients remains a controlling factor in their distributions. When agricultural fields are abandoned, crop species, primarily grasses, are replaced by the tree and shrub species that occupied the different soil types prior to agriculture. The distributions change only when the underlying substrate is fundamentally altered, such as the lowering of groundwater in the riparian areas. Herbaceous species dominate the exotic flora. Species planted along streets and in parks represent artificial situations that require maintenance in order to survive. All of the distributions noted in this chapter occur under a particular climate regime. With climate change, we can expect that these distributions will change depending on how a change in climate, drier or wetter, warmer or colder, will affect water and nutrient availability as well as ability to adapt to different temperatures and changing seasons.

Acknowledgments We thank Geoffrey Buckley, David Nowak, and Guy Hager for information they kindly provided for this chapter.

CHAPTER FOURTEEN

Principles of Urban Ecological Science

INSIGHTS FROM THE BALTIMORE SCHOOL
OF URBAN ECOLOGY

Mary L. Cadenasso and Steward T. A. Pickett

IN BRIEF

- Urban ecological science as an integrative approach to social-ecological systems is a young discipline.
- The Baltimore Ecosystem Study (BES) has made discoveries that help identify general principles for this growing science.
- This chapter presents a system of thirteen principles to represent the breadth of findings from BES.
- The principles are general tools for understanding and serve to frame social-ecological research and communication, summarize key empirical insights, point to methodological approaches, and suggest comparisons within and among urbanized regions.
- Principles are part of nested hierarchical conceptual structures. Above the level of principles still larger generalizations appear, while below the principles are found the detailed empirical results and specific models discussed throughout this book.

INTRODUCTION

Urban ecology is a rapidly developing discipline. As an interdisciplinary arena of scholarship, urban ecology incorporates theoretical, conceptual, empirical, and methodological elements from a variety of sources beyond biophysical ecology, including engineering, geography, hydrology, urban design, and a vast array of the social sciences. The city as a focus of study has gained traction among ecologists more recently than in many other disciplines. In fact, urban ecology is considered a frontier science, meaning that it is still in flux as a science; it is incomplete and contains ideas that may be controversial. These characteristics are in contrast to a textbook science made up of a well-accepted body of understanding that is relatively well integrated, complete, and uncontroversial. Urban ecology is based on principles developed by the field of ecology but shares boundaries with many other established disciplines and integrates relevant principles from these disciplines in an effort to build its own understandings as it moves toward becoming a textbook science.

The goal of this chapter is not to provide a history of the development of the field of urban ecology but to propose a set of general understandings, or principles, which can assist in the formulation of a comprehensive urban ecological theory (box 14.1). Principles can be distilled from empirical generalizations or conceptual models, and they can represent different stages of acceptance from tentative to firm. Developing a comprehensive theory is an overarching goal of urban ecology, but no such theory has yet emerged, leaving the field well within the realm of frontier science. Steady progress is being made, however, on various components that would contribute to a complete theory of urban ecology. Empirical research, model development and testing, comparative analyses, framework generation, and theoretical advancement have all played a role in urban ecological science. This book focuses on long-term social-ecological research in Baltimore, Maryland. We will draw on insights gleaned from some two decades of work by a large number of investigators who make up BES, collaboratively gaining understanding of Baltimore as a social-ecological system. Where possible, we illustrate each principle with research conducted within BES. These illustrations may not be a complete representation of the principle, as principles emerged from knowledge gained through various research efforts as opposed to the principles serving as an a priori guide to research. In addition, some of these generalizations have emerged as a result of comparing Baltimore to other cities and processes of urbanization, so they cannot necessarily be exemplified only by referring to research conducted in Baltimore. This range of principles reflects the three fundamental questions focused on structure, flux, and understanding that form the foundation of BES (chapter 1).

BOX 14.1. KEY TERMS

Anoxic: Lacking oxygen; a condition in soils required for denitrification, by which nitrate is converted to nitrogen gas.

Continuum of urbanity: A conceptual framework indicating that any site in an extensive urban region is both urban and rural to some degree. The sites differ in the mix of livelihoods each supports, the lifestyles each engenders, the connectivities each participates in, and the effects on and local constraints of the ecosystem at each place.

CSE: City-suburban-exurban systems are spatially extensive mosaics that contain patches representing different degrees and kinds of urbanization. They typically cross various jurisdictional boundaries and can achieve regional extent.

Environmental justice: The ethical concern that environmental benefits and burdens are fairly shared among different social groups and that the processes by which benefits and burdens are apportioned is open and fair.

Framework: A network of models and concepts representing the structure of explanations, predictions, and generalizations in a study area. Frameworks organize existing empirical findings and point to gaps in knowledge. Frameworks are a major structuring element of theories.

GF: The Gwynns Falls stream or the associated watershed.

HERCULES: A novel land cover classification for cities, suburbs, towns, and exurbs that resolves those areas into patches described as hybrids of elements that originate from biological, physical, and social processes.

HOLC: Home Owners' Loan Corporation, a New Deal–era federal agency charged with rating neighborhoods by creditworthiness.

Human ecosystem: A concept that recognizes that ecosystems that contain humans or reflect human influence can be considered to consist of biological, physical, social, and constructed or built components, all interacting with each other.

Legacy: A social, built, or ecological pattern existing at a particular time whose effects can exist or appear at later times.

LST: Land surface temperature, a metric useful for quantifying and modeling urban heat island effects (see chapter 10).

Metaprinciple: A very general principle that draws the insights from

several more focused principles together into an overarching generalization.

Phylogenetic: The lineages of evolutionary descent among organisms.

Principle: A conceptually oriented statement, generalizing across empirical results or indicating a general expectation about the structure
or function of a system.

Social-Ecological System (SES): A system composed of interacting social, biological, and physical components. This concept emphasizes
that urban areas as an SES reflect feedbacks across both social and
biophysical components and processes. The SES concept is similar
to the human ecosystem idea.

Theory: A conceptual device to support understanding of a specified
domain of interest. Theories consist of stated scope, assumptions,
definitions, concepts, generalizations (sometimes as laws), models,
hypotheses, all tied together by a framework.

Urban form: The identity and spatial patterns of constructed and biological components of urban areas.

WUI: The wildland–urban interface, a conception that emphasizes the
interaction of urbanized lands, including suburbs and exurbs, with
wildlands such as large regional parks, production forests, and grazing lands.

Here, we will discuss how BES research contributes to each principle
as well as its implications for conceiving of, gaining understanding of, and
managing a metropolitan region as an integrated social-ecological system.
Each principle is a more specific and refined articulation of broad understandings about urban systems than we have presented elsewhere. Following our discussion of each of the thirteen principles in turn, we synthesize
them into a network in order to reinforce the idea that they are not intended
to be isolated building blocks that can be ordered in a particular progression. In addition, the principles are complemented by crosscutting themes.
We expect that our list of principles may prove to be incomplete. The field
of urban ecology is actively developing, and there are sure to be future insights that will lead to the refinement, expansion, or even replacement of

the principles proposed here. Many of the empirical results reflect several principles, and so we apply some examples to multiple principles.

PRINCIPLE 1. CITIES AND URBAN AREAS ARE HUMAN ECOSYSTEMS IN WHICH SOCIAL-ECONOMIC AND ECOLOGICAL PROCESSES FEED BACK TO ONE ANOTHER

BES conceives of the entire city-suburb-exurb (CSE) system of Baltimore as an integrated, social-ecological system. A fundamental assumption of urban ecology is that a metropolis and its component areas are ecosystems. The ecosystem was eloquently defined in 1935 by Arthur Tansley as a specific physical space that contains biotic and abiotic components that interact with each other. There are also interactions within the biotic and abiotic components, and they can be referred to as complexes—abiotic and biotic complexes—to highlight those interactions.

Tansley's abstract definition of an ecosystem is powerful because it can be applied to a vast array of specific cases wherever organisms and physical processes interact in a defined area. Urban areas can therefore be considered ecosystems because they contain organisms such as plants, animals, microbes, and people, and they contain physical features such as air, water, soil, nutrient cycles, buildings, and pavement. Cities can be viewed as hybrid social-ecological systems and be referred to as human ecosystems (figure 14.1). As human ecosystems, they incorporate not only the traditional biological and physical features ecologists are used to considering but also features of social structures, such as demographic, institutional, behavioral, economic, and technological complexity, as well as features of the built environment, such as highly modified or covered soils, buildings, utility and transportation infrastructure, and various kinds of paved surfaces. These new features can be seen as extensions of the basic ecosystem concept, not violations of it. In sum, cities are human ecosystems that contain biotic, social, physical, and built components interacting with each other. By explicitly considering this more inclusive realm of biological and physical features, we recognize the reciprocal feedbacks between social features and ecological processes.

Because BES has adopted the perspective of the Baltimore metropolitan region as a human ecosystem, researchers have investigated reciprocal links among ecological processes, the built environment, and social fea-

The Human Ecosystem

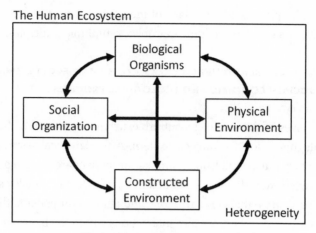

FIGURE 14.1 The human ecosystem containing the four complexes of biological organisms, physical environment, constructed
environment, and social organization. All of these complexes
reciprocally interact with each other. The interaction of complexes
generates and responds to heterogeneity, which can be represented by the background of the entire figure.

tures. For example, important feedbacks exist among socioeconomic characteristics, legacies of the built infrastructure, and biodiversity in the urban
environment. Based in metacommunity theory, it was hypothesized that
the degree of social investment in plant community composition, from low
investment to high investment, would result in plant communities that
could be characterized by either self-assembly or human-facilitated assembly (chapter 11). These hypotheses were tested on vacant residential lots by
comparing the biodiversity of plants found within the footprint of the demolished building to those found in the remnant backyards. The plant
community within the building footprint was a self-assembled community
that was characterized by a subset of species from the regional species pool
able to disperse to, and persist in, the parcel without human intervention.
In contrast, species found in the remnant garden or yard included those
from the regional pool that could disperse into the plant community following abandonment of the property as well as species that were selected for
the garden by the past residents. The previous social investment influenced
which species were found within the boundaries of the former structure's
footprint and which were found in the remnant backyard.

The feedbacks among ecological processes, the built environment, and social features were also quantified at scales coarser than the residential parcel. Patterns of land surface temperature (LST) were investigated to determine the influence of the composition and distribution of vegetation and paved surfaces on heat exposure to human communities. Not surprisingly, LST was higher in areas with greater building coverage and lower in areas with more tree canopy cover. The amounts of physical and biological components of the landscape were important for determining LST, and, in addition, the way in which the components were arrayed in the landscape, or their configuration, influenced LST. Greater shape complexity and edge density of patches of trees, for example, were related to lower LST in areas adjacent to the patches. Because LSTs varied widely across the city, the human population may be differentially impacted by heat exposure. Census block groups with higher LST also experienced higher levels of social stressors, including high poverty, less formal education, lower income, and greater risk of crime, as did more vulnerable populations, such as ethnic minorities and elderly residents. These residents may be less able to advocate for green space in their neighborhoods or less able to have the space or ownership to maintain green space on private property. The differential vulnerability to excess heat exposure across the city gave rise, in part, to efforts to increase the urban tree canopy cover to 40 percent by the year 2030.

PRINCIPLE 2. URBAN AREAS CONTAIN REMNANT OR NEWLY EMERGING VEGETATED AND STREAM PATCHES THAT EXHIBIT ECOLOGICAL FUNCTIONS

Ecologists have historically placed urban areas in opposition to pristine or natural areas and have assumed that ecological processes are either greatly diminished or highly altered by human interaction with the system. Yet ecological phenomena occur in urban areas because biotically dominated areas exist either as remnant patches of vegetated or aquatic habitat or patches that have newly emerged through planning and restoration. Remnant patches provide the basis for much ecological research. A guiding motivation is frequently to determine whether the ecological processes occurring in these patches are similar to or different from those occurring in nonurban settings and to investigate the mechanisms of how the urban setting has altered a specific ecological phenomenon. Emergent patches are often the result of human alteration of the system, including disturbance, management, and restoration. These patches may represent novel systems because they are characterized by a unique combination of species or environmental conditions or both. We provide two examples from BES research; the first addresses the ecological functioning of remnant riparian zones, and the second presents the ecological functioning of lawns as an example of newly emerged vegetation.

Streams and their associated riparian zones exist as remnants of past continuous habitat, and these zones can perform biogeochemical functions. Borrowing from basic ecosystem science, BES applied the watershed concept and research approaches to study the capacity of these remnant riparian zones to buffer streams from nitrogen leaving the terrestrial system. High levels of nitrogen are a pollutant of concern in waterways worldwide, and sources of nitrogen are abundant in the urban environment, including atmospheric deposition, historical fertilization practices on agricultural fields, contemporary fertilization of residential lawns, pet waste, and human waste entering the system from leaky infrastructure. Riparian vegetation can take up nitrogen that enters the riparian zone through atmospheric deposition and from overland, subsurface, and groundwater flow. In addition, under saturated, anoxic conditions nitrate in riparian soil can undergo the process of denitrification and be released to the atmosphere as nitrogen gas. These ecological processes, however, may be altered in an urban riparian zone because of changes to the hydrology of the system that are caused by urbanization of the surrounding landscape. Urban streams can experience "flashy" flow that increases dramatically after a storm due to the abundance of pipes and paved surfaces that move the water quickly from the terrestrial system into the streams. These fast flows cause bank scour and a deeply incised channel that may result in a drying of the riparian zone soil. This change leads to less denitrification, less interaction between runoff and riparian soils, and therefore less opportunity for the riparian zone plants to take up nitrogen and, eventually, a change in species from riparian obligate species to upland species. This complex of characteristics has been termed "hydrologic drought," and research investigating these ecological functions along the Gwynns Falls watershed confirmed the existence of the characteristics in the urban portions of the watershed.

Lawns are a dominant emergent vegetation type when land is urbanized. In the state of Maryland lawns occupy more than 10 percent of the terrestrial land cover. Lawns are also thought to contribute to N pollution of receiving waterways because of management decisions to add fertilizer and water. BES research, however, showed that while the pools and rates of production of reactive N in residential lawn soil may be higher than in the surrounding deciduous forests, they were remarkably low. Surprisingly, residential lawns retained high amounts of nitrate.

PRINCIPLE 3. URBAN FLORA AND FAUNA ARE DIVERSE, AND THIS DIVERSITY HAS MULTIPLE DIMENSIONS

Organisms in the urban environment are found within remnant patches, as discussed in Principle 2, but they are also found dispersed across the entire metropolis. In general, species that can tolerate the stresses of the urban environment tend to exist in higher numbers there because of the subsidies to resources or release from predators that also characterize the urban environment. But some species cannot maintain viable populations in the urban environment and are lost from it. This may result in an overall decline in species richness but an increase in the density of those species that can tolerate the urban environment.

Organisms found in the urban environment consist of species native to the biome and species that have been introduced from outside the system. Plants and animals can be introduced to the urban environment intentionally, as many horticultural species are, or they may be unintentionally introduced via soil, ballast, lumber, packing materials, or other transported media. Native and nonnative species contribute to the biodiversity of the system. However, urbanization may have resulted in the loss of some native species that performed crucial functions, and following urbanization those functions may be performed by nonnative species. Therefore, management considerations must shift from a concern for origin of species toward a focus on the function that a species provides in the landscape so as to avoid leaving important ecological roles unfilled.

Bird counts were conducted in the City of Baltimore at fifty randomly selected points. Despite the assumption that nonnative bird species exist everywhere, it was found that the urban environment is not homogeneously occupied and that abundances within the city vary widely. Thus the quintessential urban bird species—house sparrows, pigeons, and European starlings—did not overlap significantly in location, instead being associated with distinct urban habitats. Using vegetation and land use/land cover characterizations, four distinct clusters of birds were identified in Baltimore, each associated with particular habitat features. The first cluster included birds that occupied the urban shoreline and were associated with water while the second consisted of birds that were most frequently found in the inner city and can successfully deal with highly urbanized areas. The other two clusters were characterized by bird species that can exist in low-density residential housing, and these two clusters differed by the vegetation stature—shrub cover versus tree cover—that the birds preferred.

Patches of remnant forests may serve as refugia for urban fauna, particularly resident and migratory birds. The utility of a forest fragment as bird habitat may be determined both by internal patch features and by the characteristics of the landscape matrix surrounding the patch. Breeding-bird communities in fifteen forest fragments (2–8 ha) in the Gwynns Falls watershed were surveyed during the 2005 breeding season. In residential areas management of properties surrounding forest patches may augment the resources available to forest birds. Models containing different subsets of variables describing the forest patch and the surrounding residential matrix, including neighborhood age, land-cover composition, and the density of bird feeders, baths, and nest boxes within one hundred meters of the forest edge were tested. Bird species richness was greater in forest patches with high levels of tree and shrub cover within the one-hundred-meter buffer, which tended to occur in intermediate-aged neighborhoods (figure 14.2). These relationships were observed within a narrow buffer width, indicating that individual landowners can manage their property to enhance avian diversity in adjacent forest patches. This research demonstrates not only the importance of remnant habitat for maintaining urban biodiversity but also the crucial role that management of the landscape surrounding these remnants can play.

Biodiversity, however, extends beyond taxonomic diversity or species richness. In fact, there is increasing interest in exploring the interactions among the three primary dimensions of diversity: taxonomic or phylogenetic, functional, and genetic. BES research that investigated plant communities on vacant lots as described under Principle 1 explored differences in taxonomic or phylogenetic diversity compared to functional diversity. Beta diversity among plant communities in the remnant backyard portions of the vacant lots was found to be significantly higher than among plant communities within the footprint of the demolished structure for both metrics of diversity.

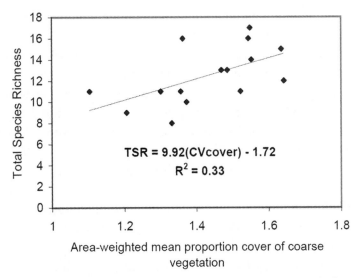

FIGURE 14.2 The relationship of bird biodiversity to spatial structure of residential neighborhoods. Bird diversity was quantified in remnant forest patches surrounding by residential neighborhoods that differed in the amount of coarse (woody) vegetation present. (From Chrissa Carlson, 2006, used by permission)

PRINCIPLE 4. HUMAN VALUES AND PERCEPTIONS ARE A KEY LINK TO MEDIATING THE FEEDBACKS BETWEEN SOCIAL AND ECOLOGICAL COMPONENTS OF HUMAN ECOSYSTEMS

The feedbacks between ecological and social components of human systems are mediated by human values and perceptions. Ecosystem services are benefits conveyed to humans by components and processes that comprise the ecosystem. Because ecosystem services depend on the relationships between human values and perceptions and environmental structures and processes, they are an expression of this principle.

An obvious example of this principle—one that has received much research attention from BES—is the urban tree canopy and the perceptions residents hold about trees and the value they place on them. For example, there is a feedback between tree canopy cover in the city and heat exposure to humans; the urban tree canopy can provide climate regulation of the urban environment by cooling through shading and transpiration. Urban residents, however, need to recognize and value that climate regulation before they can evaluate it as a service of the ecosystem.

There are other benefits provided by an urban tree canopy, including aesthetic appeal, habitat for birds and other wildlife, and potential moderation of water and air quality. However, there are costs associated with the urban tree canopy, including tree maintenance such as watering and pruning, cleaning cars and sidewalks of leaves and bird droppings, and hazards to property and persons of falling branches during storms or damage to aboveground and belowground infrastructure. Residents may find that the services conveyed by the tree do not outweigh the disservices they experience by having responsibility for or being in proximity to a tree. These trade-offs between services and disservices will vary across groups of residents, will likely change as residents age, and will be experienced differently depending on whether that tree is located on the street, in a yard, or in the neighborhood park.

In addition, the species of the tree will influence its growth form, size at maturity, phenology, resource needs, and therefore perceptions of its beauty and value. Incorporating human values and perceptions in an understanding of the causal links between ecological and social components of human ecosystems is complex. A crucial but often overlooked step is to appropriately quantify and evaluate the services and disservices provided by the urban tree canopy and not assume that these exist equally to all residents simply because the urban tree canopy exists.

As discussed in Principle 1, the City of Baltimore aims to double its urban tree canopy to 40 percent cover by 2030. In order to achieve this goal, private property owners within the city must be involved because BES research has estimated that approximately 85 percent of the land on which trees could be planted is private property. Achieving this goal requires that residents value trees and perceive them as net benefits in the environment. Historically as well as currently, some residents may resist the push to plant trees. Residents maintain different values and perceptions of trees in the urban environment for many reasons. One may be the perception that greater numbers of trees in the urban landscape can reduce visibility and provide places for criminal elements to hide or hide their goods. In contrast to these perceptions, BES research found a strong *negative* relationship between crimes such as burglary, robbery, theft, and shooting and the amount of tree canopy. A myriad of potential confounding variables were controlled for, such as socioeconomic status, defined by race and income, as well as variables of housing age, construction type, land tenure, and the natural environment. A significant conclusion was that, all other things being equal, a 10 percent increase in tree cover should be associated with a 12 percent reduction in crime rate and that this effect of increased tree cover would be even more dramatic on publicly owned land. The belief by some residents that vegetation in the urban environment may lead to higher crime was not borne out by reality.

PRINCIPLE 5. ECOLOGICAL PROCESSES ARE DIFFERENTIALLY DISTRIBUTED ACROSS THE METROPOLIS, AND THE LIMITATION OF SERVICES AND EXCESS OF HAZARDS IS OFTEN ASSOCIATED WITH THE LOCATION OF POPULATIONS THAT ARE POOR, DISCRIMINATED AGAINST, OR OTHERWISE DISEMPOWERED

A critical aspect of urban ecosystem services is how equitably they and associated disservices or hazards are distributed in the landscape. Concern with equitable distribution of environmental risk or burden is referred to as environmental justice, and this concern first gained traction in communities that were located adjacent to contaminated sites such as landfills or industrial locations that produced hazardous pollution. These communities are typically disadvantaged economically or are communities of color. Environmental justice, as a social movement, battles the burdens of contamination these communities are exposed to at a greater relative proportion than others and works to participate in land use or regulatory decisions affecting their communities. More recently, environmental justice concerns have expanded to include recognizing differential access to environmental benefits as an injustice and seeking environment-based amenities for all communities, as discussed below.

Environmental justice concerns centered on proximity to environmental burdens and access to amenities have been the focus of research in BES. The relationship between the distribution of Toxics Release Inventory (TRI) sites within Baltimore and the racial demographics of neighborhoods in close proximity to these sites was investigated. Surprisingly, the distribution of TRI sites was found to be in primarily white neighborhoods, not African American neighborhoods. However, this finding is a result of long-standing discrimination keeping African Americans from living close to places of work. In the nineteenth century, when Baltimore first became an industrial city, white residents occupying working-class neighborhoods were advantaged by living near the factories where they worked. Consequently, they had shorter walks to work than most African American workers. The persistence of neighborhood boundaries in Baltimore and the history of racial segregation have left an unexpected legacy of white residents living in closer proximity to TRI sites than African American residents.

The other side of the coin from environmental burdens is access to environmental amenities. Parks in urban landscapes are viewed as an amenity, and they may provide a suite of ecosystem services, including escape from urban heat, a place for leisure and recreation, reduction in air pollution and noise, absorption of rainwater and flood reduction, and, potentially, a place to grow food. The distribution of parks in the City of Baltimore relative to demographic factors of surrounding neighborhoods was investigated. A park was deemed accessible if it was within four hundred meters of a residence because this distance is recognized as a standard reasonable walking distance. It was found that African American and high-need populations had greater walking access to parks than white and low-need populations. However, park acreage per capita was lower for African American and high-need populations. Crucially, access as an outcome measure is not the only variable to consider because it does not include the needs of the neighborhoods. Neighborhoods with more multifamily structures, smaller yards, and more children or elderly people may also merit access to more parks and places to experience leisure or recreation.

An urban park, however, may be an amenity or a disamenity depending on the coarser-scale context within which it is situated. In neighborhoods experiencing less crime, housing values for homes within four hundred meters of a park were higher, indicating that the park was viewed as an amenity. In contrast, in neighborhoods experiencing more crime, housing values for homes within four hundred meters of a park were lower, indicating the perception of parks as disamenities. A threshold within which this shift from higher to lower home value occurs was identified at approximately 406 to 484 percent greater than the national average crime rate. In other words, in neighborhoods experiencing crime rates 484 percent or higher than the national average, being close to a park had negative effects on the value of homes.

Two important messages emerge out of Principles 4 and 5. First, amenities (or disamenities) are not the same as services (or disservices). Services are rooted in measurable ecological processes and evaluated by the human beneficiaries to determine whether or not a service has been received. Characteristics of the human population and the amenity may shift through time and space, which may result in a shift in the evaluation of whether or not a service is conferred. Second, equal distribution of an amenity across the landscape may not necessarily mean equitable distribution of a service. This may be because the amenity provides the service differentially based on the landscape context or because evaluation by the human beneficiaries is not homogenously distributed among groups.

PRINCIPLE 6. URBAN FORM IS HETEROGENEOUS ON MANY SCALES, AND FINE-SCALE HETEROGENEITY IS ESPECIALLY NOTABLE IN CITIES AND OLDER SUBURBS

Heterogeneity in ecological systems has become a primary theoretical and research concern in ecology because it is a driver of spatial differentiation in natural processes. Spatial heterogeneity must also be a concern in urban systems. Indeed, urbanists and urban sociologists have noted social and architectural heterogeneity as a key feature of urban areas, compared to villages and rural settlements. Heterogeneity may be created and maintained by natural features, social differentiation, management decisions, or the links and feedbacks between ecological and social features. The heterogeneity found in cities becomes apparent as soon as one walks through city streets. Often spatial shifts in the social or ecological structure or function of the system are abrupt. This sudden change in the urban fabric may be generated by social processes that can occur at very fine spatial scales, including aesthetic and management choices made at the scale of the individual residential parcel. At a coarser scale, shifts in characteristics may be made between neighborhoods that differ in development history, land use activities, or occupancy, for example.

Cities can be viewed from many different perspectives and scales. Interest in the growth and spread of a city over time may lead to a view of the city at coarse, regional scales where the city itself is depicted as an amorphous splotch on the regional map (figure 14.3). Alternatively, interest in the spatial distribution of vegetation and the density of built structures in the city would require a very different perspective, likely at a finer scale and a different resolution of categorization. The first example may employ a simple categorization of urban versus nonurban land use while the second example would need greater disaggregation within the urban land use to include some categorization of vegetation and built structures at different

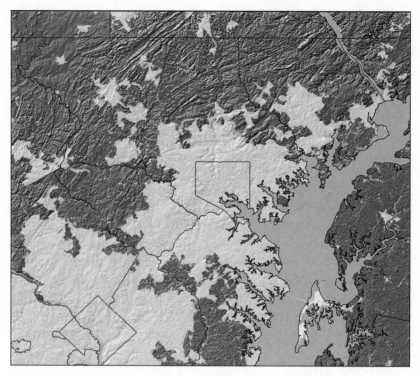

FIGURE 14.3 Baltimore-Washington metropolitan areas: urbanization in 2000 census represented as an urban splotch. This representation simplifies the heterogeneity of the system into urban versus nonurban.

magnitudes or intensities. In contrast, social scientists may be interested in the differential distribution of people and may use classifications based on demographic factors and the consumer choices made by residents. For example, the PRIZM classification hierarchically aggregates the human population into five classes based on population density, fifteen classes when socioeconomic status is added, and the addition of "social characteristics," which includes such things as magazine subscriptions, charitable giving, and frequent purchases, creates sixty-two "lifestyle clusters." These three kinds of clustering represent different social theories. The five-class model represents a theory that urban structure is based on density, the fifteen-class model represents a theory of human and social capital focused on cities as structures for industrial production, and the sixty-two-class model acknowledges a theory of urbanization based on consumption.

Because spatial heterogeneity is a fundamental characteristic of cities and because it is a primary concern of ecology theoretically and empirically, it is crucial that the spatial heterogeneity be accurately quantified and described at spatial and temporal scales relevant to the specific research question. Frequently in urban ecology, heterogeneity in land use/land cover is selected as the primary spatial descriptor of the system because it is readily available through national databases and it is assumed to be the most appropriate depiction of spatial heterogeneity. We argue, however, that this assumption should be tested through empirical research.

In BES we have developed a new land cover classification motivated by the desire to more accurately capture the fine-scale heterogeneity we were observing by using not only finer resolution imagery but also, more important, a classification with a finer resolution categorization of patch types. We also hypothesized that relationships between the heterogeneity of the system and ecosystem function would be improved if the heterogeneity was determined by the physical attributes on the landscape, or land cover, because these attributes would be more tightly linked to ecosystem processes. The new classification, named HERCULES (High Ecological Resolution Classification for Urban Landscapes and Environmental Systems), was developed to quantify urban heterogeneity by focusing on the elements of land cover: buildings, woody and herbaceous vegetation, and paved and bare soil surfaces. The resulting patch mosaic generated by variation in the distribution of land cover elements can serve as a structural template on which additional heterogeneities can be layered depending on the research question. For example, watershed boundaries, topography, and water infrastructure layers can be added if the research question focuses on factors that may influence the dynamics of water quality and quantity, or social structures of demographics, ethnicity, education, and lifestyle can be superimposed on the structural template to ask questions about the relationship between social differentiation and physical differentiation.

Through our application of the HERCULES classification we have discovered that heterogeneity in land cover is more fine-scaled in the urban core and older suburbs compared to more recently developed areas. Two reasons may offer some explanation of this pattern: (1) the development practices and how they have changed over time and (2) the amount of turnover and reworking of the urban fabric. Both of these explanations are predicated on the assumption that the inner urban areas are older than their surroundings. When Baltimore was first developing, land was purchased and developed in small parcels by many developers, each with a different style and intention. Even when one developer built several blocks, the cellar holes were dug by hand, and walls were built brick by brick. Following World War II, with the rise of long-distance transportation infrastructure and the use of prefabricated architectural components, the size of developments grew substantially, and suburbs were valued by how efficiently they achieved a single land use type—single-family homes. Therefore, the development pattern was repeated across the landscape in large swaths, diminishing the granular variation in land cover that was typical of the inner urban developments of an earlier time. In addition, developments near the core of the city have existed for a longer period of time and have gone through many transitions in use. Each stage leaves a legacy resulting in very fine scale heterogeneity of land cover.

PRINCIPLE 7. URBAN FORM REFLECTS PLANNING, INCIDENTAL, AND INDIRECT EFFECTS OF SOCIAL AND ENVIRONMENTAL DECISIONS

Urban form refers to the physical layout of a city-suburban-exurban system across space. This topic has engaged urban researchers, planners, designers, and residents for centuries. Much attention has been paid to top-down, direct control of urban form. Such an approach to planning may suggest that urban areas are fundamentally rational products. For instance, the deep history of cities identifies cosmological significance in the layout of the cities of the Middle East, the Aztec Empire, or dynastic China. Likewise, in the United States famous examples of formal layout have intellectual, commercial, and lived significance. For example, James Oglethorpe's Savannah, Georgia, or William Penn's Philadelphia are touted as the epitome of formal, gridded layout. The gridding of the borough of Manhattan north of Houston Street in 1811 was a vehicle for enhancing development. However, scholars note that even in such iconic cases as Philadelphia, the actual development of the town and its relationship to the bounding streams were much less regulated than the grid would suggest. The idea that cities are rationally laid out and modified runs afoul of the actual complexity, incremental change, and multiple motivations of urban form and change.

Baltimore illustrates the principle that cities contain incidental and indirect effects in addition to rational planning. Baltimore, like the iconic

Savannah and Philadelphia, also exhibits gridded layout, but that layout in fact represents many clashing grids. The various grids reflect the multiple orientations to three major northeast-trending streams and their roughly parallel ridges. It also reflects the complicated shape of the harbor and the boundaries of early plantations (chapter 3). Initially, spatial complexity of the Baltimore layout shifted when the originally separate grids grew together. The diverse, clashing grids originated in the era of the walking city. Even in that era key market roads outside of the city followed the high ground of the watershed divides (for example, Reisterstown Road) or cut northeastward across both the local grids and the strike of the stream valleys (for example, Belair Road). Over time the sutures of Baltimore's grids changed as advances in transportation technology suggested to the city and regional planners that more efficient long-distance mobility could be achieved. Taming the diversity of grids became all the more important as new transportation modes developed. Trolley lines first brought contrasting parts of Baltimore's city grid together in the late nineteenth century. Later, in the early twentieth century, wide boulevards and surface arterials suitable for private automobiles were established. Finally, the divided, high-speed interstate or expressway system of the 1960s and 1970s became a predominant factor in transportation and the shaping of the urban fabric.

These changes in the transportation arteries were intended to knit the city and its region together for different purposes. The walking city placed workers in easy pedestrian distance of factories and the docks. It was the origin of the racial segregation based on easy access to factory work that underlies the unexpected finding of the association of TRI sites with contemporary white, working-class neighborhoods discussed in Principle 5. This gridded layout would have had significant ecological effects, as gutters and curbs became the de facto headwaters of the drainage network (chapter 9).

Social processes inhabit the grid. For example, even in the era of the fine-scale, walking grid, social segregation existed. Slaves and freepersons of color along with unskilled white laborers lived in the inner blocks in tiny two-story houses facing alleys (figure 14.4). Managers, merchants, and those afforded higher social status lived on the outer block faces. Generally, the grandest of the row houses fronted onto formal public squares furnished with fountains and attractive landscaping. In the late nineteenth century Baltimore was remarkably integrated at the ward level, but within wards this fine-scale, alley-based differentiation held sway. This arrangement en-

FIGURE 14.4 Fine-scale heterogeneity of Baltimore neighborhoods shown by inner block housing on alleys (left), and block face housing on main streets (right). The structures in both of these images are of similar age. The alley houses are viewed through a void created by demolition of two block face houses. The vacant lot has been landscaped with new trees, grass, and a mulched path. (Photographs, S. T. A. Pickett)

sured that domestic, factory, and dockworkers all had easy access to their employment.

Such living arrangements were not permanent, however, and from the 1880s to the 1910s large changes in Baltimore's formal structure took place. One of the most significant was the result of the cholera epidemics of the 1880s. These dynamics have been studied extensively by BES, and we have documented that the infant mortality recorded by the 1880 census was both high—one in four children died in their first year—and concentrated in Baltimore's low-lying areas. This staggering loss of life stimulated those who had the financial resources or who were not excluded by racial classification to move to higher ground. Thus began Baltimore's first round of suburban spread.

At the time of this demographic shift, racial attitudes that had loosened somewhat during Reconstruction after the Civil War began to harden again. In 1890 racial segregation, which had been absent at the level of Baltimore City's twenty wards, began to take on coarser-scale patterns, with African Americans increasingly concentrated in certain neighborhoods. In 1911 Baltimore passed the first municipal ordinance in the United States requiring racial segregation by neighborhood. Although this law was struck down by the Supreme Court in 1917, additional measures were put in place to ensure racial segregation. For example, the comprehensive zoning plan of 1923 slated alley districts and the immigrant tenement neighborhoods of

South and Southeast Baltimore as industrial zones, promoting displacement of existing residents. Neighborhood associations, the real estate industry, and perhaps most notably the Federal Home Owners' Loan Corporation (HOLC) maps of mortgage-worthiness, all contributed to enforcing and even exacerbating racial segregation. The HOLC maps in Baltimore are closely congruent with today's building vacancy, vacant lots, and lack of tree canopy. These legacies in the urban form of Baltimore were indirect effects of social decisions but have profound ecological consequences.

"Pull factors" add complexity to Baltimore's urban form, as is common among American cities. Suburban spread, first begun in earnest with the flight from the cholera risk of the 1880s discussed above, became a tsunami in the second half of the twentieth century. Sometimes referred to simply as white flight, suburban spread has a history and dynamics that are considerably more subtle than that. First, it wasn't entirely white, as members of the African American middle class were among those leaving the older city as well. Second, it was also a matter of pull focused on the new suburban districts, not just push from the older city. Pull consisted, for instance, of tax credits for mortgage interest, the availability of larger houses and yards, the construction of federally subsidized interstate highways, and the development of industrial and business parks. Some of these pull factors were the result of the GI Bill, which effectively advantaged the white population because of the segregation of the armed forces that continued through World War II. Push factors included the disinvestment in schools, the continued pressure of racial bias, the social isolation and discontent of disenfranchised populations, and the crowded, unsanitary conditions in old row-house neighborhoods. For example, the sanitary infrastructure of many of the alley blocks was substandard. Although the motivation for suburban expansion may be simply described in terms of a middle class, structural ideal of suburban form, many strands of push and pull, infrastructure, lifestyle, and perception are intertwined in the dynamic. Baltimore is typical in this regard among American urban regions.

The suburban expansion of Baltimore and the differential investment in existing neighborhoods have a number of lasting ecological consequences. Although some are intentional, such as the installation of aboveground stormwater retention or detention basins in newer suburban developments, an unintended ecological difference appears in vegetation between neighborhoods. Neighborhoods having wealthier, more-empowered residents are characterized by horticultural plantings and active vegetation management. The vegetation in poorer, less-empowered neighborhoods is more likely to be of volunteer origin and to experience scant management. Differential neighborhood investment therefore results in social and built heterogeneities that affect vegetation cover and ecosystem fluxes. Tree canopy cover and management of trees, shrubs, and herbaceous vegetation are major drivers of ecosystem function in urban areas.

The difference in vegetation resulting from unintentional interactions also exists on finer scales. For example, in lots on which an abandoned row house has been demolished, the contrast in vegetation composition and layering in the former building footprint and the remnant backyard differ substantially. Such row-house lots in central Baltimore typically measure only 13 × 75 ft. (circa 4 × 23 m). The differences in biodiversity, functional diversity, and phylogenetic diversity are large across these fine-scale contrasts. Demolition of vacant houses is intended to reduce risks of fire and illegal activities, but it accidentally determines local patterns of biodiversity.

Although this principle would seem to be of local significance only, complexity also results from global connections of any given city with other cities, resource-generating regions, and hinterlands for food and waste processing around the globe. Globalization has many unintended consequences for Baltimore and other cities. Baltimore's industrial zenith occurred with the steel industry and defense mobilization associated with World War II. The significance of that industrial base declined for two reasons. First was the federal policy decision to move military-oriented industries to locations seen as less vulnerable to coastal threats. Second was the decision by corporations themselves to save costs by consolidation and by sending lower-skilled jobs offshore. Although the rationality of security, efficiency, and profit is unarguable, such objectives had vast negative and presumably unintentional effects on industrial cities like Baltimore and other traditional industrial powerhouses. Further corporate consolidation in the financial industry following deregulation robbed Baltimore and other metropolitan areas of access to charitable and civic commitment by local or metropolis-based firms. The concentration of corporate headquarters in continental or global centers like New York, London, and Shanghai had unintended civic consequences on many American cities.

PRINCIPLE 8. URBAN FORM IS A DYNAMIC PHENOMENON AND EXHIBITS CONTRASTS THROUGH TIME AND ACROSS REGIONS THAT EXPRESS DIFFERENT CULTURAL AND ECONOMIC CONTEXTS OF URBANIZATION

A key characteristic of the urban landscape is its spatial heterogeneity, as described in Principle 6. Cities, like all ecosystems, change in form across space and are also constantly undergoing change in that spatial form through time. The urban form of Baltimore has certainly undergone slow, gradual change over time. For example, neighborhoods may experience change as young trees planted at the time of development grow to form arching tunnels of greenery above the streets, and houses undergo modifications to fit the needs and desires of young versus aging residents. These gradual changes are a result of plant succession and a myriad of social choices and management decisions. In Baltimore we found that temporal lags exist between social and ecological drivers of the spatial structure of the system such that the socio-demographic factors of the 1960s explained the heterogeneity in tree cover in 1999 better than the socio-demographic factors of residents in 1999.

Urban form may also experience abrupt changes. For example, a planning decision may be made to replace housing and light commercial with an expressway, as was often done in the urban renewal era. Natural disasters that dramatically and suddenly alter urban form by damaging structures are another source. Baltimore has undergone several abrupt changes over time and was altered by the impacts of the Great Baltimore Fire of 1904, Hurricane Agnes in 1972, and Hurricane Isabel in 2003, for example. Each event itself altered the urban form but so did the response to the events, including changes to building codes and zoning intended to diminish risk to life and property in future fires and storms. These responses leave lasting changes in the urban form.

This principle has emerged from our study of the changes in urban form in Baltimore as well as our efforts to place those changes within the context of urbanization globally. Cities embedded within different biophysical, cultural, and historical contexts may have alternative trajectories of changes in urban form. Early cities were frequently walled to make a distinction, both socially and physically, between the city and the surrounding hinterlands. The rise of industrialization made it necessary for cities to look beyond their immediate surroundings for labor, raw materials, and markets for manufactured goods so that walls became porous and cities became more regularly connected through transportation routes. Following World War II, American cities purposefully expanded outward. The Federal Aid Highway Act of 1956 facilitated this spread by creating a network of interstates and thereby making it easier for people to live in surrounding residential suburbs and commute to work in the core of the city. Commercial districts became more dispersed as the zone of travel for the average consumer increased and the population density in the urban core decreased. This linear progression of change in urban form is highly generalized and does not reflect the development of urban areas in all parts of the United States or the world. Some cities experienced no industrial period and are characterized by rapid, sprawling growth as migrants from the surrounding areas come to cities in search of employment and other economic and social opportunities.

PRINCIPLE 9. URBAN DESIGNS AND DEVELOPMENT PROJECTS AT VARIOUS SCALES CAN BE TREATED AS EXPERIMENTS TO EXPOSE THE ECOLOGICAL AND SOCIAL EFFECTS OF DIFFERENT DESIGN AND MANAGEMENT STRATEGIES

Urban design is a process of physical modification of urban fabric for purposes of social benefit, commercial profit, or civic good, among other objectives. Rarely is urban design thought of as an exercise in constructing or modifying an ecosystem. However, the idea of urban areas as ecosystems (Principle 1) suggests that urban designers actually create, restore, or otherwise modify social-ecological systems—they design and plan ecosystems. From this perspective, urban design becomes a practical process with potentially measurable ecological outcomes. Hence, designs can be considered to be experiments in exactly the same sense that forestry, fisheries, restoration, and conservation are. These interventions are all experiments because they intentionally manipulate natural, built, and hybrid natural-built ecosystems, and if the phenomena that are altered are documented and the outcomes of the alteration assessed are compared to unmanipulated sites, rigorous conclusions can be drawn about ecosystem processes. Even the unintentional effects of designs (Principle 7) can be assessed with careful, creative measurements. Design as experiment becomes an important tool for both learning and ecological engagement in urban ecosystems, just as adaptive management reflects learning in natural resource management.

An example in Baltimore is the stormwater management interventions in the 907-acre (367 ha) Watershed 263. Defined by a common storm drain infrastructure and outlet, Watershed 263 is a dense, old row-house neighborhood to the west of downtown Baltimore. A collaboration between BES, the city Department of Public Works, and the Parks & People Foundation installed various stormwater management interventions in comparable subwatersheds where the water quality in the storm drains exiting each subwatershed was instrumented and sampled. Interventions included (1) removal of impervious surfaces in schoolyards, (2) curbside rain gardens, (3) tree planting on streets and in vacant lots, and (4) curb cuts into vacant lots. In addition to the water quality measurements residential surveys were conducted to determine social responses and conditions associated with the interventions.

Research documented that dense, old residential neighborhoods could be hotspots for both nitrogen and phosphorus in storm drains. Although the declines in nitrogen and phosphorus after installation of stormwater best management practices in one of the subwatersheds were substantial, they far exceeded the expected impact of the interventions. The mismatch is attributable to other documented factors (described below) that changed at the same time. Not only were stormwater best management practices installed, but the city trash pickup strategy and schedule were altered and an unauthorized horse stable was closed. However, the interventions did have notable educational and social benefits. In the elementary schoolyard where asphalt removal began, and the project planned with student involvement, environmental science and literacy improved. Social perceptions throughout the watershed improved, with increases in resident satisfaction concerning the neighborhood, watershed knowledge, and social capital. However, as in many urban interventions, multiple factors were altered simultaneously. Community engagement in planning processes may have contributed to the enhanced neighborhood satisfaction along with any perceived environmental benefits of trees and greening. There remains substantial opportunity to employ urban designs of various kinds as experimental entry points for understanding urban ecosystem function. Both biophysical and social outcomes should be measured in urban design as experiments.

PRINCIPLE 10. DEFINITION OF THE BOUNDARIES AND CONTENT OF AN URBAN SYSTEM IS SET BY THE RESEARCHERS BASED ON THEIR RESEARCH QUESTIONS OR THE SPATIAL SCOPE OF ITS INTENDED APPLICATION

Thus far we have focused on the interactions among biological, social, physical, and constructed complexes of a human ecosystem that are encapsulated in Principles 1–9. Principle 10 suggests that the second key feature of the definition of an ecosystem is methodological and requires specifying the spatial extent of the ecosystem or establishing system boundaries within which the internal- and among-complex interactions occur. The ecosystem concept is scale neutral, meaning that no single spatial scale for an ecosystem is set a priori. Instead, the ecosystem can be of any size. For example, an entire city can be an ecosystem but so can subregions within the city, such as a watershed, a neighborhood, or even a single residential parcel (figure 14.5). The boundaries of the ecosystem demarcate what is in and what is out of the target system, which is necessary for determining the spatial extent of system measurements and comparisons among systems. Boundaries are determined by specific research questions. In nonurban systems, ecosystem boundaries are often established around areas that have similar physical or biological structure—for example, a lake or a forest patch—or they may be established to include an area integrated by flows, such as a watershed. The same approach can be taken in urban systems, but additional features may need to be incorporated, such as municipal or neighborhood boundaries.

BES uses many different ecosystem boundaries depending on the particular research question. Much of the BES long-term data collection is focused on the Gwynns Falls watershed (chapter 1). This watershed extends from its headwaters in Baltimore County to its confluence with Baltimore Harbor near downtown in the City of Baltimore. Measurements of stream water quality and flow integrate the social, ecological, and built complexes that are interacting within the watershed boundaries. Measurements within the watershed boundaries also integrate across the two jurisdictions—Baltimore County and the City of Baltimore. Notably, the city and county have entered into a formal agreement to consolidate management of this watershed. Bounding a watershed inclusively is required if the research focus is at the scale of the entire watershed. However, if a research goal is to assist managers in Baltimore City to define tree-planting goals and iden-

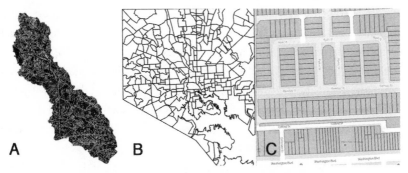

FIGURE 14.5 Application of the ecosystem concept across scales. Ecosystems can be defined at any scale, as shown in these three panels. Panel A is a watershed defined by a hydrological boundary and shown here with land cover patches; Panel B is the City of Baltimore defined by the municipal boundary and shown here with the recognized neighborhoods; and Panel C is an urban block defined by streets and shown here with property parcels. The flows of materials, energy, information, and people, for example, can be assessed for each of these three instances of the ecosystem. (Source for B and C is a public data repository, https://data.baltimorecity.gov/api/geospatial/r3qj-2ifh?method =export&format=shapefile, and https://data.baltimorecity.gov/Geographic/Parcels/rb22 -mgti, respectively)

tify priority areas for planting, measurements with the boundaries of the Gwynns Falls watershed would be inappropriate. Instead, the municipal boundaries that were established for political reasons may be an appropriate ecosystem boundary for this research goal. It is the responsibility of the investigators to state what the boundaries of their study units are and what the criteria for inclusion in the urban system are.

PRINCIPLE 11. URBAN COMPARISONS CAN BE FRAMED AS LINEAR TRANSECTS OR AS ABSTRACT GRADIENTS, AND THE ABSTRACT COMPARISONS ACKNOWLEDGE THE SPATIAL COMPLEXITY OF URBAN HETEROGENEITY

This principle also expresses a methodological concern: how to establish comparisons between urbanized and nonurbanized, or less urbanized, areas. In ecology a gradient is gradual change in a variable of interest across space. This change can be linear, meaning that the change in the variable is unidirectional—always increasing or decreasing across space. This type of spatial change is referred to as a direct gradient, and the change in the variable can be quantified using a transect approach such that quantifying the variable at some regular interval across space would capture the gradual change in the variable. Applying this approach to urban areas suggests that the amount of "urban-ness," measured by some indicator such as human population density or the amount of impervious surfaces, declines steadily from the urban core out into the hinterlands. In such cases the distance from city center may be meaningful and may offer a useful way to make comparisons at very coarse scales.

More commonly, contrasts in the system are not neatly arrayed along linear transects; instead, the spatial heterogeneity exists in patches. In addition, differences in any one variable of interest do not necessarily overlap in space with differences in other variables of interest resulting in a complex spatial mosaic. In this case, the spatial complexity is better depicted as an abstract gradient, also referred to in ecology as an indirect gradient. Thus linear transects may be appropriate bases for some comparisons, but, more generally, ordinated gradients are appropriate for most comparisons. These methodological concerns apply to the biophysical components of urban systems and the social interactions and processes, including such things as sense of place, livelihood, and lifestyle that characterize urban regions.

In BES we recognize that urban systems are hybrid systems consisting of human and natural components that are heterogeneously arranged across space. Within this complexity there are many axes upon which to frame comparisons within the urban system. Criteria for comparison may include the proportion of cover by building, vegetation, and surfaces, the lifestyle characteristics of neighborhoods, land use impacts on stream hydrology and nutrients, history of social investment in residential yards, or the sizes and eras of subdivision. This list highlights only some of the research activity in BES and in no way exhausts the list of research within BES or elsewhere.

PRINCIPLE 12. URBAN LAND COVERS AND LAND USES EXTEND INTO AND INTERDIGITATE WITH RURAL OR WILDLAND COVERS AND USES

Rapidly growing cities expand outward into adjacent agricultural production lands, forests, deserts, or whatever landscape type predominates in the surroundings. In the United States first the streetcar and then the private car facilitated the movement of urban residents out of the noisy, dirty industrial center of the city toward seemingly more idyllic landscapes in the surrounding rural areas. This expansion increased in both population and distance as transportation and the infrastructure to support it became increasingly available. Eventually, light industrial, commercial centers, and even corporate headquarters followed, established at major crossroads and in suburban communities. These new nodes of business could draw from a larger, spatially dispersed workforce that could travel on fast, limited-access highways and avoid the congestion of commuting into the city center. The suburban fringe interdigitated with surrounding rural or wildlands in spatially convoluted and complex ways.

Increased interdigitation of urban lands with wild or rural lands results in loss of land for agricultural production, fragmentation of habitat and interruption of dispersal corridors for organisms, and potential disruption of natural disturbance processes as humans manage land in an attempt to prevent such events or minimize their impact. The wild–urban interface (WUI) is increasing dramatically, and it is estimated that 10 percent of all land that contains housing in the conterminous United States is in the WUI. In some locations, particularly the western United States, this increased interdigitation of urban lands with rural or wildlands can be costly. For example, the annual cost of fighting wildfires has tripled since 1990 in the United States and as of 2015 consumed almost half of the annual budget of the U.S. Forest Service. The more severe fire seasons account for some of this increase, but increased development on the WUI adds tremendous expense to fighting fires, especially for state and local agencies.

Cities have adopted a variety of strategies to stem the spread of urban development. In 1967 Baltimore County, which surrounds the City of Baltimore, established the urban-rural demarcation line (URDL) in an effort to prevent development out into the county. One-third of the land in the county is inside the URDL, and, forty-five years after establishment, 85 percent of the county's human population lives within the URDL. The adjacent counties of Carroll and Harford have not instituted a spatial limit to urban development. An analysis of the distribution of subdivisions in the three counties of Baltimore, Carroll, and Harford from 1960 to 2007 found that the vast majority in Baltimore County were located within the URDL, whereas in Carroll and Harford the subdivisions were spread throughout the counties and numbered many more. In addition, the abundance of smaller subdivisions was dramatically higher in these two counties, indicating that the interdigitation of urban development with rural lands is dramatically higher in the counties that have not implemented a strategy to contain sprawl. This example reinforces Principle 7, which states that design and decision making can have unintended ecological consequences.

Note that not all urban change is on the fringes, in contrast to familiar coarse-scale images of amoeba-like urban spread. Infill can occur in existing suburbs, exurban nodes, or core cities. Demolition and replacement of buildings or upgrading of infrastructure can change the composition and configuration of city, suburban, and exurban lands. For example, in the inner suburbs of Baltimore County smaller, postwar houses are sometimes demolished, and a larger structure occupying a much larger footprint on the same lot is built to replace it. This practice is sufficiently common to have elicited the local term "scrape off." Urbanization in reality is an ongoing process in an extensive and dynamic landscape mosaic consisting of built, natural, social, and physical components.

PRINCIPLE 13. THE FLUX OF WATER, INCLUDING BOTH CLEAN WATER SUPPLY AND STORMWATER MANAGEMENT, IS OF CONCERN TO URBAN AND URBANIZING AREAS WORLDWIDE AND CONNECTS THEM EXPLICITLY TO LARGER REGIONS

Access to clean drinking water and the management of stormwater flow are key concerns for all cities, regardless of whether they are in moist or dry climates. Not only is water a flux that links cities to their upstream and downstream neighbors, but also the quantity, location, timing, and quality of water flow integrate ecological processes in the system. Research in BES recognizes this crucial integrative role of water and has applied the watershed concept from ecosystem science as a fundamental approach to understanding the structure and function of Baltimore as an ecosystem. The Gwynns Falls watershed has so far served as the primary research watershed of BES, and the watershed and several tributaries are instrumented to provide long-term weekly data on water quality and quantity in the GF stream. The GF watershed encompasses a rich variety of land use types as it flows from Baltimore County through the City of Baltimore before draining into the Chesapeake Bay. Land uses include agricultural lands in Baltimore County, postwar suburban developments outside the city boundaries, and early-twentieth-century streetcar suburbs and older row-house neighborhoods within the city. The quality and quantity of water in the GF stream are compared along the stream from reaches that drain different portions of the watershed, compared across seasons within one year and across years to investigate cumulative effects of ecological and social processes occurring in the watershed. This multifaceted approach has provided numerous insights into the functioning of riparian zones, the prevalence of pollution from fertilizers, personal care products, and road salt, and the impact of land cover/land use on runoff.

Many U.S. coastal watersheds are urbanized. Because urban systems contain vast amounts of impervious surfaces, the primary concern for the management of stormwater in cities has been to remove the water from the system as fast as possible in order to avoid damage to infrastructure and property. This management focus results in little water absorption by the soils and plants in the system and rapid flow of water in the receiving streams, which is often characterized as flashy. The altered magnitude and duration of peak flows in streams following a storm have consequences for stream morphology and riparian zone structure and function. But these flows also deliver sediments and pollutants into receiving water bodies. In the case of Baltimore, water from the GF watershed makes its way into the Chesapeake Bay. Therefore, water flowing into the bay carries with it all the impacts, including heat, sediments, nutrient pollution, and organic chemical contaminants, that the urban system has had on water quality, thereby linking the urban system with the larger region.

CONCLUSION: BUILDING NETWORKS OF UNDERSTANDING

These thirteen principles have emerged out of the collective thinking and research focused on Baltimore as a complex social-ecological system. Principles 1–9 focus on empirical or conceptual content; 10 and 11 highlight methodological concerns; 12 and 13 focus on practical issues confronting urban systems based on their patterns of growth and the need everywhere to manage drinking, waste, and storm water. Each principle captures a crucial concept or generalization needed to build an inclusive theory of urban ecological science. However, the principles should not be thought of as individual building blocks that can be ordered into a particular progression. Nor are they intended to be distinct. Instead, the principles, taken together, are a network of ideas that overlap and reinforce each other. They can be loosely clustered into five initial categories, each codifying an important dimension of cities, suburbs, and exurbs. These dimensions may be considered to be metaprinciples—still higher-level generalities about the structure and function of urban ecosystems. Each principle supports one or more of the larger dimensions; the dimensions are not mutually exclusive. The five dimensions can be summarized as follows:

1. Cities are ecosystems that contain ecological and social processes that feed back and affect each other;
2. Cities are spatially and temporally heterogeneous;
3. Cities are dynamic across space and time;
4. Cities are distinct but are linked to other areas that may not be adjacent;
5. City form and dynamics are influenced by their physical, biological, social, economic, historic, and cultural context.

The first three dimensions are all fundamental characteristics of cities. Cities are ecosystems, the extent of which must be defined by establishing boundaries in space and time to set the limits of what is being studied. Like all ecosystems, urban ecosystems contain biological and physical complexes that interact within the area defined by the chosen boundary. Urban ecosystems also contain social and built complexes that interact with each other and interact with the biological and physical complexes. The feedbacks that occur between the different complexes of an urban ecosystem do so within a heterogeneous matrix. This heterogeneous matrix is created and maintained by a myriad of actors and is situated within the biogeophysical, social, and historical context of the city and geographic place. The rate of change in heterogeneity of the matrix will vary across time and space. Some periods of time experience rapid change, while at other times the system is relatively stable. Within cities, some neighborhoods or districts may change rapidly while others maintain consistency, with residents occupying neighborhoods for generations. The spatial and temporal dynamics of cities across scales is crucial because feedbacks are rarely instantaneous, and interactions at one scale may have implications at coarser or finer scales. This gives rise to an overarching principle:

6. Cities are complex adaptive systems.

We have used the principles as tools to summarize insights from social-ecological research in BES. However, they are intended to apply across urban systems. Indeed, several of them are fundamentally comparative in nature. At the same time, not all comparisons that might ultimately be pursued across urban areas around the globe are implied in our discussion. Indeed, many comparisons will be suggested by other disciplines, such as political ecology, critical geography, urban design, economics, and physics. This partial list of framing disciplines suggests some of the richness of dimensions along which urban areas, including Baltimore, might ultimately be compared. We are keenly aware that the developmental models used to explain the history of Baltimore as a postindustrial, increasingly service-based urban region will fail in many global comparisons, or even in comparison to urban areas in the United States, such as Phoenix or Los Angeles, that began their urban growth spurts after World War II. Still, we hope that the principles that are relevant to Baltimore will not only help put Baltimore research in a larger perspective, but also improve the ecological understanding of other regionally extensive urban ecosystems (box 14.2; figure 14.6). This broader understanding will rest upon joint social and biophysical perspectives, approaches, and models.

One of the most conspicuous changes in the Earth's urban estate is its expansion and connectivity. Baltimore has exhibited medium-scale expansion based on suburban and exurban spread, on connections with the northeastern U.S. urban megalopolis, and on deepening and accelerating connections with urban regions and disjunct hinterlands across the globe. The fact that Baltimore boasts a neighborhood called Canton and is sometimes called the Clipper City acknowledges its long history of global integration through maritime trade. While our work in Baltimore shows the deep roots of these changes, the shifts in demographics, manufacturing, shipping, economic investment, and racial segregation continue to shape the regional metropolis and its relationship with larger territories. These linkages will continue to shape Baltimore's ecology as well. The continuum of urbanity, a new theoretical tool, helps to place Baltimore in the fast-changing global patterns and processes of urban dynamics. The continuum of urbanity highlights both connectivity across space at different scales and the fact that livelihoods and lifestyles are key dimensions of change and feedback in and among urban social-ecological systems. This theoretical framework suggests the richness of social-ecological processes and patterns that the

BOX 14.2. PRINCIPLES OF URBAN ECOLOGY

1. Cities and urban areas are human ecosystems in which social-economic and ecological processes feed back into one another.

2. Urban areas contain remnant or newly emerging vegetated and stream patches that exhibit ecological functions.

3. Urban flora and fauna are diverse, and this diversity has multiple dimensions

4. Human values and perceptions are a key link to mediating the feedbacks between social and ecological components of human ecosystems.

5. Ecological processes are differentially distributed across the metropolis, and the limitation of services and excess of hazards are often associated with the location of populations that are poor, discriminated against, or otherwise disempowered.

6. Urban form is heterogeneous on many scales, and fine-scale heterogeneity is especially notable in cities and older suburbs.

7. Urban form reflects planning, incidental, and indirect effects of social and environmental decisions.

8. Urban form is a dynamic phenomenon and exhibits contrasts through time and across regions that express different cultural and economic contexts of urbanization.

9. Urban designs and development projects at various scales can be treated as experiments to expose the ecological and social effects of different design and management strategies.

10. Definition of the boundaries and content of an urban system is set by the researchers based on their research questions or the spatial scope of its intended application.

11. Urban comparisons can be framed as linear transects or as abstract gradients, and the abstract comparisons acknowledge the spatial complexity of urban heterogeneity.

12. Urban land covers and land uses extend into and interdigitate with rural or wildland covers and uses.

13. The flux of water, including both clean water supply and storm-water management, is of concern to urban and urbanizing areas worldwide and connects them explicitly to larger regions.

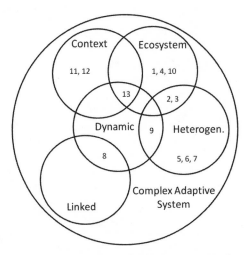

FIGURE 14.6 Urban ecological science principles organized into metaprinciples, which are high-level generalizations. Each metaprinciple is represented as a sphere and contains one or more principles keyed by the numbers in the text and summarized in box 14.2. An overarching metaprinciple emerges: urban systems are complex, adaptive systems. This emergent principle encompasses the remaining conceptual content of the framework. The principles of urban ecological science are thus a nested conceptual hierarchy. The term "heterogeneity" is shortened in the figure.

principles of urban ecology as expressed in Baltimore must link to. It is too soon to say how this new theory or, indeed, the variety of other disciplinary perspectives noted earlier will ultimately relate to the insights of BES. But it is clear that together they point the way toward a new kind of urban ecological science.

PART THREE THE GROWTH AND APPLICATIONS
OF URBAN ECOLOGY

Lessons Learned from the Origin and Design of Long-Term Social-Ecological Research

J. Morgan Grove and Jacqueline M. Carrera

IN BRIEF

- The transfer of perspectives and practices from rural, social forestry programs in developing countries was important to the development of the Baltimore School.
- Sociologists with exposure to and expertise in ecology and forestry were involved in the formation of the Baltimore School.
- The Baltimore School began with social and ecological concepts, concerns, and data.
- Linking science and decision making was critical to advancing both science and practice in the Baltimore School.
- The Baltimore School would not have happened without substantial long-term resources from the LTER program of the National Science Foundation and USDA Forest Service.
- In order for it to continue to grow into the future and to be a relevant model for other cities throughout the world, it will require an expanded and diversified consortium of participants, investments, and outcomes.

INTRODUCTION

Origin stories and their founding effects can be important to many groups. Origin stories can have long-lasting influences on a group's perspectives and practices. The origins of a group do not necessarily occur in a moment, but they may coalesce over a period of time. These stories can be helpful for understanding a group and sustaining it over time. In this chapter we examine how the Baltimore School came to be and lessons learned for designing long-term social-ecological research programs.

The origin of the Baltimore School of urban ecology is part of Kingsland's long march to cross-disciplinary research (chapter 2). Although the roots of the story were established in the early 1970s, its main stem appears in the late 1980s and early 1990s with the Urban Resources Initiative (URI) of the Yale School of Forestry & Environmental Studies and the Urban-Rural Gradient Ecology (URGE) project of the Cary Institute of Ecosystem Studies (box 15.1). In this chapter we describe the history of how and why we came to establish a program of urban long-term social-ecological research in Baltimore, key perspectives and practices, and conclude with lessons learned. Our experience suggests that these lessons are relevant to other cities and situations.

WHY BALTIMORE?

The Baltimore Urban Resources Initiative
Yale's involvement in Baltimore began in 1988, when Ralph Jones and Bill Burch met while participating in a Blue Ribbon commission on the U.S. National Parks System. Jones and Burch were members of a seventeen-member committee to develop recommendations for research and resource management policy in the National Parks. At the time, Jones had just been recruited from his position at the University of Baltimore as a professor of urban recreation to be the director of Baltimore City's Department of Recreation and Parks (R&P). Burch had recreation experience too, working as a research social scientist for the Forest Service and studying recreation uses in Forest Service wilderness areas early in his career. Since then, he had joined the faculty of the Yale School of Forestry & Environmental Studies (F&ES) as a sociologist and professor of natural resource management.

Jones and Burch found themselves seeking each other out during commission breaks for lunch, discussing their work and exploring other issues. Jones shared his worries about some of the key issues Baltimore faced, including decaying homes, falling property values and a declining tax base, rising crime rates, illegal dumping, and other environmental problems such

BOX 15.1. KEY TERMS

Abiotic: The nonliving or physical components of ecosystems.

Biotic: Refers to living organisms.

F&ES: The School of Forestry & Environmental Studies at Yale University.

PPF or P&PF: The Parks & People Foundation, also refered to as Parks & People, originated as a nonprofit shadow organization to support the Department of Recreation and Parks. It emerged into the premier organization linking neighborhood social revitalization and environmental quality in Baltimore.

Property regime: The control of land, not necessarily via ownership. Land can be controlled, that is, occupied or managed by the state, by private entities, or by communities.

RB: Revitalizing Baltimore, a partnership to bring communities, agencies, and policy makers together in facilitating social forestry as an environmental and social benefit.

URGE: The Urban-Rural Gradient Ecology project, conducted in metropolitan New York and surrounding areas, initially to compare the ecosystem function of similar forests ranging from the Bronx to rural northwestern Connecticut on the same bedrock.

URI: Urban Resources Initiative, a program to improve social and environmental quality and resources in neighborhoods in cities. In Baltimore, the URI was a partnership between the USDA Forest Service and the Parks & People Foundation.

VSA: Variable Source Area is originally an approach in watersheds to determine the controls of water flow across the landscape.

as air and water quality. Burch talked about his forestry work through F&ES in Nepal as the director of the Institute of Forestry and in tropical areas as the director of the Tropical Resources Institute. Burch spoke of what he had learned in addressing community-centered rural development issues through his efforts to link natural resource management to community development. Critical to these efforts was the importance of including local people in research and decision making and the need to transform traditional agencies to better meet local needs. Burch talked about how these programs had created important training opportunities for F&ES professors and students and how these opportunities made abundantly clear the need for academics to develop an applied orientation and produce useful knowledge.

Listening to Burch, Jones immediately saw clearly that he wanted to transfer these lessons from rural areas in other countries to his own city and agency. Although surprising, many of the needs and opportunities that Burch described sounded very relevant to the issues Jones faced as director of his R&P agency. Jones challenged Burch to apply these much-needed practices back home in U.S. cities. Jones wanted to know how he could transform his R&P by linking the environment with community development. Burch had no immediate answer except to say, "I don't know, let's hire an intern to figure it out."

John Gordon had been the chair of the National Park Service committee that had included Jones and Burch. He was also the dean of F&ES. Burch approached Gordon with the idea of inviting Jones to Yale to talk about the issues and opportunities his agency faced in Baltimore. Jones had impressed Gordon with his opinions and committee work, and Gordon agreed to support Jones's visit. In the spring of 1989 Jones visited Yale and encouraged the faculty and students to develop a program of mutual learning and service that would, in effect, apply the lessons they had learned from their rural, international work to urban areas at home. During Jones's visit, Burch recruited Morgan Grove, a master's student at Yale, to be the first intern to work with Jones. Burch believed that Grove would be a good choice to assess what opportunities might exist in Baltimore for developing environmentally based, community-centered urban development through the R&P. As an undergraduate, Grove had taken several of Burch's classes, including social forestry and parks management. Grove had also earned dual degrees in architecture and environmental studies, with a focus on urban planning. Grove agreed to work with Jones for the summer and prepare a report for Jones, Burch, and Gordon on the possibilities of initiating a new partnership program between Baltimore and Yale.

Several weeks later Jones died suddenly of a heart attack. Burch traveled to Baltimore that weekend to attend Jones's funeral. While Burch was in Baltimore, several members of Jones's staff told Burch about how Jones had been so excited about working with Yale and his vision for reinvigorating the department by focusing it on community development issues through recreation, parks, and forestry programs. Burch returned to Yale even more committed to working in Baltimore and honoring the many memorable conversations he had had with Jones.

Burch made plans with Jones's staff for the summer's work in 1989. Grove would shadow staff from each of the department's divisions to learn about their capacities and needs. At the end of the summer, Grove would prepare a report describing how Yale and the department could work together to realize Jones's vision of creating capacity within the department to link environmental initiatives with community development. Burch's instructions to Grove were to use the perspective of social forestry programs, which had developed internationally in rural areas since the mid-1970s.

Social forestry projects had tended to be small-scale activities that addressed local needs and provided benefits to local residents. The concept of *society* in social forestry signified a broader agenda than growing trees. The goals of social forestry also included group formation and collective action, institutional development, and the establishment of sustainable social structures and value systems to mobilize and organize individuals. An underlying assumption of the social forestry approach was that forestry could be more effective in meeting existing needs if communities were integrally involved in the planning, decision making, and implementation of forestry projects. Thus social forestry projects were highly participatory (Burch and Grove 1993; Grove et al. 1993).

Burch visited Grove and department staff several times during the summer. Grove took Burch on driving and walking tours throughout the city to share what he had seen and learned. Both Grove and Burch were struck by the extent of abandoned lots and homes in Baltimore, which numbered nearly sixty thousand at the time. They were also surprised to discover that Baltimore had clearly recognizable and distinct watersheds that could serve as an organizing framework for linking the city to the Chesapeake Bay both ecologically and culturally. These two observations suggested the opportunity to link environmental rehabilitation and urban revitalization at local and regional scales.

Burch remembered taking walks with Herb Bormann, his ecologist friend and colleague at Yale. Burch and Bormann had worked together at Yale since the late 1960s to rethink the relationship between ecology and social sciences within an ecosystem framework. They had been involved since the early 1970s in efforts by the Ecological Society of America and the

Institute of Ecology to advance interest and work on urban ecosystems (chapter 2). This work had led to a symposium and the publication of Stearns and Montag's book *The Urban Ecosystem: A Holistic Approach* in 1974. Bormann had also cofounded the Hubbard Brook long-term watershed studies in New Hampshire with Gene Likens, who would later direct the Cary Institute of Ecosystem Studies. Those studies had helped pioneer the development of ecosystem studies within a watershed approach.

Considering these experiences, it is not surprising that Burch said to Grove during one of their walks, "You know, we really should start thinking about this whole effort as the first urban long-term study. But instead of studying forest clear-cuts like Hubbard Brook and its impacts on water, we could study urban forest restoration to improve water quality and revitalize neighborhoods."

The results from Grove's summer in Baltimore led to a joint program between the city's R&P, the Parks & People Foundation (PPF; a city-based NGO), and Yale. The program was called the Urban Resources Initiative (URI) and was established with five goals: First, promote urban revitalization and environmental restoration using a social forestry framework. This framework included a participatory approach to both practice and research in order to build strong linkages between science and decision making. Second, promote professional training for students from Yale and staff from the department. This focus on training would be critical to developing a cadre of professionals equipped to address the ever-expanding needs of cities and their environments, increasing the capacity of current departmental staff and preparing students from Yale to work in rural areas as well. Third, use Baltimore as an outdoor classroom to educate local students and Yale students about cities and the environment. Fourth, develop the capacity to use Baltimore as a "living laboratory" for applied and basic research as a long-term, urban watershed research program. Fifth and finally, create demonstration projects to promote learning within the city as well as exchange with other urban areas and with social forestry projects in rural areas. With support from Dean Gordon, Grove took on responsibilities for staffing the URI program and developing internship opportunities for students to work in Baltimore.

In 1993 the USDA Forest Service launched a program in Baltimore that built on the activities and partnerships that the URI program had developed. This program was called Revitalizing Baltimore (RB) and was funded by the Forest Service through the PPF. This would be Jackie Carrera's first project as the new executive director of PPF and take advantage both of Carrera's background as a community organizer with Maryland Save Our Streams and of the emerging community of residents and neighborhood associations working to improve the city's streams and watersheds. Also new to the effort would be scientists from the Forest Service's Urban Forestry Research Work Unit, located in Syracuse, New York: Bob Neville, Wayne Zipperer, and Rich Pouyat. Neville, Zipperer, and Pouyat contributed experience in urban hydrology, forestry, landscape ecology, and soil ecosystems. Pouyat had an appointment at the Cary Institute of Ecosystem Studies (Cary). Pouyat would introduce Grove and Burch to Mark McDonnell, Steward Pickett, Peter Groffman, Mary Cadenasso, and other members of the URGE project at Cary.

The Urban-Rural Gradient Ecology Project
The ecological roots of the Baltimore Ecosystem Study (BES) are in comparative ecological science conducted in the New York metropolitan region. Mark McDonnell, then of the Cary Institute of Ecosystem Studies in his role as forest ecologist for the New York Botanical Garden (NYBG) in the Bronx, New York, planted a seed which grew into BES and which we believe was crucial in establishing contemporary urban ecological science in the United States. When he and Carl White attempted to measure the nitrogen metabolism in the old growth forest on the grounds of the NYBG in 1985, they discovered the soils to be hydrophobic. Although this phenomenon was known from other cities, the finding stimulated McDonnell to compare the urban forest with other oak forests on similar substrates but located at greater distances from the New York City urban core. Ultimately, this comparison became known as the Urban-Rural Gradient Ecology (URGE) Project and was advanced by interactions with Pouyat and Groffman and an increasingly broad group of ecological researchers, including Margaret Carreiro. With support from the Cary Institute of Ecosystem Studies under the leadership of Likens, Kimberly Medley, a postdoctoral researcher in geography, was hired as the first expert in social structures and processes to join the collaboration. The URGE project could soon claim extensive findings concerning, for example, soil contamination, nitrogen loading, denitrification, the role of exotic earthworms and different fungi, and forest structure along the gradient. Attempts to increase the scope of the study by adding

social science and economic collaborators beyond the expertise provided by Medley met with little success, due perhaps to limited interactions at the time between the social sciences and ecological sciences in general and to the high level of prior commitment that characterized the social scientists McDonnell and his colleagues approached.

In 1993 McDonnell became the director of the Bartlett Arboretum, the Connecticut State Arboretum. Research in the New York metropolitan URGE project began to be carried out in the diverse institutional homes of the established collaborators as graduate students and postdoctoral associates moved on to other positions. Continued efforts to establish working relationships with social scientists bore fruit when Pouyat introduced McDonnell and Pickett to Grove and Burch and their decade-long project in Baltimore. The desire of Grove and Burch to familiarize the ecologists with their social science research and community engagement resulted in a field trip to Baltimore. It became clear that these social scientists had established significant social capital, including "street cred" in Baltimore. Their social networks with communities, action-oriented NGOs such as the PPF, and key environmentally relevant Baltimore city agencies were clearly a precious and site-specific resource. If the desire to increase integration between biophysical science and social science was to be fulfilled, Baltimore seemed to be an ideal place to realize that goal. Pickett's position permitted him the freedom to pursue funding opportunities, using Baltimore as a research arena. Over the next several years, colleagues interested in Baltimore, some from Baltimore area institutions like the University of Maryland, Baltimore County, and Johns Hopkins University, were courted and became contributors to an emerging intellectual framework to support integrated biophysical and social research and outreach in Baltimore.

This historical process of recruitment established a pattern that persists today. Specifically, BES adopted an inclusive process in which participants join because they are self-nominated or recruited. BES is not constrained by a singular academic institution or academic departments. And our teams perform more like a jazz jam session than a regimented orchestra. Our collaborations build lines of research based on riffs of listening, trust, and field-walkabouts: visits to locations and sharing observations, questions, and interpretations about the history, dynamics, and future of a place from social, economic, and ecological perspectives. The collaborators became used to playing off of each other organically yet with direction.

FOUNDING EFFECTS: KEY PERSPECTIVES AND PRACTICES

The origins of the Baltimore School continue to affect its ongoing perspectives and practices. Many of these perspectives and practices are part of the key building structures of BES.

Interdisciplinary Orientation

The rural, social forestry approach that Yale brought from developing countries to Baltimore provided an interdisciplinary orientation. As we noted earlier, the concept of *society* in social forestry signifies a broader agenda than growing trees. The goals of social forestry can also be social, economic, and cultural, including group formation and collective action, institutional development, and the establishment of sustainable social structures and value systems to mobilize and organize individuals and households. Social forestry is problem-oriented and addresses social and ecological concerns. Social forestry solutions may incorporate both social and ecological concepts and data (in contrast, see chapter 2 for issues related to the metaphorical appropriation and use of ecological concepts).

Solutions occupy a middle ground that neither blames humans for their destructive impacts nor places too much faith in technological solutions. Further, solutions do not assume that individuals and social groups can be counted on to behave as economic-ecological actors. Certainly, material improvements and well-being are important: skills learned, jobs accomplished, levels of living made better. Shelter, food, gainful employment, health, and welfare are crucial domains of civil society. But far too often it is clear that a great deal of human behavior is motivated by additional things: the quality and development of our relations with other human beings; an improved understanding of ourselves and others; a gift of compassion or the exercise of civility in our daily lives; a life saved, an idea gained, a creative turning point made; a habitat saved or restored.

A Human Ecosystem Approach

Burch's interest in a human ecosystem approach dated back to the late sixties and his work with Bormann. In 1976 he worked with several of his graduate students to produce *A Handbook for Assessing Energy-Society Relations,* which included a conceptual diagram of the elements in a human resource system. Burch would later refine and apply this diagram to social forestry practices and natural resource management in general (figure 15.1).

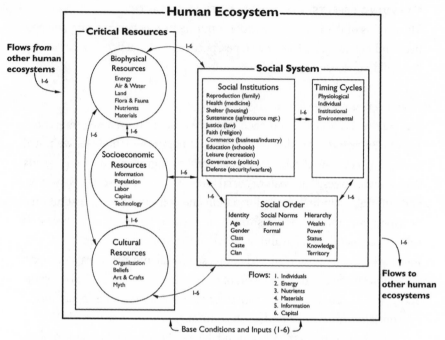

FIGURE 15.1 The Human Ecosystem Framework: critical resources, social system, and flows.

Burch stressed to Grove the need to apply a human ecosystem approach to the social forestry projects and watershed issues in Baltimore.

Some elements of the diagram are familiar, such as critical biophysical resources of energy, air, and water. Other elements are less familiar, such as critical cultural resources that include organizations, beliefs, and myths. Social institutions are strategies for solving human needs by allocating critical resources. Social order is important for understanding who receives critical resources and who does not.

Burch and Grove used the human ecosystem framework and experiences from rural social forestry projects to guide the efforts in Baltimore. Some of the initial assessments and research concentrated on the distribution of critical resources over time and how social institutions affected the allocation of critical resources. Some critical areas of focus included trees and parks; stormwater; social stratification and power associated with social class, race, ethnicity, and gender; property rights and land abandonment; and organizational change and social networks.

Watershed Approach

Burch and Bormann encouraged Grove to develop a strategy for adapting existing ecosystem and watershed approaches from rural to urban areas by assigning the topic to Grove for his term paper in Bormann's ecosystem ecology course. The premise of this approach was to consider how watersheds could be thought of spatially and functionally as heterogeneous landscapes and described and analyzed in terms of physical, biological, and social differentiation and regulation. In other words, just as ecologists seek to understand the physical and biological mechanisms that regulate the flows and cycles of critical ecosystem resources in rural watersheds, Grove, Burch, and Bormann wanted to include the social mechanisms that affect the flows and cycles of critical ecosystem processes in urban watersheds. The intellectual building blocks for this approach came from the hydrologists' Variable Source Area (VSA) approach, the ecologists' Shifting Steady State Mosaic concept from Hubbard Brook, and the sociologists' and geographers' Social Area Analysis approach (figure 15.2). Grove initially thought of this conceptual combination as an integrated VSA approach (box 15.2).

As Grove struggled with his idea of an integrated VSA approach, he searched for other relevant approaches that could be adapted to urban areas. Grove read about an emerging approach in ecology called patch dynamics. Pickett seemed to be a leading developer of patch dynamics. Pickett's contact information was included in several of the book chapters that Grove read. Grove called Pickett and described his idea of combining social area analysis with a Shifting Steady State Mosaic and VSA approach. Grove asked Pickett if this would be an example of patch dynamics. Pickett may have been taken by surprise by this unexpected call and barrage of ideas. Grove thought Pickett seemed intrigued and encouraging when Pickett replied, "Well, in a special sense, I guess so." Grove thanked Pickett for his time, hung up the phone, and returned to his term paper for Bormann describing this idea and its usefulness to linking environmentally based, community-centered urban revitalization. The use of a phone might sound quaint, but this was before the widespread use of email. Grove would later meet Pickett at the Cary Institute and learn about the URGE project through Rich Pouyat.

Ecology of Cities

The transition from an ecology *in* cities to an ecology *of* cities is a fundamental idea to BES and the Central Arizona–Phoenix LTERs (chapter 1). Pickett

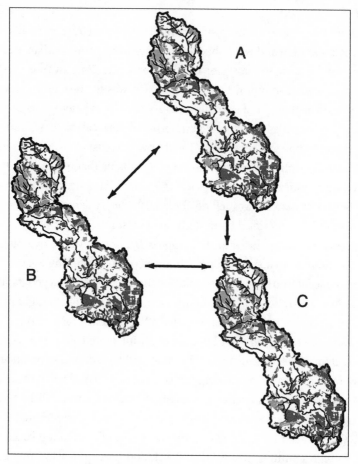

FIGURE 15.2 Gywnns Falls watershed and a Social-Ecological Variable Source Area (VSA) approach for integrating physical, biological, and social patch dynamics: (A) social differentiation; (B) abiotic differentiation; and (C) biotic differentiation. (See box 15.2 for enumeration of the aspects of differentiation for each type)

and others first articulated this idea in the prologue to a special issue of *Urban Ecosystems* in 1997. Some of the antecedents of this transition come from earlier, practical experiences in Baltimore.

Jones and Burch had identified the topic of environmentally oriented, community-based activities as a core opportunity for reshaping the city's R&P. This would require a change in the department's business model from a service-based organization in which the city was the "owner" of the

BOX 15.2. ASPECTS OF DIFFERENTIATION IN EACH COVER TYPE IN AN URBAN ECOSYSTEM, USING A VSA APPROACH

Social Differentiation (e.g., in residential areas):
    Forest and vegetation management
    Income
    Race
    Property regime (communities, landowners, tenants)
Abiotic Differentiation
    Topography
        Slope
        Aspect
    Soils
        Infiltration
        Percolation
        Water storage
        Permeability
        Antecedent soil moisture
    Hydrology
        Water table elevation
        Discharge
        Sediment loss
    Nutrients
        Nitrogen dynamics
        Phosphorus dynamics
Biotic Differentiation
    Vegetation
        Extent
        Structure
        Species composition
        Growth rates
        Distribution

public trees, parks, and open spaces to a "client-driven" organization in which the city's residents were the owners and beneficiaries of the city's natural resources across all types of ownerships—public, private, community—and uses—residential, commercial, industrial, and open space. From this new perspective, the department's roles would include manager, extension agent, educator, advocate, and cheerleader for the entire mosaic of land uses and management in the city, not just the "analog" parks and streams from the ecology *in* cities. This shift in perspective—both practical and scientific—to serving and understanding all land uses, landowners, and residents required new, interdisciplinary approaches.

Data Collection Strategies

Academic disciplines—both basic and applied—may preferentially use a limited set of methods, tools, and techniques. Social forestry as a scientific practice tends not to privilege one method, tool, or technique over another because the goals and constraints associated with many social forestry projects involve social, economic, cultural, and environmental drivers at multiple scales.

In Baltimore, Burch and Grove employed a variety of methods, tools, and techniques that were often used in social forestry projects; specifically, what was called rapid rural appraisal. Some of these methods included ethnographies and case studies, experimentation or direct manipulations, after-the-fact analyses, cross-sectional studies, and longitudinal studies to collect data. Specific tools included remote sensing, key informant and focus group surveys, observational studies, and analyses of "social scats"—administrative studies of documents, records, and operational data from government agencies, private businesses, nonprofits, and neighborhood associations. Finally, these tools could be applied through a variety of techniques, including social area analyses, maps, transects, point surveys, seasonal calendars, flow diagrams of critical resources, decision-making trees, and Venn diagrams of organizational relationships and networks.

These empirical approaches from social forestry were inclusive of social and ecological data. A challenge was how to strategically combine extensive data such as remote sensing or census data with intensive data such as field plots or open-ended interviews. These extensive and intensive data were physical, biological, or social and were associated with different scales.

The early URI approaches to social forestry created a local awareness of, openness to, preparedness for, and expectation of interdisciplinary research. Eventually, BES adapted and formalized many of the methods, tools, and techniques from URI's social forestry efforts for organizing long-term monitoring, comparisons, experiments, and modeling; methods such as transects and patch analyses; and extensive-intensive data frameworks.

Geographic Information Systems: Interdisciplinary, Spatial Data Framework
URI staff developed the first Geographic Information System (GIS) for the city in order to advance social forestry activities and parks management. There were several reasons for developing a GIS. First, because social forestry projects address social, cultural, and ecological needs and constraints, the initial motivation for developing a GIS was to organize, integrate, and coanalyze different types of data from different sources and at different scales in a unified framework. The unifying element in this case was that any type of data could be connected to a specific place and time.

For example, working with the City of Baltimore's Departments of Recreation and Parks, Public Works, Planning, and the Division of Forestry, URI staff linked biophysical and socioeconomic information to geographical points such as houses and trees, lines such as streets and streams, and boundaries such as neighborhoods and forested areas. GIS layers of land use, vegetative cover, streams, watersheds, soils, roads, and census tracts were combined spatially to form a single, comprehensive map. Different spatial data layers were selected and grouped together for a variety of uses: local community resource maps, citywide forestry inventories and greenway assessments, and regional land use analyses such as modeling the impact of land use and land cover change on nonpoint source runoff.

A second reason to develop a GIS was that it enabled URI staff to share information among different groups in a multidirectional way. For example, as the GIS was developed, the entire GIS could be shared back with each of the agency contributors, allowing them to empirically include other sources of data and perspectives in their planning and management. The GIS was also used to empower local community members by using it in meetings with residents to create maps, promote discussion, and incorporate their knowledge in planning, managing, and evaluating social forestry projects.

A final reason for developing a GIS was to promote discovery and shared understanding of social forestry opportunities and constraints. GIS was used to bring together diverse information and create maps to spark conversation about social-ecological patterns and the processes that create those patterns. Thus GIS-produced maps were objects that focused mutual education about a stream, a neighborhood, or a region. These conversations about the past, present, and future of any place were often interdisciplinary and invoked space, time, and organizational scales of explanation and learning.

Linking Science and Decision Making

An important goal of the URI partnership was to generate new and useful knowledge that connected environmental rehabilitation and urban revitalization. Each of the major partners in URI—Yale, the department, and the PPF—had a distinct role in reaching this goal.

Yale provided students and staff to summarize current research and analysis of existing data, conduct new research, and develop applications. The need to analyze existing data led to the development of the GIS, discussed above. New research involved the collection of new data for assessments, monitoring, and evaluation of pilot social forestry projects. These analyses were particularly relevant to adaptive management: what worked, what did not work, and how to improve? Finally, Yale worked to develop watershed modeling applications to estimate the hydrologic impacts of trees and pervious and impervious surfaces on stormwater quality and quantity.

The department and the PPF identified knowledge needs and developed pilot social forestry projects with Yale. PPF took on the initial role of implementing pilot projects and developing prototypes as the pilots were evaluated and tested in different types of neighborhoods. The department supported the projects with resources and committed to eventually implementing the prototypes as a full-fledged program. The three partners worked together to evaluate the social forestry projects and identify adaptations that were needed to improve them.

This process of joining decision making and research (figure 15.3) led to a dynamic process that created cumulative cycles of adaptive management and research that advanced both decision making and science.

The PPF played an additional networking role to connect decision making and research. Staff from PPF facilitated connections with managers from local and state agencies, nonprofit organizations, community groups,

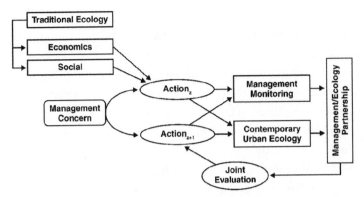

FIGURE 15.3 Dynamic feedbacks between decision makers and scientists: a generic example of the dynamic interaction between research and management. The cycle begins with the separate disciplines of ecology, economics, and social sciences interacting with a management or policy concern. A management or policy action results (Action$_z$). Management monitors the results of the action to determine whether the motivating concern was satisfied. Contemporary urban ecology, which integrates with economics and social sciences, is now available to conduct research that recognizes the meshing of natural processes with management and policy actions. Combining this broad, human ecosystem and landscape perspective with the concerns of managers can generate a partnership to enhance the evaluation of management actions. New or alternative management actions can result (Actions$_{z+1}$).

businesses, and researchers from government agencies, universities, and institutes. This network became formalized as the RB Technical Committee (1994–2006). Meeting quarterly, the RB Technical Committee provided a regular forum for the identification of knowledge needs and exchange of information.

The long-term relationships between researchers and decision makers in Baltimore has created a culture and dynamic capacity for what the political scientist Donald Stokes defines as Use-Inspired Basic Research, or Pasteur's Quadrant. Stokes describes this quadrant as science that is designed to both enhance fundamental understanding and address a practical issue. The quadrant is labeled Pasteur's Quadrant because the biologist Louis Pasteur's work on immunology and vaccination both advanced our fundamental understanding of biology and saved countless lives.

CONCLUSION

BES developed from early experiences that preceded it, principally the URI of the Yale School of Forestry & Environmental Studies, the Urban-Rural Gradient Ecology (URGE) project of the Cary Institute of Ecosystem Studies, and local initiatives through Maryland Save Our Streams to restore streams and watersheds in Baltimore City and County. These projects provided many of the building blocks of perspectives and practices for what we now think of as the Baltimore School of Urban Ecology. Social scientists with ecological exposure and expertise from URI were involved from the beginning, as they were familiar with ecological concepts, concerns, data, and long-term ecological research programs. Several practices and concepts were important for linking science and decision making. Social forestry is a practice that depends on social and ecological concepts, methods, and data and is oriented toward community-based concerns and solutions. An ecology *of* cities is a concept that is inclusive of all land uses and owners, and social-ecological patterns and processes. An important goal of linking science and decision making in BES is the mutual benefit, whereby both science and practice advance.

Collaborative research is not for everyone. To return to the music metaphor used earlier, scientists who require solos or prefer duets might not do well in our BES jazz ensemble. Likewise, scientists who prefer to play in a symphony with clear instrument sections, sheet music, and notes to play may not do well either. Playing in an interdisciplinary ensemble takes time and commitment. It can be expensive in terms of time and energy yet intellectually intriguing and rewarding. The success and progress made through BES would not have been possible without significant, long-term funds and resources from the National Science Foundation and USDA Forest Service and the commitment and interest of local agencies, NGOs, and neighborhood associations.

Acknowledgments   The authors are grateful to Gordon Geballe, Herb Bormann, John Gordon, J. K. Parker, Joyce Berry, Mark McDonnell, Sally Michel, Lynne Durbin, and the Honorable Kurt Schmoke.

CHAPTER SIXTEEN

Ecological Urban Design

THEORY, RESEARCH, AND PRAXIS

Brian McGrath, Victoria Marshall, Steward T. A. Pickett, Mary L. Cadenasso, and J. Morgan Grove

IN BRIEF

- The Baltimore Ecosystem Study (BES) provides a cogent political and geographical context in which to locate a critical evaluation of the intersections between ecology, social justice, and urban design.
- This critical intersection provides a basis for a new theory on the metacity that integrates ecosystem, social justice, and design thinking.
- The metacity is motivated to understand the functional role of urban heterogeneity as a linked network of neighborhood patches that work together to achieve urban resilience and social equity at regional scales.
- Academic urban design studios establish a foundation for incorporating interdisciplinary, project- and action-based urban design research into ecological science, where patch dynamics provided a metaphorical and practical tool shared by ecologists and designers to integrate built, ecological, and social criteria into research projects.
- Research on new forms of critical ecological urban design praxis employed a novel urban land cover classification system called HERCULES as a way to understand hybrid built and vegetated urban systems.

- In conclusion, the metacity is presented as a framework for an integrative field of urban design praxis, which provides theories, research, and new forms of action for integrating ecological knowledge and social justice.

SOCIAL JUSTICE AND THE POLITICAL GEOGRAPHY OF BALTIMORE: THE CONTEXT FOR DESIGN

Based on an analysis of Baltimore beginning in the 1970s, David Harvey and Sherry Olson, at the time geographers at Johns Hopkins University, developed an important body of empirical evidence of the persistence of environmental injustice in North American cities. To understand spatial disparities that result from social injustice, Harvey began with the nineteenth-century industrial metropolis and the Marxist theory of a metabolic rift (box 16.1). The metabolic rift describes the rupture or estrangement between the urban and the rural as well as the worker and the land, manifest in the depletion of soil nutrients as a direct result of the exportation of food and fiber to expanding cities. Olson identified and mapped this rift spatially in eight eighteen-year building boom cycles over two centuries in Baltimore, resulting in rings, wedges, and spider web patterns of segregation. This history of industrial exploitation of land and labor, and the subsequent increasing speed of urbanization, resulted in a simultaneous urban expansion and internal implosion. Olson's mapping also represents increasing flows associated with the metabolic transformation of Baltimore from a compact city within an agricultural hinterland to an industrial metropolis to a node within the sprawling Boston–Washington megalopolis.

The argument presented in Harvey's highly influential book *Social Justice and the City* is structured around his detailed studies of housing and rental markets in Baltimore neighborhoods. As opposed to seeing residential differentiation in urban areas simply in terms of social-ecological processes, consumer preferences, utility-maximizing behaviors on the part of individuals, and the like, for him, evidence in Baltimore suggests that "financial and governmental institutions play an active role in shaping residential differentiation and that the active agent in the process is an investor seeking to realize a class monopoly rent." U.S. government intervention in housing finance, construction, and markets fueled economic growth in these markets in the 1930s through the Federal Housing Authority (FHA), which "created a large wedge of middle income people who are debt en-

BOX 16.1. KEY TERMS

Adaptive cycle: A conceptual framework indicating how systems change as they accumulate social and biological resources, how they embed those resources into their natural, social, constructed, and infrastructural components over time after disturbance, and how those resources may be released and reorganized by subsequent disturbance. Key to the continued existence of a system in a given general form or to adjust to disturbance and internal change is its ability to release but not lose resources and to reorganize those resources into structures that can respond quickly after disturbance. A system is not adaptive over the course of disturbance, change, and accumulation if resources become locked into forms that cannot support reorganization or if the accumulation mechanisms after disturbance are not nimble enough to capture the resources released by disturbance. See, for example, Wu and Wu 2013 and Pickett et al. 2014 for background.

HERCULES: High Ecological Resolution Classification for Urban Landscapes and Environmental Systems is a conceptually nuanced approach recognizing that spatial areas in urban areas comprise structures resulting from interacting natural and social processes. Rather than basing the classification on the presumed contrast between natural and human products, the system emphasizes urban areas as integrated social-ecological systems.

Megalopolis: A large regional aggregation of core cities, urban nodes, suburban areas intimately connected, initially by heavy rail and, later, by air, express highway, and media infrastructure. Megalopolises can extend across state or provincial and sometimes national boundaries. Megalopolis contrasts with relatively isolated metropolises centered in a supportive hinterland, which were the predominant earlier form of regional urban structure.

Metabolic rift: The disconnection of labor from the land via the industrial revolution. Modernized by Marxist theorists of social science in the twentieth century (e.g., David Harvey 1973), the term now refers to the disconnection between biogeochemical flows (supply and return) between cities and suburbs, and the systems that supply resources and provide waste processing (Foster 1999)—what would now be called external ecosystem services. The metabolic metaphor of the nineteenth century essentially refers to the biogeochemical cycling that is now embodied in ecosystem ecological theory, which was not available to the scholars of that time. Hence a metabolic rift, in ecological terms, is the breaking of nutrient and energy flows between urban and nonurban systems, including agricultural, forest, pastoral, and wildlands and waters.

Patch: A three-dimensional volume of the Earth's surface or of a human settlement that differs from adjacent areas in structure, composition, or processes. Patches may interact across boundaries with adjacent or remote patches that are similar or different.

Patch dynamics: The conception that ecological systems consist of discrete patches that may change through time, and which are differentially connected to contrasting neighboring patches as well as to similar and dissimilar patches at a distance. Patches are arrayed in mosaics, which themselves change as individual patches or types of patches change with biophysical processes such as succession, disturbance, matter flow, and energy flux. Patches also change as a result of human and social decisions, and the actions of institutions and networks of decision making. Design is a significant driver of patch dynamics in urban systems consisting of cities, suburbs, exurbs, and hinterlands.

Planning or master planning: Based on the modernist idea that urban functions should be distinct and separate. Master planning is a top-down process, spatially arranging contrasting but uniform areas of land uses, such as residential, commercial, industrial, transportation, green space, and so on, as nonoverlapping uses throughout a jurisdiction. This usually coarse-scale, expert-driven process contrasts with urban design.

Metacity (pronounced meta-city): An application of the patch dynamic concept to settlements of any size and extent, not just to very large cities as originally introduced by the United Nations as part of an urban size ranking. The metacity idea suggests that urban systems in the broadest sense are patchworks of social-biophysical patches, and these patches experience various kinds of local and long-distance connections. Furthermore, they experience intentional and incidental social and biophysical changes that alter patch structure, identity, function, and connectivity. This view of urban systems emphasizes not only that they consist of different kinds of landscapes, reflecting social and biophysical dynamism, but that these systems are manifestly heterogeneous, differentially connected, and far from static. See McGrath and Pickett 2011, McGrath 2016, and Pickett et al. 2013 for background.

Praxis: A reflective approach to practice of a discipline or profession.

Urban design: The collaborative process of shaping urban form by bringing together the perspectives of such disciplines as engineering, architecture, landscape architecture, ecological science, and social science to create vital, livable human settlements. Focus on changing form of individual sites and projects is balanced by considering their relations to social, environment, and economic contexts and must include active engagement with local communities in the design process.

Urban Design Concept Team (UDCT): A consortium of architectural firms charged by the Maryland Department of Transportation in 1970 to manage community criticism of the plans to route a major interstate highway complex through Baltimore. The community concerns for the damage to the city's largest parks, some of its most historic neighborhoods, and disadvantaged neighborhoods in the center city eventually led to the adoption of a tunnel-based route.

cumbered homeowners." Harvey analyzes how these economic forces and government policies in the 1960s isolated poor residents in the inner city and arranged those who could afford to move through locational choices (chapter 7) into geographical submarkets.

House sales were sorted by financing structure, with the most expensive homes purchased by upper-income groups through savings and commercial bank mortgages in the north of the city. Middle-income people mostly purchased homes through 1930s FHA programs in the northeast and southwest neighborhoods as well as in Baltimore County. As Harvey wrote, "FHA programs and policies create a plateau of house prices between the 'disinvestment sink' of the inner city and the stable middle-income area." Mortgage bankers took advantage of new FHA guarantees in the high turnover areas with racial change following the social unrest of the 1960s, the

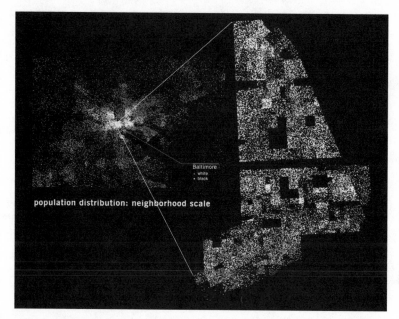

FIGURE 16.1   U.S. Census data by race for 2000. (Justin Moore, 2004)

vitality of community Savings and Loans in European ethnic areas of east and south Baltimore, and inner-city cash and private transactions for the lowest-income areas. These themes connect to previous chapters in this book, both historically (chapter 3), in terms of redlining and racial injustice (chapter 4), and as continued in more recent building cycles in increasingly distant suburbs (chapters 6 and 7; figure 16.1)

Olson portrays Baltimore's building cycles as a spider web pattern of successive new transport systems from rail to automobile as well as neighborhood street patterns, consisting first of dense row-house grids and later of dispersed single-family homes on cul-de-sacs. The result was rings of row-house neighborhoods inhabited by a new immigrant labor force employed in new industries surrounded by car-based suburbs. Each generation inherited a new city form as well as a new metabolism, as increasingly extensive spatial webs of natural resources were exploited and new exchanges developed with world markets. This metabolic and morphological change was intertwined with the mobility of people, as those with the means to, sought social status and an escape from environmental and health problems. According to Olson, "The long-swing rhythm of urban investment has produced distinctive spaces and forms in the city. The timing and spatial structure of city building also created a tempo and a spatial matrix for income redistribution, social conflict, and social pathology."

The Baltimore spider web, for Olson, was a mechanism of social organization and redistribution that in each generation served to maintain structural inequality based on differentials in investment, environmental quality, and mobility. What began as racial discrimination within households between master and servants later became a separation between alley and street houses and today has become starkly racially segregated neighborhoods. By the twentieth century a black ghetto emerged in Baltimore, extending to the northwest, and a green wedge of wealth to the north interrupted the ring pattern of neighborhoods. With the arrival of massive suburbanization after World War II, segregation was at the scale of the metropolitan region. For Olson, poverty, race, and structural dependency are core segregators and take the form of a brick circle in the spider's web of the Baltimore region, while wealth is a green wedge. The rapid movement of capital into new construction in the suburbs with very little money flowing into maintenance and renovation in the city resulted in the abandonment of one thousand houses per year between 1960 and 1970 in Baltimore. Rings of population decline now grew around the center while homeownership declined in the outer ring of the city. Also evident was a ring of zoning exceptions for conversion of single-family row houses to multifamily dwellings where "race discrimination and inequalities of income are played off against each other, reinforcing the very structure of inequality."

URBAN DESIGN ORIGINS AND PROVISIONAL PRACTICES

Urban design was created as an academic and professional field in the United States shortly after the end of World War II, as national attention and unprecedented levels of funding were directed toward renewing the nation's overcrowded, neglected cities. This intense interest was focused on both the crowded, old city centers, which had not seen infrastructural investment since before the Great Depression in 1929, and the rapid growth of new suburbs fed by the FHA housing loan guarantees, homeowner tax incentives, and the new Interstate Highway System. Architects, landscape architects, and planners hoped that the creation of a new discipline of urban design located in between these established professional fields could face the challenges of the regional scale of the car-based city as well as the rehabilitation of historical centers. However, in both the urban center and periphery, economic development interests and highway engineering came to dominate the postwar American city, and the disciplines remained split in their scholarly and professional pursuits and limited in their impact. Fine-grained, architecturally based design processes were too slow and expensive to compete with the rush of highway construction and urban renewal, landscape architecture often provided merely a green veil to suburban sprawl, and normative planning remained focused on rezoning for economic growth and further land use separation.

As a result of such disciplinary factionalism, two mutually exclusive forms of urban design practice emerged in postwar North America: on the one hand, the design of pedestrian-scaled city fragments both in old city centers and new suburban enclaves; and on the other, the provision of environmentally sensitive and attractive, yet increasingly segregated, new greenfield settlements, carefully masked by vegetation from the high-speed commercial strips of the new, car-dominated urban regions. In the 1960s, however, urban social justice, civil rights movements, and the environmental movement were fueled by the destruction unleashed through urban renewal and highway construction and the exclusion from new suburban developments of the urban poor, racial and sexual minorities, single women, and the elderly, among others. Urban design, therefore, was born within but was ill-equipped to address the polarizing debates between diverse urban stakeholders, government policies, powerful financial interests, neighborhood preservationists, social justice advocates, and environmental conservationists.

Baltimore is situated near the southern end of the northeast U.S. seaboard megalopolis, the conurbation within which much of the early thinking and experimentation in the new field of urban design were located. First Harvard University in 1956 and then the University of Pennsylvania in 1958 held conferences and established new graduate programs situating a new field of urban design between architecture, landscape architecture, and urban planning programs. At MIT, Kevin Lynch established the cognitive rules for pedestrian-based "good urban form." However, his proposal for the preservation of the old markets of Faneuil Hall in Boston occurred just as the Central Artery and City Hall urban renewal projects were replacing Scully Square, the public space at the heart of the city. In New York, following Jane Jacobs's critique of modernist planning and urban renewal,

a more neighborhood- and community-based approach to city redevelopment emerged, and in 1965 Mayor John Lindsay established the Urban Design Group, headed by Jonathan Barnett. Barnett focused on microzoning
policies for special districts, such as Times Square, rather than master planning. In Philadelphia, Edmund Bacon took on a more imperial role as mod-
ernizer of William Penn's City and focused on monumental projects in the
center city such as Independence Mall and the Galleria. At the University
of Pennsylvania, Ian McHarg outlined the landscape-planning rules for
designing with natural systems in mind on the city periphery and, as a
member of the office of Wallace, McHarg, Roberts and Todd, designed many
greenfield suburban new towns around Baltimore and elsewhere in the
Northeast Corridor.

Baltimore was the site of various experiments in postwar suburban expansion and urban renewal. In January 1944 a group of Baltimore business
leaders formed the Downtown Committee, which recommended rehabilitation of the old central city, an outer ring road collecting radial highways,
a nonstop highway between Washington and Philadelphia, and a modern
internal highway system. By 1949 the Baltimore County Planning Commission, a political jurisdiction separate from Baltimore City, developed a
beltway plan completely bypassing the existing city (figure 16.2). The belt-

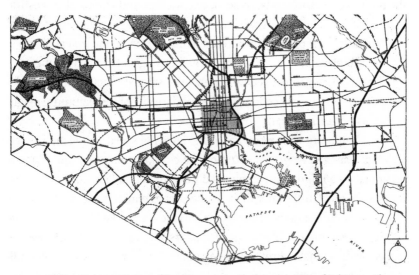

FIGURE 16.2 Baltimore's East–West Expressway Study, 1960. (City of Baltimore Planning Commission)

way was begun by the county, taken over by the state, and eventually completed in 1961 with federal interstate highway funds. Large amounts of federal funds were available not only for this new network for highways but also for slum clearance and the construction of new public housing. According to the urban planner Sydney Wong, Lafayette Courts in Baltimore City concentrated over eight hundred families in six high-rise and seventeen low-rise buildings on a 21.5-acre site created through the demolition of blocks of old row houses. Families living in public high-rise projects faced many unconsidered problems, and the projects reinforced segregation, as 89 percent of public housing residents were African American.

In 1962 the Greater Baltimore Committee, concerned with the modern viability of the old business district, unveiled the Charles Center plan. The plan resulted in a modernist superblock composed of towers and plazas with strong architectural controls maintained through design competitions. This was strategized to catalyze a sequence of renewals that succeeded in bringing suburbanites back into the city, bypassing the older neighborhoods: the Inner Harbor renewal, planned by Wallace-McHarg, was the second phase, and, later, Camden Yards became a model for new center city sports stadiums. The tall ships bicentennial celebrations of 1976 attracted huge crowds to the waterfront, and the Inner Harbor became an international model of leisure- and consumption-led development, followed in 1980 by the National Aquarium, Maryland Science Center, and Harborplace Festival Marketplace. Olson was an early critic of such development and questioned the expenditure of public funds in remaking the city in the image of a "festival marketplace" for tourists rather than focusing on the needs of the city's inhabitants. The gap between the new professional practice of urban design focused on the city center and the social reality of the Baltimore neighborhoods is clearly evident.

The Baltimore City region has a long legacy of social segregation (chapter 4), as waves of migrants occupied the city's row-house neighborhoods separated by a rolling topography and three major streams: Gwynns Falls, Jones Falls, and Gunpowder Falls. As the stream names suggest, Baltimore, the most inland of the megalopolis coastal cities, is marked by a topographic fall line within the city boundaries (chapters 3, 13). Parks and parkways designed by the Olmsted Brothers were built in the river valleys and outer hilltops, and people with means moved to wedges of healthier, greener places. According to Olson, the "rolling terrain, geologic variety of the vast semicircle of countryside north and west of US 1 and a number of unspoiled necks and rivers, peninsulas and tidewater inlets" provided a perfect setting for new developments that "feature streams, private lakes or marinas, woods, and pasture for horses."

For Olson, the names—Fox Meadows, Camelot, Quiet Inheritance, Tanager Forest—hint at the premium combination of prestige and outdoorsy vigor recalling the Maryland plantations, the hunts, and the English aristocracy. Most postwar urbanization took place in Baltimore County, a separate political entity outside the city's boundaries. The interest in planning based in landscape ecology principles was embraced in Baltimore County with the establishment of an urban-rural demarcation line (URDL; figure 7.2) in 1967, restricting suburban developments with water and sewer services to the ring road and new arterials outside the political boundaries of the city and including conservation zones for farmland and natural resources. According to Donald C. Outen, the URDL closely traces the beltway and radial interstate corridors, successfully containing 90 percent of the county's population within a boundary served by water supply and sewers on one-third of the land. The management of sprawl also led to innovative sustainable landscape designs, such as Wallace-McHarg Associates' plan for Green Spring and Worthington, Maryland.

Design strategies for the aging ring between the renewed downtown enclaves and greenfield suburbs were more difficult given the great social inequality of postwar renewal projects. According to Olson, 80–95 percent of the people relocated by urban renewal and highway construction in Baltimore were African American. The Baltimore Plan of 1952 was one of the first in the country to focus on rehabilitation. In one innovative project, in the West Baltimore neighborhood of Harlem Park, alley dwellings were demolished and block clubs were formed to participate in the design of inner-block parks. The Urban Design Concept Team (UDCT) was another unique experiment in Baltimore in response to community criticism and conflict between rehabilitation of the old residential districts and the city's ambitious highway plans. Created in December 1970, UDCT was composed of the architectural firm Skidmore, Owings & Merrill, J. E. Greiner Company, Parsons, Brinckerhoff, Quade & Douglas, and Wilber Smith & Associates. Their client was the state of Maryland's Interstate Highway Division, which sought federal money for their ambitious plans. The urban designers' mandate was paradoxical in that they were charged with harmonizing conflicting views and providing for the social, economic, and aesthetic needs of the city but without deviating from the highway route.

According to Wong, the team suggested alternate routes and helped to redesign a highway plan that originally would have run eastward through Rosemont, Harlem Park, and Federal Hill, cross the Inner Harbor, and landing in Fells Point and include a giant, sixteen-lane interchange over the Inner Harbor. Competing in the court of public opinion, the UDCT opened a field office that worked closely with community groups and provided them with valuable inside information. Twenty-four disparate organizations struggling against the highway proposals coalesced into MAD, the Movement Against Destruction. Environmental protection groups such as VOLPE—the Volunteers Opposing Leakin Park Expressway—added environmental conservation to the neighborhood-based social injustice protests as the highway was planned to cut through Leakin and Gwynns Falls Parks, the largest green spaces in the city. As a result of UDCT efforts outside of their clients' mandate, MAD and VOLPE coalesced to prevent the east–west expressway across central Baltimore, and Interstate 95 was diverted south via a harbor tunnel below historic Fort McHenry, sparing the Rosemont, Harlem Park, Federal Hill, Inner Harbor, and Fells Point neighborhoods.

William Burch's 1991 *Strategic Plan for Action* for the Baltimore City Department of Recreation and Parks was another innovative praxis model emanating from Baltimore. In the face of the reduction of public parks funding that accompanied the shrinking of the city's population and tax base, Burch and the team from the Yale Urban Resources Initiative connected parks management within a social–natural resource framework (chapter 15), improving connections to government agencies, nonprofit organizations, and community groups. In the wake of this effort, in 1994 the U.S. Forest Service Revitalizing Baltimore (RB) project developed a national community forestry and watershed restoration plan managed by the Parks & People Foundation in cooperation with the Maryland state forester. Working with local and state nongovernmental organizations, local, state, and federal agencies, and universities, RB carried out community forestry, watershed restoration, and educational projects centered on conservation stewardship and outdoor experiences in underserved neighborhoods. This work linked natural resource management and social capital in Baltimore, forming the foundation of ecological urban design as a social praxis.

The Baltimore region was the site of many experiments in new urban design theories, but only recently are the conflicts between highway construction and urban renewal finally being addressed. Lafayette Courts was demolished and replaced by less dense townhouses in 1995; Charles Center was redesigned to include more activities in 2005; and Baltimore's first limited-access highway, the Jones Falls Expressway, reaching its fifty-year expiration date, may be slated for removal in the face of costly rehabilitation. While Baltimore County's growth boundary and strong zoning culture successfully contained sprawl in planned communities just outside the city's boundaries, scattered subdivisions have leapfrogged beyond the boundaries of Baltimore County. As a result, the problem of historical social and racial inequities has only worsened. The urban poor who were caught in the massive relocations to facilitate construction of the expressways built to link the new residential and commercial suburbs to the central core remain isolated in an inner ring where the majority of the region's structural dependents are pooled. Citizen activism was able to stop expressway construction but could not spare the neighborhoods of West Baltimore from a submerged "highway to nowhere" nor the large presence of vacant lots and abandoned buildings.

PATCH DYNAMICS AND URBAN DESIGN DISPARITIES

The Urban Design Working Group (UDWG) was formed to conduct experimental academic design studios in 2003 as a novel branch of BES research. The studios, illustrated in this section of the chapter, were hosted at the Graduate School of Architecture, Planning and Preservation (GSAPP) at Columbia University between 2003 and 2006 and at Parsons School of Design from 2007 to 2010. The urban design studios remain of current interest, as they have acted as forums for a deep interdisciplinary dialogue between the ecological theory of patch dynamics and new, expansive notions of urban design praxis engendering a synthesis of ecological theory and social action. The UDWG applied ecological and social justice theory in project-based research for neighborhoods across the Baltimore region. The studios included field trips, student projects, presentations, seminars, lectures, and conferences with Baltimore residents, local architects, neighborhood groups, and planning officials, primarily through Baltimore's Parks & People Foundation, neighborhood organizations, and the city's Office of Sustainability.

The design studios also took advantage of the new computer platform of the "paperless studios" at Columbia. Digital tools introduced into architectural and urban design education in the 1990s expanded urban designers' ability to represent, visualize, and communicate complex urban adaptive cycles in space and time. More important, digital platforms allowed for students to work collaboratively in research teams, to efficiently include the various partners outside the academic setting, to pool the resources offered by different studio design instructors and BES partners, and to innovate with novel communication techniques for neighborhood residents. The students worked in design teams comprising social networks and affinity groups that went beyond individual student knowledge and singular design authorship. The structure of the studio sought to uncouple urban design from the professional limits and resultant top-down model that so hampered the examples cited above, such as the work of UDCT in West Baltimore and similar structures elsewhere. Complementing digital tools was the commitment to extensive fieldwork in Baltimore's neighborhoods in dialogue with multiple neighborhood, nonprofit, and government organizations. Paperless studios allowed the UDWG to both understand the emerging form of what we call the metacity and to create new ways of designing with communities.

J. Morgan Grove, the leader of the Demographic and Social Working Group in BES, initiated the idea of an urban design studio. Grove approached the architect Brian McGrath, coordinator of the urban design studio at GSAPP in 2002. McGrath's work had engaged the adaptation of New York to financial cycles and technological change and explored the possibility of creating a collaborative and interdisciplinary urban design method based on diverse urban actors consensually reshaping existing cities. The basic premise of the social scientists working in Baltimore is that ecological knowledge in the absence of the development of social capital will not result in sustainable environmental change; social models need to be developed alongside ecological models in order to implement change.

BES's then-director Steward Pickett and co-principal investigator Mary Cadenasso framed the work of the urban design studios around the ecological theory of patch dynamics. Patch dynamic theory is based in disturbance rather than equilibrium ecology and therefore emphasizes ecosystem heterogeneity and flux rather than homogeneity and stasis. Cadenasso, an ecologist, enriched the theoretical framework of patch dynamics with her work on ecological boundaries as landscape components that manage the flows of organisms, nutrients, and information in cities. The landscape architect and urban designer Victoria Marshall provided urban landscape mosaic as a lens from the first studio in 2003, while the architect Joel Towers introduced the concept of resilience to the second year of the design studio in 2004 and brought David Harvey to speak to the students. The architects Petia Morozov and Sandro Marpillero at Columbia University and Mateo Pinto and Eugene Kwak at Parsons School of Design contributed to establishing a foundation for incorporating interdisciplinary, project-based urban design research into BES in collaboration with BES scientists. Also instrumental were president Jackie Carrera of the Parks & People Foundation and senior director of Great Parks, Clean Streams & Green Communities Guy Hager. More than one hundred students from Columbia and Parsons participated in the studios, as did visiting students from Aalborg University in Denmark in 2005 and Katholieke Universiteit Leuven in Belgium in 2010.

Grove introduced the Olmsted Brothers' Plan for Baltimore of 1904 as evidence of his theory of design and social capital. The 1904 plan remained an incomplete vision, but it anticipated the growth of the city of Baltimore, and it advocated structuring that growth upon the preservation of the city's three stream corridors (chapter 9) and linking these corridors via parkways. The first group of students, led by Marshall, updated the symbolic power of the Olmsted legacy in American urbanism to an era inspired by Burch's 1991 Parks Plan of community-based action. The design studios examined the city with the same breadth of vision as the 1904 plan but worked with a social-ecological imagination that broke the prevalent cognitive separation between nature and city, respectively, into parks and neighborhoods. Additionally, Olson's work demonstrates how these parks become buffers in the social segregation of neighborhoods. Rather than marking wilderness boundaries and preservation territories and imagining nature as the other, the design studios acknowledged that humans and nature are inseparable, and a just city is an integrated one. It rains on every street corner, making every urban block part of a watershed ecosystem, and all neighbors share a watershed. The work of the students maximized intertwined relationships between city and nature as a tool to bring neighborhoods in contact with the newly completed Gwynns Falls Trail and with each other (figure 16.3).

Geographically, the first studio studied the Gwynns Falls watershed, a sixty-six-square-mile (17,150 km²) stream catchment beginning at the headwaters in the Glyndon neighborhood of Baltimore County and ending at

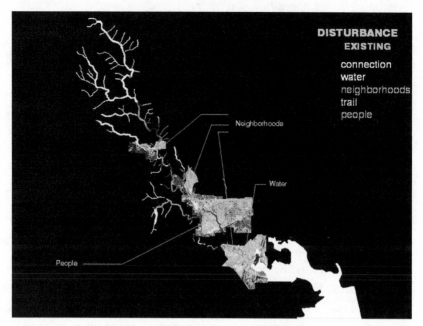

FIGURE 16.3 Gwynns Falls Watershed Study. (Rennie Tang, 2002)

the Middle Branch of the Patapsco River in Baltimore City. Divided into eleven subwatersheds, this huge catchment area crosses the urban-rural demarcation line and the socioeconomic and political divide between Baltimore County and Baltimore City. Rather than as a nature corridor separated from neighborhoods, the urban design studios considered the Gwynns Falls watershed as an integral part of the city region's green infrastructure, on the one hand, and, on the other, as intimately linked to the social integration of Baltimore neighborhoods. Small hills enclose the city grids, opening out into valleys and ridges, further diversifying and intensifying ecological niches (chapters 1 and 13) and diverse pockets of social life (chapters 3 and 4). The team entered the city first by walking on trails through woods to witness riparian systems, storm runoff erosion, and sanitary sewer leakage, then by riding bikes through the previously impenetrable river corridor, and then by paddling around the mirrorlike Middle Branch, with its muddy, trashy sediment below and highway flyovers above. The UDWG developed a deep knowledge of the mosaic of the neighborhoods and social life of West Baltimore with its street-corner drains as well as its suburban-edge city parking lots and strip malls over the ridge.

The first studio work coincided with the completion of phase one of the Gwynns Falls Trail, a fourteen-mile-long (23 km) hiking and biking trail that begins at the city boundary. Subsequent studios engaged in the later phases in connecting this trail system to the center of the city via the B&O rail spur and along the Gwynns Falls outlet to the Middle Branch waterfront, where an existing waterfront park can connect inland to the state park system. Gwynns Falls had become disconnected both ecologically and socially from its adjacent neighborhoods, and design projects were developed to reconnect Gwynns Falls watershed to Baltimore's neighborhoods and to provide linkages between communities that have been historically divided. Newly connected neighborhoods could provide civic support for ecosystem restoration at multiple geographical scales encompassing regional forestry, stream valleys, large protected areas, abandoned industrial sites, and neighborhoods.

The academic design studios focused on translating the watershed, patch dynamic, and social capital frameworks (chapter 1) at the scale of the river basin to the neighborhood and the block. As an analytical tool geared toward experimental and exploratory design methodologies, the studios established understanding of the structure and flux of the watershed as a social–natural system across space and time; and from those understandings empowered residents to alter the watershed structure, flux, and meaning of their community. Components included brownfields, post-industrial sites, fragmented urban blocks, the watershed, infrastructure connections and disjunctions, and patterns, including neural, organic, and constructed ones.

In subsequent years the urban design studio focused its efforts in collaboration with the very neighborhoods caught in between the renewed center city and expanding suburb, just east of the Gwynns Falls watershed, the urban subcatchment referred to as Watershed 263. This subwatershed area became the primary focus for the UDWG as the issues of social capital and patch dynamics within the watershed framework were tested with several neighborhood groups in West Baltimore (figures 16.4 and 16.5). These neighborhoods were considered part of the entire hybrid patch structure of the Gwynns Falls regional watershed and a critique of the limited, uneven spatial model of the ecological corridor.

The 2010 design studio at Parsons looked at the internal logic of multiple blocks in Harlem Park in partnership with the Harlem Park Neighbor-

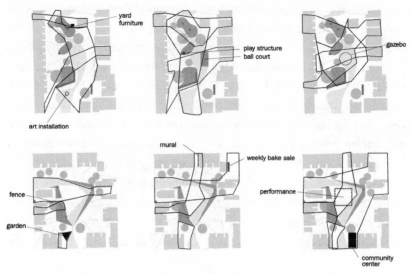

FIGURE 16.4   Harlem Park backyard share proposal. (Justin Moore, 2004)

hood Association, Parks & People Foundation, and the Baltimore Office of Sustainability. Urban blocks were seen as ensembles of buildings, vacant lots, and inner-block parks. Taken together, they are unique ecological and social units. Each is its own miniwatershed with roofs and backyards, high points and low outlets, and various household types. Instead of seeing the house, neighborhood, and city as the scalar units of urban design, the block is seen as a cohesive social support structure and ecological patch. Educa-tion, recreation, agriculture, work, housing, and community are pooled within a block providing social support beyond the family unit (figure 16.6).

The East Baltimore studio included students from Katholieke Univer-siteit Leuven, Belgium, as part of a transatlantic academic exchange on urban inclusion in partnership with the Historic East Baltimore Commu-nity Action Coalition (HEBCAP) and Morgan State University. While many blocks of East Baltimore have been razed for an entrepreneurial develop-ment extending from Johns Hopkins Medical Center, HEBCAP acquired a string of vacant buildings and lots along the Northeast Corridor rail line. Projects engaged water management, ecological and community resto-ration, and health and educational service for jobs creation and income-generating resources (figure 16.7).

The UDWG introduced transdisciplinary collaboration as a method to

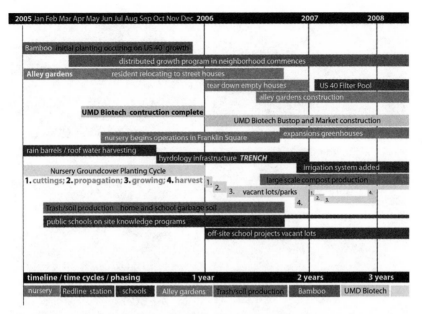

FIGURE 16.5 Watershed of Fortune (detail): timeline for implementation of West Baltimore neighborhood plant nursery cooperative business. (Phanat Xanamane, 2005)

advance knowledge, regardless of the anxieties generated by the guardians of disciplinary boundaries. Ecologists and designers linked and shared models, frameworks, and theories and saw each other as fellow urban actors looking for research questions. The UDWG works as a network in order to undertake significant research assisting neighborhood residents to initiate, leverage, or adjust to change. The design studios engaged drawing and visioning as a tool to trigger discussion and to communicate how urban form can shape and express the incredibly rich spatial language of ecology. Drawings communicated urban models, understood as mental images that can be shared by all decision makers in envisioning possible futures. The UDWG imagined a newly critical role for urban design in relation to the innate human ability to adjust to complex change, given the right access to information leading to better integration of our social and biological web of relationships. Adaptation and adjustment were key drivers of design solutions, which searched for resilient urban form and processes that encompassed heterogeneity, spontaneity, vitality, and equity. Themes that emerged across the student projects included altered everyday practices ingrained in

FIGURE 16.6 Inner-block proposals for Harlem Park. (Parsons architectural design studio, 2010)

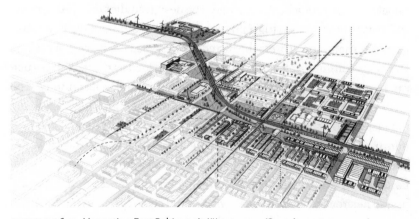

FIGURE 16.7 Harvesting East Baltimore's Waterscape. (Sven Augusteyns, 2011)

material existence through actions and rituals and the production of new ways of life, civic imagination, and social connections. The studios were imbued with a sense of possibility for urban design praxis as a space for new processual action and new ways of thinking, working, and building that are collaborative, improvisational, and experimental and always based in a concern for community action and quality design.

RESEARCH ON NEW FORMS OF ECOLOGICAL URBAN DESIGN PRAXIS

In addition to the academic design studios, the UDWG engaged in developing new forms of ecological urban design praxis in a research setting through projects coordinated by the design firm urban-interface. Two research projects are described below, first, the mapping and analysis of a novel classification system of spatially heterogeneous urban land cover patches that can be used for a more equitable diffusion of urban design activity and, second, a set of neighborhood-based participatory design scenarios for urban watershed nitrogen retention. Both projects shift the agency of urban design to networked bottom-up local initiatives within the meta-understanding of regional-scale ecosystem processes. Common cause is found between residents in outer exurban, inner-ring suburban, and center city neighborhoods, thereby reducing urban design disparities. In these two collaborative projects the research team comprised a wide array of collaborators and interests: urban designers, scientists, policy makers, and urban residents aimed to provide scenarios for local, neighborhood-based action. The results suggest a new mode of metapraxis that goes beyond the tradition of site-specific, problem-based, top-down or client-initiated projects, but that provide frameworks for alternative pathways to collective citizen action.

Integrated Land Cover Classification of Urban Spatial Heterogeneity

The U.S. national land use planning system separates urban from agriculture and forested areas, and in urban areas it separates built from nonbuilt components in the landscape at a continental scale. Urban areas are typically coarsely mapped by land use classifications of residential, commercial, manufacturing and other uses. Additionally, people are geographically classified according to census questions about race, age, education, employment, and income. Based on the logic of the industrial city, land use and demographic classifications are the basis for planning functionally differentiated areas zoned by use and density, political constituency, and the provision of public services rather than structurally as ecological systems. Urban designers are typically brought into a project of limited scope of damage control after most large-scale infrastructural and planning decisions have been determined or are responsible for the design of privately developed enclaves.

In contrast to such top-down and coarse-grained *land use* planning, high ecological resolution *land cover* classification is important for design action that links spatial heterogeneity, social justice, and ecological function in urban patches. Social-ecological participant analysis reveals the types of processes that are actually occurring in various socially stratified neighborhoods and seeks to leverage those processes through design action. Although both are focused on spatial heterogeneity, land cover is a physical pattern, and social preferences reveal human ones. Together, land cover and social preference classification allow social-ecological data to be cast independently for explaining ecologically functional variables such as biodiversity, nutrient retention, and carbon storage, compared to social variables based on lifestyle and prestige. The development of collective interest rather than self-interest is important because it helps communities to see problems and solutions beyond property lines. As physical descriptors of spatial heterogeneity, land cover and social patterns may be better lenses through which to view novel ecological processes as tools for equitable urban design.

Cadenasso and colleagues created High Ecological Resolution Classification of Urban Landscapes and Environmental Systems (HERCULES) to link spatial heterogeneity to the social-ecological functioning of urban systems. Urban regions are integrated social-ecological systems that are characteristically heterogeneous. They comprise built and nonbuilt elements that interact and jointly influence ecological processes across fine-grained spatial scales. HERCULES uses the amount and type of five land cover elements— woody vegetation (shrubs and trees), herbaceous vegetation, bare soil, pavement, and buildings—as criteria to determine patches containing different combinations of land cover. In HERCULES the proportional cover of each of the five features is used to classify patches. Different patches, meaning those areas with different types and amounts of land cover elements, can be compared to determine whether they also differ in a specified ecological process, such as nutrient cycling. Furthermore, patch identity reveals social disparity in access to quality urban design services and can be a tool for implementing community-based design action.

Through the process of developing a visual framework for the HER-
CULES system covering the Gwynns Falls watershed in Baltimore, some
important new discoveries were made about land cover heterogeneity and
uneven urban design patterns. The various combinations of different per-
centages of the five base land cover types suggest hundreds of possible
patches. In organizing a chart of these possible patches, various families
of patch types were uncovered, some dominated by a single type, such as
forest-dominated residential neighborhoods. Other patches might have two
land cover types that together dominate the patch, such as buildings and
pavement that characterize urban commercial strips. Equal mixtures of
more than two types appear more rarely. These families of patch types can
be mapped to understand where they are located in the city, such as forest-
dominated neighborhoods surrounding the planned parkways in the first-
ring suburbs of Baltimore (figure 16.8). Finally, while the HERCULES chart
shows all the possible patches that can appear in a given urban watershed
or politically bounded area, any watershed or city will have a specific signature that is determined by which patches actually appear.

HERCULES is relevant to an inclusive ecological urban design praxis
because the physical and biological elements that influence ecological
processes can be understood within neighborhood-scaled, self-determined
patches. Designers, ecologists, and residents together can evaluate the spe-

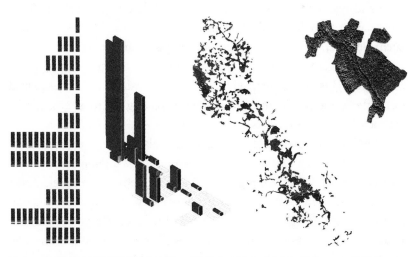

FIGURE 16.8 HERCULES Patch Atlas. (Victoria Marshall, M. L. Cadenasso, S. T. A.
Pickett, Brian McGrath, 2007)

cific land cover signature of a neighborhood as a baseline for projected change. HERCULES patches elucidate the remarkable heterogeneity of the city within the variable hybrid mixture of built and vegetated surfaces. Small-scale design choices alter the movement of water, nutrients, and organisms across patches and affect parameters such as temperature and moisture within patches. Community understanding of the links between design choices and ecological processes can aid in producing more democratic and just processes in designing resilient, ecologically sound cities.

Neighborhood Scenarios for Ecological Urban Design

In the second research project, a matrix of design scenarios was created within the HERCULES land cover framework. The design options play out on three widely different test subwatersheds: in a recently constructed low-density exclusive suburb in the watershed of Baisman Run; in Dead Run, a first-ring suburb repopulated by new immigrant families; and in Harlem Park, an African American row-house neighborhood within Watershed 263 (figure 16.9). Additionally, the preferences of the residents of three focal neighborhoods in these subwatersheds were evaluated according to the relative social capital and cooperation by and between various actors: Private Property Parcel Change (Me), Cooperative Change Between Neighbors (Us), Public Infrastructure Change (Ours). Individual households (Me) would benefit from understanding their gardening habits within neighborhood patch and regional watershed frameworks, while the public right-of-way (Ours) could be redesigned according to the specifics of a neighborhood patch. However, it is the Us scenario that suggests an ecological urban design praxis creating a new kind of commons between private properties and public rights-of-way. Shared backyard fences, hedges, and underutilized areas offered the possibility of neighbors cocreating the most ecologically and socially productive zones.

The three neighborhoods have different balances between individual, neighborhood, and public infrastructure reflecting the uneven development inscribed in the Baltimore region as described above. Springfield Farm Court Road is a private subdivision of estates on five-acre lots draining into a forested, open-stream tributary to Baisman Run above Oregon Ridge Park in Cockeysville. Gilston Park Road is a street of older ranch houses on small lots that drains through Westview Park into Dead Run in Catonsville, just outside the Beltway. This area represents a condition of private homes with

FIGURE 16.9   Three neighborhood watersheds: (from top) Baisman Run, Dead Run, and Harlem Park. (Brian McGrath, 2009)

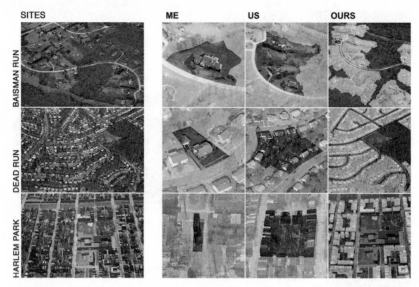

FIGURE 16.10 Three social arenas for intervention: me, us, or ours. (Brian McGrath, 2009)

front yards and backyards draining into a public street filtered through an open stream in a public park. Lanvale Street Headwater Catchment contains a few blocks in the Harlem Park neighborhood in Watershed 263 in West Baltimore. This area contains a much higher percentage of public land, including the unusual condition of inner-block public parks created in an urban renewal project of the 1970s. The design scenarios provide different ways for a diversity of residents within a regional watershed to take action that comes from local preferences yet achieves shared regional goals (figure 16.10).

In the two research projects discussed above, ecological urban design can serve social justice aims rather than being unevenly distributed through individual client-based or local, public initiatives within wealthy political jurisdictions. Instead of having technical experts redesign only those neighborhoods that can afford it, scenarios allow all neighborhood residents to create change in their own property, cooperate with neighbors, or advocate investment in public works. In such a process, issues of social justice and inclusion can be integrated with ecosystem design by learning how residents' small-scale decisions affect a whole watershed. Normative urban design practice has tended to reinforce top-down patterns and processes of social stratification as well as the release of ecological degradation down-stream. A strategy of the commons allows people to make design decisions across private property lines, in cooperation with neighbors, or within pub-lic right of ways but always within a feedback loop of ecological and social cooperation. Local decisions are part of a regional patch matrix that can be understood through the HERCULES system and within a watershed framework. The networked possibilities of such a complex adaptive system comprise a new ecological urban design praxis that we have termed the metacity.

CONCLUSION

Ecologically informed and socially just urban design is defined here as a synthetic discipline within an expanded field of architectural and landscape design, ecological theory, social research, and community activism. This crit-ical ecological urban design praxis finds its richest potential as a new urban model: the metacity. We have coined the term "metacity" to both critically evaluate the new form of diffuse urbanism worldwide that has accompa-nied digital and financial globalization and to create a theoretical and prac-tical bridge between ecology, social justice, and urban design. The word "praxis" is used to convey theory and research synthesized within a broader spectrum of embodied design activities beyond standard forms of academic research and professional practice. The Urban Design Working Group, as part of BES, has taken on this larger systemic context by engaging the inner city, older suburbs, and outer periphery as interrelated ecological localities in which community-engaged design is key. Such engagement, empha-sizing inclusiveness, connectivity, and community, is crucial to achieving sustainability as a social goal and resilience as a model of adaptation and change. Ecological and socially motivated urban design is founded on a new, integrated understanding of cities within key contextual factors: the muta-bility of cities, the failure of the modernist development model, the explo-sive growth and increasing disparities of urbanization worldwide, and the shifting vulnerabilities resulting from rapid climate change.

The metacity implies a nonhierarchical spatial and ecological matrix model. Irene Guida, in her 2012 dissertation based on fieldwork in Balti-more, criticizes ecology's metaphorical use of the architectural term "the corridor." She describes a corridor as a spatial device to organize vast urban territory around networks for movement. The corridor is the most basic and elementary piece of a network, and the more we live in between the connections of a network society, the more we are subject to the logic of corridors. As a result, our bodies, perceptions, behaviors, and thinking are all shaped through this very device, and access to this infrastructure, especially via the automobile, has greatly increased social segregation. If in Victorian society the architectural corridor was introduced to separate functions and social actors, such as servants from the served, in postwar urban American, ecological corridors often masked the large highway infrastructure that fundamentally altered social relations in both cities and countryside. The neighborhood struggles against this network of highway corridors in 1970s Baltimore opened up a new space for socially activist ecological urban design that sought to link immediate ecological interactions to human well-being.

Corridors sever local space by unifying regional flows. They discipline access and govern priorities, disconnecting some flows and connecting others along linear paths. It can be seen as a connecting device, if we look at structures such as Interstate 95 or the Northeast rail corridor, as well as a disconnecting one, if we look at the many poor immigrant and African American neighborhoods razed to build these infrastructural lines through cities. Ecological corridors have often had the same result, as was seen in Sherry Olson's mappings of segregation patterns in Baltimore, buffered by parks and green spaces. In an urban region, a highway, river, or even a linear park can separate neighborhoods from each other just as an architectural corridor separates a line of rooms. Even green corridors can block immediate connections while providing the possibilities for further connections beyond. Corridor space privileges a space of flows and extension, against an intensive space–body of living matter in place, within a heterogeneous patch matrix. Therefore, the ecological corridor as a metaphor and device makes linkages and connectivity but also reinforces social divisions.

The urban design work at BES can be seen as a revision of the ecological metaphor of the corridor, providing a new and different description of the Gwynns Falls watershed within what the architect Susana Torre refers to as a spatial matrix. Biologically, a stream system limited to a linear corridor reduces the whole system's capacity to react to stress. In the Gwynns Falls watershed, the Olmsted Brothers' park corridors are biologically fragile, flashy, and polluted by leaking sanitary infrastructure laid out to exploit the stream valley's descent. Also, socially, they separate neighborhoods by their

lack of cross-connections. Biodiversity's reduction, paradoxically due to ecological corridors, is linked to social segregation and results in a more fragile
local environment.

Releasing nature from the limits of ecological corridors and focusing
on land cover as a common matrix rather than land use as regulated private property opens the door to integrating our ecological understanding of
cities with the collective design of those cities. Neighborhood design determines the type, amount, and specific arrangements of land cover elements,
so it plays a critical role in establishing the heterogeneity of a city. The
choices of neighborhood residents are frequently driven by visual variety
and the experiences desired for using these spaces. These choices may,
however, also influence ecological processes like the movement of water,
nutrients, and organisms and affect parameters like temperature and moisture. Understanding links between design choices and how heterogeneity
influences ecological processes can aid us in designing sustainable, ecologically sound cities. This evolution from the social dynamics of ecological
corridors to the patch dynamics of urban neighborhoods within a matrix
of choice is also a switch in power dynamics and a change from top-down,
metropolitan regulation to a dispersed metacity imagination.

The metacity theory radically differs from the metropolitan tradition,
providing a close-up reading of the interactions between different ecological actors and including the feedback between humans and biotic activity.
The patch dynamic, watershed, and social justice frameworks have inspired
the formulation of a new ecological urban design praxis based on this model.
BES provides evidence of cities as a patchy matrix with a greater degree of
connection between social action in relation to water, vegetative, and built
systems, demonstrating that urban design can foster social inclusion as well
as ecosystem function. As further demonstration of the contemporary relevance of the design studios, former students are currently enacting these
frameworks in communities beyond Baltimore in order to provide a more
inclusive social space and a more interconnected biotic ecosystem (figures
16.11 and 16.12). This is achieved through an enhanced social–natural matrix and a reduction of the spatial hierarchy and isolation brought by large-
scale infrastructures, whether rail, highways, or corridor park systems.

In ecology, "meta-" indicates something more inclusive than the ecological structure of a certain place. Hence, a metapopulation comprises a
number of isolated, discrete populations of a species that are connected by

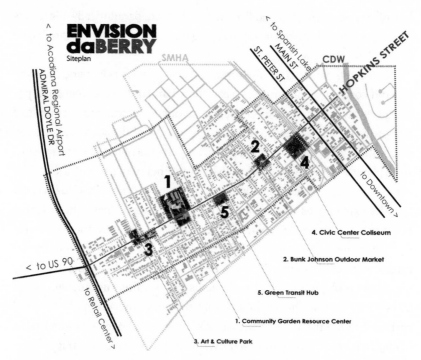

FIGURE 16.11   Envision Da Berry, New Iberia, Louisiana. (© Phanat Xanamane, 2017, used with permission)

migration (chapter 11). Individual populations can grow, persist, decline, or become extinct. New populations are established, sometimes in new patches and sometimes in a patch that had been vacated. Thus a metapopulation is a spatially heterogeneous, dynamic, differentially connected system. A similar concept applies to metacommunities of various species. A community may rise and fall at a site, be obliterated, or become established in a new location. Exchanges of species, information, and resources across the matrix of patches will affect and be affected by the spatial changes in the communities. In other words, "meta-" in ecology is about dynamics and flux across heterogeneous space.

Metapopulation and metacommunity concepts from ecology reinforce the use of the term "metacity" in social theory and design contexts, and it is manifestly dynamic and process oriented. Metacity theory assists one in moving beyond the specific form of an immediate community at a particular point in time and allows one to consider the whole urban matrix in a

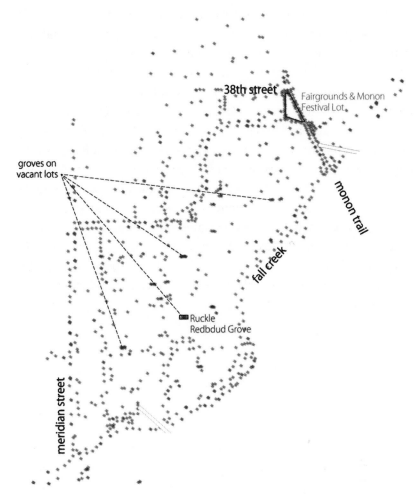

FIGURE 16.12   Urban Patch, Indianapolis, Indiana. (© Justin Moore, 2017, used with permission)

more inclusive way than modernist master planning. Normative urban design's focus on specific elite urban fragments can now be redirected to larger social and ecological contexts, both physically and virtually. The ecologists of the UDWG initially challenged the designers to create urban designs as models of patch dynamics. This involved a switch in agency in order to provide a space for the creation of a critical ecological urban design praxis, not just with the goal of greening the city but to address systemic social stratification through collective action.

The metacity concept questions the politics of urban design practice tied to social forces that continue legacies of spatial segregation and environmental destruction. New theories and modes of praxis are required if we are to deploy our technical, theoretical, and practical skills in an effort to resituate urban ecology in relation to social justice movements. Today, the metabolic rift theorized by Marx is no longer between the urban and the rural, the worker and the land. The estrangement is between the ecological impact of an urban locality within a globalized system of flows, redefining what we mean by "environment." Therefore, new forms and subjects of struggle have emerged, necessitating a meta-methodology for multiple patchy urban design tactics within larger ecosystem and social justice frameworks. The metacity provides a multiscalar framework for the ecological design of cities as complex, adaptive systems and can articulate a way forward for democratic societies to contribute to improved global urban ecological change through local initiatives, regional integration, and global cooperation.

CHAPTER SEVENTEEN

Teaching and Learning the Baltimore Ecosystem

Alan R. Berkowitz, Bess Caplan, Susan Blunck, Janet Coffey, and Janice L. Gordon

IN BRIEF

- Baltimore provides a rich context for teaching and learning core disciplinary ideas, science practices, and cross-cutting themes embraced by school standards.
- The Baltimore Ecosystem Study's (BES) work with teachers and students supports a shift from the use of generic concepts and distant places to education based on discovery and application in the real environments in the local metropolis.
- Baltimore's teachers are enthusiastic about and responsive to professional development opportunities that help them focus on teaching in, about, and for local ecosystems through direct investigation, data exploration, and systems thinking.
- BES has provided a diversity of young, aspiring scientists with first-hand opportunities to do science, often in the very neighborhoods BES studies.
- Baltimore has been a laboratory for research about teaching and learning, contributing to our knowledge of effective teaching strategies and of the pathways toward mechanistic, principle-based understanding of the environment.

INTRODUCTION

BES is unusual among Long-Term Ecological Research (LTER) projects because, in addition to having requirements for ecological research, it was charged with working in the local school system. The project accepted this charge as an opportunity for engaging with the rich diversity of communities in Baltimore. It also has been a way to share what the project has learned about the metropolitan area and particular neighborhoods with students, teachers, and administrators. *Fostering an understanding of the metropolis as a social-ecological system or ecosystem* (chapter 1) lies at the core of the BES education mission, most recently in the context of contributing to the movement toward sustainability in the city. BES education strives for a seamless interplay between its social-ecological research and (1) teaching, learning, and information flow; (2) teacher professional development programs; and (3) collaborations among scientists, educators, and students. Novel research seeks to understand what supports teachers in mainstream school settings in embracing this mission, and what supports students in developing principle-based reasoning about key facets of their urban landscape.

In this chapter we share insights from our twenty years of work with Baltimore City and surrounding schools, painting a picture of our approaches, programs, and accomplishments and ending with reflections on the implications for the field of urban ecosystem education in general. After an opening vignette that highlights core aspects of the BES education approach (box 17.1), we (1) summarize the principles that guide our work and then describe our efforts to (2) infuse BES science into middle and high school classroom instruction; (3) support students in considering and pursuing environmental science career pathways, and (4) advance scholarship about urban ecosystem education.

THEORIES, RATIONALE, AND MOTIVATIONS FOR BES EDUCATION

A number of theories and motivations guide BES education practice. These start with core ideas about the ultimate goals for BES-supported instruction—environmental citizenship and urban ecosystem literacy—then move to those that shape our vision of how students acquire this knowledge and proficiency and culminate in those that identify core features of effective professional development practice to help teachers develop the knowledge, skills, and attitudes they need for successful teaching. An understanding of the ecology education landscape in Baltimore and its social

BOX 17.1. A MAGIC MOMENT ABOUT TRANSPIRATION

"That means each tree loses one hundred of these every day," the teacher said, waving a plastic bottle of water in the air. He had just followed the calculations we made, starting with the amount of water we'd collected in plastic bags holding a few leaves from each of several trees in the schoolyard: multiplying the volume of water per area of leaf per day by estimates of leaf area per ground area and of ground area per tree. This was one of those magic, light bulb moments every educator lives for: everything coming together for a learner.

The speaker in this case was a high school engineering teacher in BES's Comp Hydro (computer science + hydrology) education project. The goal is to develop ways to integrate hydrology, data, and computation for high school students to learn about local water issues. This moment, and the project, exemplify the ambitious aspirations and key strategies of BES Education. Transpiration is a big idea in urban ecology, something most teachers and virtually no students know about. Learning about transpiration is a celebration of ecology IN the city, a key understanding for ecology OF the city, and, given its implications for application, important for ecology FOR the city. It also exemplified 3-Dimensional Teaching at its best, bringing together a big idea (transpiration); powerful science practices (investigating real phenomena, arguments from evidence, and modeling); and crosscutting concepts or ways of thinking: quantitative reasoning, conservation of mass, and scale. And it demonstrates our professional development practices of engaging teachers as genuine learners, reflecting on how to make learning engaging and accessible to their students and empowered by lessons, supplies, in-class support from BES educators and from a sustained professional community of their peers, educators, and scientists.

dynamics provides valuable context and motivation for key features of our practice.

Environmental Citizenship

Environmental citizenship (box 17.2) is the overarching goal for students and teachers. Students need the knowledge and abilities to act from their

BOX 17.2. KEY TERMS

Constructivist learning: Constructivism entails recognition that learning takes place as people add to and modify what they already know about something. Rather than seeing learners as empty vessels passively receiving new knowledge, it enjoins us to see learners as actively constructing their knowledge via new experiences, insights, guidance, and support.

Data Jam: A competition where middle and high school students analyze authentic local datasets, make a claim supported by their evidence and reasoning, and then use their creativity to explain their claim to a general audience. First developed in 2012 by Asombro Institute educators as part of the Jornada Long Term Ecological Research program, there now are Data Jams in Puerto Rico, New York, and Baltimore.

Ecosystem literacy: Berkowitz and colleagues at Cary Institute coined this term for the knowledge and skills needed to "read and converse" about ecosystems for a useful purpose. It embraces the broadened definition of "ecosystem" that acknowledges humans as integral parts of every place on Earth, and purposes that include both utilitarian understanding and appreciation of the beauty of ecosystems.

Environmental citizenship model: This model was proposed by Berkowitz, Ford, and Brewer in 2005 as an alternative to "environmental literacy" as the primary goal of environmental education. They defined it as "having the motivation, self-confidence and awareness of one's values, and the practical wisdom and ability to put one's civics and ecological literacy into action." Environmental citizenship involves empowering people to have the knowledge, skills, and attitudes needed to identify their values and goals with respect to the environment and to act accordingly, based on the best knowledge of choices and consequences. We use "citizenship" to suggest civic engagement; it implies nothing concerning immigration status.

KidsGrow: A signature after-school and summer program of the Parks & People Foundation providing elementary-aged students with a rich blend of academic and homework help, character development, physical activities, and environmental enrichment.

Learning progression: Learning progressions describe the increasingly sophisticated ways students talk or write about a topic. They are usually described in a "space-for-time substitution" approach similar to chronosequence studies of plant succession, where patterns of change over time are inferred from patterns among multiple students. Complementary research describes shifts from one level to the next during the course of instruction, allowing the description of effective strategies to support student progress.

Next Generation Science Standards (NGSS): The Next Generation Science Standards were released in 2013 by a number of states that developed them. The standards lay out a comprehensive set of performance expectations for all of the sciences across grade bands from K through 12. Each standard combines a core disciplinary idea, a crosscutting concept, and a science practice, thus calling for three-dimensional (3D) teaching and learning. NGSS used a Framework for K–12 Science Education from the National Research Council as a key guiding document.

Pedagogical content knowledge: Educators have long recognized that knowledge of subject matter (content) is an essential feature of successful teachers. Pedagogical content knowledge entails understanding the key ideas within the content that students most need given their age/stage and context, along with an understanding of effective ways of both assessing their understanding and advancing it through instruction.

Socio-cognitive theory: A theory from basic psychology that highlights the importance of self-awareness and self-knowledge about one's strengths, limitations, and needs as a critical factor in shaping interest and action. It also describes the critical role played by one's social context, experiences, and relationships (including, in the case of understanding vocational interest, mentors) in influencing behavior, mediated by self-confidence and an understanding of the outcomes and requirements of different options.

will and agency when it comes to the environment. This theory guides BES education by (1) identifying ecological systems literacy, social systems literacy, and ethics literacy as three pools of knowledge and skills and the important overlaps between them, such as environmental and social ethics and ecosystem services and sustainability; (2) recognizing the importance of motivation and agency or self-efficacy in social, ecological, and ethical contexts for driving learning and bringing literacy to action; and (3) positing a central, dynamic role for action not just as an expression of one's literacy, will, and agency but also as a vital source of these things. BES education is inspired by this vision and also challenged by the gaps between it and the state of affairs among the American public in general and Baltimore youth in particular.

Urban Ecosystem Literacy

BES education focuses on the ecological system literacy component of the Environmental Citizenship model, along with its overlaps with social system and ethics literacy and with the affective and action domains, all as they pertain to urban ecosystems. Our view of urban ecosystem literacy is inspired by and evolving with the developing view from the BES research program, embracing progressions (1) from ecology *in nature* only to ecology *in the city*, (2) from ecology *in the city* to ecology in and *of* the city, (3) from ecology *in and of* the city to ecology in, of, and *for* the city.

BES Education Goals and Standards

Our vision for student learning and attainment has evolved along with the evolution of education standards at both the local (e.g., the state of Maryland) and national levels. The Next Generation Science Standards and Framework upon which they are based embrace a three-dimensional vision for science learning that BES has championed for many years. Thus we emphasize core disciplinary ideas as they apply to urban ecosystems, along with cross-cutting ways of thinking and science practices.

Constructivist Learning Theory and Learning Progressions

Our vision for how students learn the key ideas/principles, skills, habits of mind, and practices needed for ecosystem literacy is based on the constructivist notion that students actively build on prior knowledge. We have pursued research into learning progressions that describe pathways of increasing sophistication of student understanding for key dimensions of ecosystem literacy, including carbon, water, and biodiversity (figure 17.1). Several salient features of learning progressions emerge for professional and curriculum development. They (1) ensure that we anticipate and assess students' prior knowledge and experiences (they are not empty vessels) and

Learning Progressions for Urban Ecosystem Literacy

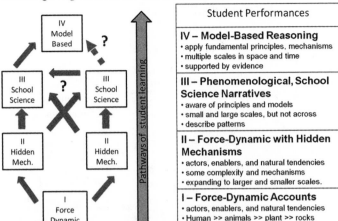

	Student Performances
	IV – Model-Based Reasoning • apply fundamental principles, mechanisms • multiple scales in space and time • supported by evidence
	III – Phenomenological, School Science Narratives • aware of principles and models • small and large scales, but not across • describe patterns
	II – Force-Dynamic with Hidden Mechanisms • actors, enablers, and natural tendencies • some complexity and mechanisms • expanding to larger and smaller scales.
	I – Force-Dynamic Accounts • actors, enablers, and natural tendencies • Human >> animals >> plant >> rocks • short time and macroscopic phenomena.

What are the pathways of student learning to principle- and evidence- based reasoning?

FIGURE 17.1 Framework for learning progressions describing how student understanding, as reflected in their performances (accounts, explanations), develops over time. The pathways shown on the left portray hypothesized differences that may obtain among students or even within students where certain understandings may be more productive than others in leading to progress to the next level. These pathways represent important questions guiding current research and practice. (Based on Gunckel et al., 2012a, b; and Jin et al., 2013)

that we help them modify and build on these as we bring new experiences and ideas to them; (2) describe patterns of student sense-making at different levels of sophistication, guiding us in the design, implementation, and interpretation of assessments used to shape and evaluate our teaching; (3) identify the requisite experiences and thinking tools needed for students to make progress between stages; (4) are developed dynamically through research and practice with students and curriculum and are thus grounded in actual student performance rather than being strictly aspirational (as in many academic standards).

Model of Teacher Knowledge and Skills for Ecosystem Literacy
BES education draws on theories of teacher knowledge and practice that emphasize the importance of strong content knowledge, pedagogical con-

Teacher Knowledge and Skills
for Urban Ecosystem Literacy

FIGURE 17.2 Framework of the three pools of knowledge
and skills needed for effective urban ecosystem teaching. The
overlaps among the three pools represent important targets for
professional development: PCK—pedagogical content knowledge, LCK—learning content knowledge, LP-PK—learning
progression pedagogical knowledge, and the three-way overlap
of L-PCK, or learning pedagogical content knowledge. Learning
progressions provide invaluable substance and resources for
each arena in its own right and, more significantly, for these
areas of overlap. Likewise, a professional learning community
in which teachers develop, test, reflect on, and refine their
knowledge and skills can play a pivotal role in success. (Based
on Gunckel et al., 2018 and Hauk et al., 2014)

tent knowledge, and knowledge of student interest and learning (figure
17.2). This practice-oriented framework suggests the importance of skills
for instruction in these same arenas. Thus we aspire to support teachers in
developing not just their own understanding of urban ecosystems but also
their ability to apply science practices such as investigating their local environment or evaluating claims using evidence and reasoning. We help them
to both elicit and understand student reasoning using learning progressions and formative assessments in the classroom. And we expose them to
a wide variety of evidence-based teaching approaches and pedagogies, sup-

porting their experimentation, adoption, and adaptation of these practices in their teaching in as many ways as possible.

Socio-Cognitive Theory of Teacher Interest and Practice

We have adapted a socio-cognitive model of vocational interest to guide our thinking about the factors that influence teachers' interest in the key dimensions of teaching for urban ecosystem literacy (table 17.1). Therefore, we seek to build their (1) *self-efficacy* ("I can use this pedagogy or teach about the Baltimore ecosystem"), (2) *outcomes expectation* ("I know what is required to teach this way and what can be accomplished in terms of student learning and my professional advancement"), and (3) *outcomes valuation* ("I value the outcomes that can be achieved"). The outcomes they achieve with each stage of adoption and experimentation feed back into their personal knowledge, skills, and attitudes, which, in turn, influence subsequent self-efficacy, motivation, interest, and identity.

Partnerships and Collaboration for Urban Ecosystem Literacy

BES is but a drop in the very large bucket of the overall education system in Baltimore and of the ecology education landscape in particular. We have

Table 17.1

Key teaching practices for fostering environmental science literacy

---

Plan instruction to focus on important **big ideas, concepts, and science practices** in the field

Develop and use **formative assessments** and other **approaches** to guide instruction and respond to student thinking during classroom discourse and in comments on student work

Have students make **explanations** (claims) **using evidence and reasoning** (applying basic principles) about environmental processes in social-ecological systems

Engage students in collecting and making sense of **evidence from real phenomena** (outside), practicing data literacy and critical thinking skills

Link **environmental science to real problems** in the local, nearby environment, anchoring students' learning in their culture and place

---

thus tried to be strategic in identifying places where our distinctive re-sources, perspectives, and expertise can have the most impact. We focus largely on middle and high school teachers and curriculum, which are best suited to the scientific substance BES brings, and on working with appli-cations to mainstream courses rather than just electives to maximize po-tential impacts. We also have sought to both enhance our impact and to better accomplish our ambitious goals through partnerships and collab-oration. Thus we have built strong collaborations both with local education partners and with leading science education scholars and practitioners from across the nation with particular interest in place-based ecology edu-cation.

ACTIVITIES TO INFUSE BES SCIENCE INTO MIDDLE AND HIGH SCHOOL CLASSROOM INSTRUCTION

Teachers are the linchpins for infusing BES science into schools. Since its inception in 1997 BES has provided professional development and other resources for middle and high school teachers in Baltimore schools. We work with mainstream teachers and schools and required core classes and curriculum, concentrating our effort on public schools in Baltimore City and County. The choice to eschew focusing on programs for gifted and tal-ented students, at private schools, or for primarily elective classes is based on our goal of developing models that can reach every teacher and student in the metropolis. In working with teachers, we have had to respond to their and their schools' needs and constraints, with the obvious payoff of ready adoption but also obvious costs in time and effort to tailor programs to changing and challenging needs. Throughout, we have maintained and built our focus on teaching in, about, and for local ecosystems through di-rect investigation, data exploration, and systems thinking. Here we describe our programs and resources for teachers, students, and schools.

Topical Workshops

Topical Workshops bring together BES scientists and educators with class-room teachers for 1 to 2½ days to explore a cutting-edge topic in urban ecosystem science and pedagogy. Teachers learn about BES research and what we are discovering about the ecosystem that they and their students are part of (box 17.3). A diversity of topics has been presented over BES's twenty years (table 17.2), supported by a large number of BES and other

---

BOX 17.3. TEACHERS LEARN ABOUT STREAMS ON DRUGS

On an overcast Saturday in March thirteen teachers, most new to BES, waded into the nondescript stream running through the University of Maryland, Baltimore County (UMBC) campus to retrieve pharmaceutical diffusing substrates (PHADs) deployed several weeks earlier. These substrates are designed to emit contaminants such as pharmaceuticals or personal care products through a small disc that is colonized by biofilm organisms in the stream (chapter 10). The teachers incubated the substrates for two hours in light and dark conditions, measured photosynthesis and respiration, and observed differences in the density of biofilms as impacted by two common drugs in the PHADs. The pathways and prevalence of drugs and personal care products in the environment were new to them, as was the significance of biofilm algae, bacteria, and fungi for in-stream processes and, ultimately, for the services stream ecosystems provide to people. The teachers took away new knowledge, a protocol they can use in streams near their schools or in classroom microcosms, powerpoints, and other resources from BES Principal Investigator Emma Rosi and postdoctoral associate A. J. Reisinger, and enthusiasm for staying in touch with BES scientists and educators during the school year.

---

scientists. Topics are drawn from the breadth of BES's research and target specific parts of the middle and high school curriculum of Baltimore City and County schools.

Workshops blend direct investigation in the local environment with sharing teaching tools and classroom applications and with insights from our research on student thinking and learning about the topic. During the workshops and through follow-up with BES educators, teachers plan concrete ways to apply what they've learned to their teaching. While we recognize that short workshops are more likely to impact "what" teachers teach rather than "how" they teach, we believe that shifting the teaching of key science concepts from textbook, distant, or generic places to real environments in Baltimore can be an important step toward fostering urban ecosystem literacy.

Table 17.2

BES education workshops, 1999–2017

Short-duration workshops (1 to 2.5 days each) offered to middle and high school teachers in the Baltimore metropolitan area, with the year(s) each was offered and the BES and other scientists involved. BES educators co-led all workshops.

MAIN TOPIC	SPECIFIC TOPIC	YEAR(S)	SCIENTISTS AND OTHER EXPERTS
Soil	Urban soil ecology	99, 01, 09	Rich Pouyat, Jonathan Russell-Aneli, Quin Holifield
	Earthworms	00, 04	Charlie Nilon, Katalin Szlavecz
Modeling	Modeling, Online simulations	00, 02, 11	Larry Band, Roel Boumans, Brian Fathe
Biodiversity	Animals in the urban landscape, Biodiversity	00, 12	Charlie Niolon, George Middendorf, Chris Swan
	Plants and people	02	Wayne Zipperer, Chris Daniel
	Inner harbor ecology	13, 14	Adam Frederick
	Vacant lots	10	Chris Swan
	Microbial ecology	12	Sujay Kaushal
	Wetlands	11	Vanessa Beauchamp
Social Science	Eco-history, Social ecology	02, 10, 14	Morgan Grove, Geoff Buckley
	Environmental justice	12, 14, 15, 16	Allison Whitmer, Glen Ross, Morgan Grove, Anna Scott, Christine Rega, Quin Holifield
Hydrology	Urban hydrology, Watersheds, waste and water treatment	02, 04, 10	Larry Band, Neely Law, Ken Belt, Gary Fisher
	Groundwater	12, 14	Aditi Bashkar
	Stream ecology, Urban watershed continuum, Drugs in streams	09, 16, 17	Sujay Kaushal, Tamara Newcomer, Ken Belt, Emma Rosi, A. J. Reisinger

MAIN TOPIC	SPECIFIC TOPIC	YEAR(S)	SCIENTISTS AND OTHER EXPERTS
Air	Urban meteorology, Air quality	02, 10	Gordon Heisler, Neil Bettez
Greening, Agriculture	Schoolyard greening	10	Mary Hardcastle
	Urban gardening, Urban agriculture	11, 15	Shaun Johnson, Peter Groffman
Disease	Disease ecology	11, 12	Shannon LaDeau
Climate Change	Climate change	13, 13	Marcus Griswold
	Phenology	11, 15	Roland Roberts, Shelly Forster
Sustainability	Sustainability	13	Abby Cocke
	Energy production	13	Ted Atwood

Teacher Institutes

Teacher Institutes are longer-term, more intensive professional development programs combining three- to ten-day summer workshops with school year sessions. The institutes nurture a professional learning community among teachers, BES scientists, and educators. In an atmosphere of cocreation of knowledge, everyone develops a deeper understanding of urban ecosystems, student learning, and teaching, with teachers often returning for more than one year. Such extended professional development can influence both what teachers teach and how they teach and help them to be more responsive to national trends and pressures for excellence in science education.

During its first phase, BES education crafted a partnership program with scientists, science educators, and teachers to develop new approaches to teaching about the urban ecosystem in their classrooms. Since this was before we had data, we shared frameworks and preliminary protocols through workshops and then supported teachers in developing activities tailored to their students and school settings. Early successes included a schoolyard ecology program in Southeast Middle School, year-long stream studies at Merganthaler Vocational Technical School and McDonogh School, and a year-long study of the lands surrounding the city's drinking water reservoirs at Western High School of Technology and Environmental Science

---

BOX 17.4. STUDENTS WRITE A BOOK ABOUT BALTIMORE CITY RESERVOIR LANDS

Students in Karen Hinson's year-long AP American History class in 1998–99 took BES's social ecology approach to build a land use history of the lands around the city reservoirs. They worked in teams that included demographics, economic structure, class/race/religion, and ecology. BES social scientist Morgan Grove worked closely with Hinson and her students and helped connect them to numerous experts from the community. Core topics in the AP curriculum were addressed in the context of the local project, and the product—an impressive book—was used by the managers of the reservoir lands in developing a forest management plan. Students benefited from the skills and knowledge they developed working with professionals and with authentic materials, plus their performance on the AP history test was actually higher than that of previous, similar classes of Hinson!

(box 17.4). With BES data, protocols, and curriculum units in hand, subsequent Teacher Institutes in such topics as eco-hydrology and investigating urban ecosystems offered a rich blend of content and BES data, teaching, and student learning knowledge and skills to participants.

From 2009 to 2014 we worked with a large community of scientists, educators, and education researchers from across the nation in the Pathways to Environmental Science Literacy Project (Pathways Project), supported by an NSF Math Science Partnership grant (http://www.pathways project.kbs.msu.edu/). Headed by John Moore at Colorado State University, this project spanned four LTER sites (BES, Kellogg Biological Station, Short Grass Steppe, and Santa Barbara Coastal) as well as schools in New York and Georgia. The Pathways Project investigated student learning progressions for some of the core ideas in environmental science, including water pathways and quality, biodiversity and its functional roles, and carbon cycling processes. In the process, three Teaching Experiments or modules were developed to address instruction in these core ideas, and six years of sustained, intensive professional development were offered to city and county middle and high school teachers. Pathways Teacher Institutes sup-

---

BOX 17.5. THE POWER OF TEACHING BASED ON THE LEARNING PROGRESSION

"The learning progressions concept was entirely a new phenomenon that surprised me. I was skeptical in the beginning but over a period of time I learned to listen to students and analyze their interpretations. Now I have stepped down from the 'commander' position in the class and learned to be a passive listener. [The BES] Pathways Project gave me more practice and confidence in teaching science with a new perspective."

Before participating in the BES professional development, this teacher would frequently gloss over details and not pay as much attention to student thinking, focusing mainly on correcting students' "mistakes." The professional development, curricula, and learning progressions motivated and helped him pay better attention to his students' ideas and to know how to help them build more sophisticated understandings.

---

ported classroom implementation of the Teaching Experiments, teacher content learning through firsthand field and lab investigations, and instruction in "learning progression teaching strategies" (box 17.5). Professional Development leaders modeled pedagogical approaches in their teaching while balancing the need for flexibility in teacher-centered instruction rather than curriculum-centered instruction. Institute teachers implemented the Teaching Experiments in their classrooms, where project collaborators collected data on student thinking and teacher implementation. Graduate students and teachers in residence supported by the grant provided classroom support to teachers while also assisting with the science education research goals of the program.

Our most mature adventure in intensive collaborative work with teachers is the Comp Hydro project, supported by an NSF Science Technology Engineering Math + Computer Science (STEM+C) grant. Part of the push by national leaders to advance the teaching of computer science and computational thinking linked to the teaching of science, Comp Hydro is working with high school earth, environmental, life, and computer science and engineering teachers in Montana, Colorado, and Arizona in addition to Baltimore. We are grappling with the many challenges of teaching across disciplines at the high school level. For example, computer science and engineering teachers have limited knowledge of hydrology content and pedagogy, while science teachers have limited computer expertise. In several Baltimore Comp Hydro schools, simply providing access to working computers for the students and teachers has proven impossible. Comp Hydro reminds us of the incredible resilience of teachers, doggedly determined to engage their students in the cutting-edge pedagogy we are developing and testing together, despite the challenges, even to the point of developing new lessons to go farther into teaching computation completely without the use of computers! Working with a small group of teachers over the course of two years, Comp Hydro is allowing us to build teacher capacity in all areas that theory suggests are required for successful science teaching, including teachers as learners, planners, innovators, and reflective practitioners.

Teacher Fellowships
Intensive research and education fellowships have been offered to teachers when resources are available. Research Experiences for Teachers fellows conduct research, working closely with a BES scientist as mentor, usually intensively over four to eight weeks during the summer, and then build on this research for a classroom application. Topics have included hydrology, stream ecology, social science, and education research. In 2009, 2010, and 2011 we were able to offer the Research Experiences for Teachers fellows, along with graduate student assistants in the Pathways Project, an intensive course in urban ecosystem ecology and education sponsored by Towson University. Other Education Fellows work closely with a BES scientist or science team and with the BES education team to learn about an important area of BES research and then develop curriculum modules to bring this area of research into the classroom. In the process, teachers gain deeper understanding of the Baltimore ecosystem and of ecological inquiry, while producing materials useful to others (see next section). Teachers in Residence is the most intensive and expensive opportunity BES has provided, and we have only had three over the twenty-year history of the project, funded by the Math Science Partnership grant. These three teachers were able to immerse themselves in BES education for a year while on leave or sabbatical from their classroom teaching positions. They were key contrib-

utors to developing the instructional materials from the Pathways Project and tailoring them to Baltimore and the city, county, and state standards. They also were key players in the research. Finally, they were instrumental in supporting other classroom teachers in implementing BES curriculum and teaching strategies. Each left profoundly changed in the way they teach and went on to assume leadership roles in the mentoring and professional development of other teachers.

Materials and Tools

Educative teaching materials and tools serve the dual purposes of providing teachers with useful resources for classroom instruction while also supporting their own learning about the content and science practices involved in the topic. We have developed materials for after-school programs for elementary school students and for middle and high school science classroom use. My City's an Ecosystem was developed for the KidsGrow program and entails an eight-unit curriculum for two years of exploration. Topics include water, phenology of people and other organisms, schoolyard habitat, birds, and decomposition of organic matter. The units have cross-cutting themes of vocational awareness, inquiry, and stewardship. The Investigating Urban Ecosystems units were developed by Education Fellow/Scientist Teams and include schoolyard hydro-ecology, salt in streams, and earthworm ecology in urban soils. Collaboratively we developed units in the Pathways Project in Carbon, Water, Biodiversity, and Citizenship. We helped develop a year-long high school curriculum called Environmental Science and Complexity. One of its four units focuses on one of our local high schools and its multidimensional ecology (soils, hydrology, plants, birds, human perceptions and use) as it pertains to a thorny decision about a football field on the campus. Currently we are working on a high school unit on urban runoff and flooding through the Comp Hydro project.

School Partnerships

BES scientists with the U.S. Forest Service and other collaborators have formed several partnerships over the years with specific schools. This allows for one-on-one engagement of students and teachers with scientists, often helping to bring equipment and research tools to the schools. An innovative blending of landscape architecture and design with ecological science and student engagement is being pursued at the Francis Scott Key

Elementary/Middle School by Mahan Rykiel Associates, a landscape architecture firm in Baltimore. It is supported by BES scientists and educators, the school's Parent-Teacher Organization and local nonprofits and businesses. Curriculum was developed to support instruction in and about the schoolyard, where mitigation money for a large construction project in the neighborhood was spent on reimagining and replanting the schoolyard space as a living, learning, outdoor classroom. The Green Street Academy is a public charter school in West Baltimore that serves students from across the city with an emphasis on sustainability. BES scientists with the U.S. Forest Service are partners in the school and provide unique opportunities for students to engage with scientists and access cutting-edge ecological protocols and research. Conservation and urban agriculture are career content tracks offered at the school and supported by the U.S. Forest Service partnership.

INNOVATIVE PROGRAMS FOR STUDENTS

While most of BES education has focused on teachers, we also have pursued several programs that connect directly to students. These have given us the opportunity to see, firsthand, how young, inquisitive students engage with the science practices and actual data of BES. An early example was the Environmental Science Summer Research Experience Program developed with support of BES by a high school science teacher at Roland Park Country School. This program engaged eleventh-grade girls in a three-week summer program investigating soil microbes on their school grounds. Students did primary, self-directed research, explored the soil chemistry of the surrounding physical environment, and determined the value of soil microbes as indicators of human impact on soils. Students and their teacher shared protocols in soil ecology with other students and teachers via their website, with significant input from BES scientists.

An opportunity to infuse BES science into an after-school program was Parks & People Foundation's KidsGrow Program. KidsGrow (box 17.6) promoted an understanding of ecological concepts, built an awareness of local environmental issues and values, developed scientific investigation and critical thinking skills, connected students to each other and their school and community, developed leadership, enhanced team-building and problem-solving skills, and demonstrated that kids can positively contribute to the community. Students also received meals, homework help, and recreational

---

BOX 17.6. CITY KIDS DISCOVER THE ECOLOGY OF THEIR PLACE

On a sunny fall day in 2008 students from Franklin Square Elementary/ Middle School waded knee deep into Dead Run at Winans Meadow in Baltimore's Leakin Park. They fished with D-nets for crayfish and macroinvertebrates. It was the first time at this park for many of these students, who live in an isolated, concrete community with low economic status and few green spaces just a mile away. The students were part of KidsGrow, an innovative, place-based ecology and reading enrichment program conceived and run by the Parks & People Foundation. The kids spent five days a week after school learning about the complexity of the urban ecosystem though the My City's an Ecosystem curriculum developed by BES, in addition to reading enrichment, homework help, and character-building activities. For instance, in one year students began the school year cataloging the living and nonliving things in their schoolyard, learned about urban habitats and the ecological history of their community, and then explored the ecology of food production, agriculture, and nutrition. They met African American scientists studying the ecology of Baltimore, planted their own vegetable gardens, and learned about water quality in field trips like the one to Dead Run. There, the students delighted in discovering aquatic species they had heard about but never seen. After a lively discussion they concluded that the stream was "fairly healthy." They were happy, but not satisfied!

---

time outside. The program provided supervised care for many children in Baltimore City at a time of day when parents are still working and children may typically be left unsupervised. Through KidsGrow, we learned that kids love outdoor, hands-on study even when they are tired after a full school day; that the environment can be a wonderful vehicle for engaging youngsters in exploration and stewardship; and that running a program where tired classroom teachers are the instructors can be challenging.

Building on BES's abundant data and in response to growing interest for data literacy and for Science Technology Engineering Art and Math (STEAM) teaching, we developed the Baltimore Data Jam, modeled after the highly successful Desert Data Jam at the Jornada LTER site in New Mexico. Students from across Baltimore enter team projects that analyze a local socioecological data set and explain their findings in a creative way that is accessible to a nonscientist audience. Data sets include stream water quality, sewage overflows, schoolyard land cover, and neighborhood social factors. Since its inception in 2015, the Baltimore Data Jam has featured three-dimensional models, videos, PowerPoint, comic strips, songs, poems, simulations, children's books, and interactive exhibits. Scientists, educators, and artists score each project, and prizes are awarded in middle and high school categories. One year the five winning teams brought their projects to the signing of the Greater Baltimore Wilderness Accord, which was attended by the mayor of Baltimore, Congressman John Sarbanes, and county executive "Dutch" Ruppersberger.

Support Students in Considering Environmental Science Career Pathways
Students of many stripes have both contributed to the BES science enterprise and cut their research teeth on mentored research projects in BES. These experiences for young and aspiring scientists give them a firsthand chance to do science, tasting its rewards and learning what is required for fulfilling work as a scientist. BES is uniquely poised to give such experiences to diverse students interested in urban ecosystems, and in some cases to students hailing from the very neighborhoods BES studies.

Since its inception we have brought undergraduates into the BES family as Research Experiences for Undergraduates (REU) students through supplements to the core LTER grant and from the Cary Institute's REU site program. Their intensive ten- to twelve-week summer projects are mentored by one or more BES scientists, with a wide diversity of mentors involved over the two decades. The twenty-odd REU students that have participated over the twenty years of BES have made important discoveries, published papers, and gone on to have impressive careers in science and related fields. Indeed, the first BES Distinguished Alumni Award was given to a person who started as an REU student, then came back to do her graduate work through BES, and now works for a federal science agency. The Baltimore Collaborative for Environmental Biology in 2000–2006 gave us the chance to work with four cohorts of diverse students from Towson and other Baltimore schools each for a two-year period that included two summers of research and academic year enrichment, team and skill building, and support from the program director and Towson professor, Jane Wolfson. The Baltimore Collaborative students all had positive experiences, with many making valuable contributions to their mentors' research and several continuing on into graduate work in the sciences. For instance, one student completed his doctorate at Princeton in biological dynamics and now works for the African Institute for Mathematical Sciences.

We also offer education and stewardship internships through the Urban Resources Initiative and other programs. Education interns contributed to many of the BES curriculum development and education research activities described elsewhere. Carol Rinke was the graduate student intern who led the Ecology Teaching Study (see next section) and now holds a Ph.D. in science education and collaborates with BES on analyzing results from the Pathways Project.

BES has supported the Parks & People Foundation's Green Career Ladder program over the years, from KidsGrow and Project BLUE (a middle school program) to their current year-round Branches program for high-school-aged students. In 2015 we launched the Young Environmental Scientists (YES) team (box 17.7) as part of Branches, with team members learning environmental science, job and life skills through authentic research with mentors, trainings, and team-building activities.

ADVANCE SCHOLARSHIP ABOUT URBAN ECOSYSTEM EDUCATION

Throughout its history BES education has pursued the dual mission of increasing people's understanding of the Baltimore urban social-ecological system through the diversity of programs described in the preceding sections and of conducting research on how such understanding is developed. During the first phase of BES we helped lay the conceptual groundwork for the nascent field of urban ecosystem education by convening a Cary Conference in 1999. This brought together leading scholars at the cutting-edge of social and ecological research on cities with researchers and practitioners at the forefront of urban, place-based, and innovative science education. This first-ever synthesis provided the field with a rich collection of perspectives from across the disciplines, helping establish the vital importance of understanding urban ecosystems for all citizens and the power of a systems-approach to education that brings together the social, biological, physical, and technological dimensions of cities. The conference produced a book that has been downloaded over sixty-five hundred times and represents an important contribution to the field. By studying the ecosystem education system, ecosystem teaching, and ecosystem learning, BES education re-

---

BOX 17.7. YOUNG ENVIRONMENTAL SCIENTISTS IN THE MAKING

It was a hot day in mid-July as five gangly teenagers exited a van after a thirty-minute drive into the suburbs to the Gwynns Falls sampling site in Owings Mills, Maryland. They struggled into chest waders, grabbed measuring tapes, vernier probes, and clipboards and slid down the steep bank to the deeply incised stream. The closed tree canopy brought welcome relief from the hot summer sun as the teens carefully made their way upstream over slick rocks and through deep pools, cataloging in-stream features and testing the water for pH, dissolved oxygen, and temperature. This was the first Branches-YES team and their stream work is part of the long-term ecological research of Peter Groffman and colleagues. The students' data will help the scientists understand the evolution of in-stream features over time. Most YES students had never spent time in a stream or much time outside of the city. After five weeks in streams, forests, and vacant lots, meeting with scientists and learning about data, research methods, and real-world applications, these students will have made connections to ecology and their local environment that most people, in or out of the city, will never make. Hopefully, this will lead at least some of them to pursue future study or a career in environmental science.

---

search addressed one of the core BES questions: How can city residents use an understanding of the metropolis as a social-ecological system to improve the environment and their daily lives?

Understanding the Baltimore Ecosystem Education System
Participants in the multi-city Urban Ecology Collaborative include Boston, New Haven, New York, Baltimore, Pittsburgh, and Washington, D.C. As part of this collaborative a survey of environmental education providers was conducted in the spring of 2005. Surveys were received from 147 organizations across the six cities, with 33 (85 percent) responding in Baltimore. Findings in the six northeast cities revealed a vibrant, diverse, and powerful community of organizations. The audiences served were quite diverse in race/ethnicity (over 50 percent African American and Hispanic), age, and

neighborhoods. Information for 118 distinct education programs in Baltimore revealed that nature, wildlife and ecosystems were the most commonly covered topics, followed by water resources and natural resources/resource management. Urban ecology, biodiversity, environmental health, and pollution prevention were next most common. Interestingly, several notable environmental topics were covered by less than 15 percent of the programs, including schoolyard ecology, energy, air quality, and sustainable development. The least-covered topics were climate change and transportation and, lowest of all, environmental justice (covered by only 2 of the 118 programs in Baltimore). Two-thirds of the programs used field activities, and 45 percent classroom or workshop instruction. These were by far the most common delivery modes listed. A smaller number (20–25 percent) involved in-class presentations, student research activities, and lab experience. Gaps in types of services provided, such as online resources, full curricula, student research, and data analysis support and the depth of program assessments used (for example, very few programs used authentic or pre/post-assessments) have helped shape the education work of BES and collaborators in Baltimore and the other cities.

Understanding Effective Urban Ecosystem Teaching

Building on our vision for effective urban ecosystem teaching from our early work with innovative teaching approaches and the Cary Conference, we sought to understand the extent to which teachers used these practices and focused on teaching about the local urban ecosystem. In 2004 we conducted an intensive study of ecology teaching in all of the public schools in Watershed 263, including thirteen teachers and seven administrators. In these schools, ecology instruction generally involved textbook use, cooperative learning, data analysis, and a connection to the local physical environment. Teachers and administrators expressed interest in teaching more about ecology, concern about connecting outdoor experiences to the curriculum, concerns about transportation, safety, and other logistics of field study, and the need for more hands-on activities and resources, including outside experts. These results inspired our work with teachers and schools in an underserved West Baltimore neighborhood as part of the multifaceted effort by BES to test the impacts of "greening" on neighborhood revitalization and satisfaction.

In 2005 we conducted a more extensive survey of ecology teaching in

Baltimore high schools, focusing on how it was being taught, for example, use of fieldwork, authentic data, community resources and action projects, and on what was being taught; are teachers teaching about local ecosystems and local issues? We received 150 responses from environmental science and biology teachers, representing most public schools (82 percent in Baltimore County and 63 percent in Baltimore City), and some of the many private schools in the city (24 percent) and county (8 percent). We found that teachers used traditional practices such as lecture, reading, and discussion much more than innovative practices like fieldwork and action projects, and they focused on the Chesapeake Bay much more than on the urban ecosystem. Indoor lab study was more frequent than study in the school-yard, and use of the surrounding neighborhood was practiced even less. Teachers' primary reasons for teaching ecology were not science focused but rather to demonstrate relevance and encourage active protection of the environment. Interestingly, outdoor-oriented activities often were independent of the curriculum and emphasized the service element rather than rigorous science learning. These findings had important implications for how we crafted our curriculum materials, programs, and professional development activities for teachers.

We have explored innovative teaching practices in more depth through a series of research projects over the past dozen years, starting with the Responsive Teaching Study in 2005–7 with David Hammer, Andy Elby, Janet Coffey, Daniel Levin, and Sandra Honda at the University of Maryland College Park along with Berkowitz and then-BES Education Coordinator Janie Gordon. This project studied how teachers modify curriculum in response to student thinking, while helping ground this attention in frameworks of "big ideas" in ecology to help focus their attention when listening for and responding to student reasoning. Eight Baltimore City and County teachers comprised the environmental science team, which, along with teams of biology and physics teachers, participated in biweekly meetings, professional development sessions on BES units, summer workshops, and in-class support. Teachers struggled to implement this practice often and well, thwarted by pressure from the testing-based curriculum that privileges definitions and breadth over depth. However, stellar examples emerged among the Baltimore teachers. For instance, one of the BES teacher's high school biology teaching was featured as a model in papers emerging from the responsive teaching study. His example highlighted the importance of teacher disciplinary *and* pedagogical content knowledge, commitment to a student- and understanding-centered pedagogy, and freedom from external constraints.

Our work on Learning Progressions (described more in Understanding Ecosystem Learning, below) has provided a deep, rich framework for helping teachers enact the practices described here. From the Pathways Project, a picture of the knowledge and skills that teachers need for effective teaching that goes beyond superficial "coverage" of topics emerged, and now guides all of our work with teachers. Assessments from hundreds of teachers across the Pathways Project, including nearly eighty in Baltimore, were used to shed light on teachers' knowledge and practices in the three arenas (see figure 17.2). In a study focusing on teaching about carbon and carbon cycling, researchers found a positive relationship between teacher knowledge and student learning. However, teachers have a hard time maintaining productive focus on principles and on interpreting student thinking. In a similar study focusing on teaching about water, BES collaborators found that professional development modestly increased teachers' content knowledge, ability to set challenging learning goals, and understanding of student thinking grounded in the learning progression, while gains in their ability to make productive responses to student thinking were limited. In a case study of sixteen teachers, including four in Baltimore, most evinced just an emergent or transitional use of productive formative assessment in their instruction. However, sophisticated classroom discourse was occasionally observed where teachers repeatedly elicited student reasoning, used the learning progression framework to interpret their ideas, and used a range of strategies to move students up levels of learning. Interestingly, such examples were not limited to classrooms where students showed the most sophisticated thinking, suggesting that the practice has broad application. Other BES collaborators found similar results in a study of teaching about water, with most teachers implementing a right/wrong approach to using the learning progressions. Videotapes and other data from four case study teachers' classrooms looked at the use of learning progression–based strategies in teaching about carbon. One of the teachers showed impressive use of these strategies, helping her students develop reasoning that is grounded in disciplinary substance and practices, similar to the findings in the earlier Responsive Teacher Study. Not surprisingly, student learning gains were higher in this teacher's classroom.

An examination of the overall pattern of implementation of the sixteen teachers in the Pathways Case Study showed that all teachers, on the basis of both their classroom practices and their reflections, benefited from the professional development. Three of the teachers limited their implementation exclusively to enacting the curriculum units shared in the professional development, while four focused on applying one specific teaching strategy. The majority (nine) showed a more integrated approach to implementation, applying the teaching strategies they learned not only in the project's modules but also across their curriculum. These teachers were distinct from the others in that all but one came to the professional development with a disposition for broader learning and professional growth, and all engaged in collaborative work with other teachers and project educators over several years of participation.

Understanding Urban Ecosystem Learning

With BES education's early emphasis on teachers, we did not conduct extensive student research until the Pathways Project and, more recently, the Branches-YES, Data Jam, and Comp Hydro projects. The main line of BES research with students has focused on understanding and learning of fundamental or generic concepts, principles, and big ideas in ecology and earth science. For the most part we ask, do they understand "ecology *in* cities?" and only to a lesser extent, "ecology *of* cities?" Through the Pathways Project we have been developing learning progressions for student understanding of biodiversity (causes and functional consequences), carbon cycling, and water pathways and quality. Learning progressions describe the patterns and pathways of student explanations about a topic, from a "lower anchor"—the least sophisticated discourse acquired through out-of-school and early school learning—to the "upper anchor"—the most sophisticated discourse attainable by graduating high school students (figure 17.2). A powerful narrative has emerged from this work for guiding our curriculum development and teaching. Most students apply a force-dynamic narrative, where actors have purposes in a storylike view (Levels 1, 2), but can develop what we view as school-science narratives (Level 3) and, ultimately, a systems-based, mechanistic narrative where principles are applied in context and with constraints (Level 4). Similar patterns of learning have been described for the three main topics of this work: carbon, water, and biodiversity. In general, middle school students give responses to our assessments at level 2, high school students at level 3, and teachers between levels 3 and 4. The "mile-wide, inch-deep" problem of science education, reinforced by teaching that focuses on preparing students for standardized tests that for the most part ask for level 3 understanding, are significant impediments to students reaching level 4 by the end of high school. However, our results show that students *can* learn key aspects of urban ecosystem literacy given the right combination of experience and support.

A second line of BES research about students explores their understanding of the outcomes and benefits of scientific study of ecosystems, using socio-cognitive theory as our guiding framework. Studies of student interest in environmental careers comparing underrepresented minority students with white students in Baltimore revealed a significant lack of knowledge about the outcomes and rewards of environmental career paths and, in the case of minority students, about an important role of perceived social support in choosing an environmental path. Opportunities offered by the Baltimore Collaborative for Environmental Biology and Research Experiences for Undergraduates had positive effects on individual students in developing an appreciation and self-efficacy for choosing environmental careers, but the sample size is too small for more than anecdotal support of our programs in this regard. We are studying high school students participating in the Branches Young Environmental Scientist program of the Parks & People Foundation to further explore the factors influencing students' choice to pursue science and ecological research. We hope to follow the students after high school to observe several cycles of choice-action-outcomes to learn under what circumstances and for which students an experience like YES can lead to success in pursuing an environmental path.

REALIZING SIX BOLD PREDICTIONS ABOUT URBAN ECOLOGY EDUCATION
In the concluding chapter of our book from the 1999 Cary Conference, Berkowitz and colleagues made six bold predictions about the future of urban ecosystem education. We will summarize our experiences and findings from twenty years of BES education by addressing each of these predictions. Nearly two decades after the conference we see tremendous evidence of progress and opportunity, while also grappling with persistent challenges.

Prediction 1. The Idea that Cities Are Ecosystems Will Be Pervasive in Society
Since the Cary Conference in 1999, the term "ecosystem" has been embraced by a broad swath of society, from the popular press's referring to the acquisition of Whole Foods by Amazon as the Amazon "ecosystem," to the STEM education "ecosystem," to software giant Apache Groovy's "fruitful ecosystem of projects," and industry's Red Hat "Ecosystem," all results of a simple Google search for the term. In all cases, the word is used to recognize a complex system of interacting components, with at least some capacity for self-organization, scalability, and sustainability. Surely if all of these can be "ecosystems," then a city can be! Perhaps even more notable is the increase in the use of the term "ecosystem services." For many years the BES social science team has collected data documenting the general public's understanding of watersheds, and we are starting to collect similar information about the understanding of ecosystems (for example, Do you live in an ecosystem?). Our informal surveys suggest that students and teachers view natural places as ecosystems more than they do cities. However, we also have found that they quickly learn to modify this view. Most of our workshops start with a simple challenge: What do you need to know to answer the question, will we run out of oxygen in this room? Teachers quickly build a conceptual model of the room as a microcosm of an urban ecosystem, including dimensions of time and space, numbers and activities of people in the room, presence of plants, ventilation, and so forth, which they readily grasp and then use in subsequent learning, including ideas about the spatial distribution of people and resources (Am I near the vent?), legacies (Did we start with low oxygen to begin with?), and context (Is the air being brought in already polluted?).

Prediction 2. Educators and Scientists Will Have a Richer, Deeper,
More Pragmatic Understanding of What Is Meant by the Concept
of an Urban Ecosystem
Through our education research and practice we have gone beyond defining the static concept of urban ecosystem to a more dynamic and layered picture. Understanding urban ecosystems is inherently a complex, interdisciplinary undertaking, involving the interplay of biophysical, engineered, and social aspects of places ranging in size from a meeting room or school to a neighborhood or entire metropolitan area. We have described how stu-

dents build pieces of this understanding, focusing on key dimensions such as carbon cycling, water flow, and the diversity and functions of organisms in the local environment. And we have found ways to support teachers in helping students see the relevance and use of their growing knowledge for addressing important problems. However, we have struggled to support more interdisciplinary teaching and learning in the face of clear separations in the curriculum and schooling between the social and natural sciences. This remains a frontier for us.

Prediction 3. The Importance of a Broad Understanding of Urban Ecosystems Will Be Recognized, Appreciated, and Well Articulated

Evidence for Baltimore's embrace of the importance of understanding urban ecosystems abounds. Maryland is one of the few states with a state-wide Environmental Literacy requirement, in this case one that is actively embraced by local districts and the general public. While the dominance of statewide education standards limits the amount of explicit attention to *urban* ecosystems in the curriculum, we have found tremendous enthusiasm on the part of teachers and administrators for teaching and learning about Baltimore. Furthermore, most city agencies have strong education programs, with the Offices of Sustainability and Recreation and Parks having extensive programs geared toward fostering an understanding of the city as an ecosystem. We continue to assert that understanding urban ecosystems is vital for citizenship—both as an end point, because these are incredibly important ecosystems, and as a pedagogy, because they are accessible for first hand, in-depth, and multidimensional investigation and learning.

Prediction 4. People Everywhere Will Have Many and Diverse Opportunities to Learn About and Apply Their Understanding of Urban Ecosystems

BES education's primary experience in this vein has been with K–12 teachers and their students. Since the urban education Cary Conference in 1999, Maryland's Environmental Literacy standards and graduation requirements were adopted, assuring that all K–12 students have Meaningful Watershed Education Experiences at least three times during their K–12 years. Our programs have provided valuable support to teachers in Baltimore City and County in providing these experiences. While for many years BES education was the primary source of learning opportunities for teachers in urban ecology, there has been a wonderful proliferation of opportunities from across the community of urban environmental education providers. Teachers and students have the opportunity to learn about the urban ecosystem through BES and other curricula specifically focused on Baltimore and also from extensive data made available through the BES Data Jam project, the BES Data Portal, the Baltimore Neighborhood Indicators Alliance, Baltimore's City Stat, and similar sources in the county and beyond. Baltimore may well be the most intensively studied social-ecological system in the world, and we are inspired and also a bit vexed by the challenge of helping its students and teachers benefit from this incredible wealth of knowledge in their formal education in Baltimore.

Prediction 5. Urban Ecosystem Understanding Will Be Developed in Many Places via Participatory Models Linking Scientists of Many Sorts, Urban Residents, Managers, and Decision Makers

BES scientists have forged very productive partnerships with managers, decision makers, and citizens, and the BES education team has built participatory projects with scientists, educators, and teachers. Starting with early successes in the Partnership Program, followed by the accomplishments of the Pathways and Comp Hydro projects, BES education has been an important nexus of collaboration for urban ecosystem education in the schools. Likewise, the Branches-YES and REU programs partner students with scientists for intensive exploration and learning. Recent partnerships between BES scientists and educators in our work with Green Street Academy and Francis Scott Key Elementary/Middle School demonstrate the tremendous potential of this approach but also the significant costs in terms of time and resources, especially for sustained productivity.

In our partnerships with teachers we have discovered productive ways to help them develop the rich, complex knowledge and skill set needed for effective teaching. Because many teachers are not from the local area, partnership with BES helped them build knowledge of local ecosystems through professional development activities. Our more intensive approach via Teacher Institutes and multiyear professional development programs helps address what otherwise is a tremendous limitation of professional development, where a very long list of desired outcomes is addressed in a very short period of time.

Prediction 6. People Will Understand the Vital Role Well-Designed and Well-Managed Cities Play in Sustaining Life—Human and Nonhuman—on Earth

While the learning goal of assuring that people understand the vital role of cities in sustaining life is fundamental to motivating our work in urban ecosystem education, we have scant evidence about how our teachers or students perceive cities or about whether our programs influence these perceptions. In Baltimore, as in many cities across the nation, we find a vexing irony: the people most impacted by the environmental ills of urban life (for example, poor people) have the least impact on the environment, the least power over environmental decisions, and often the fewest opportunities, resources, and trained teachers to learn about the environment. On the other hand, those with wealth and power are more buffered from environmental ills while still benefiting from the environmental services of the urban system, have more impact on the environment, and often have more opportunities and resources for environmental learning and exploration. Helping the full breadth of Baltimore residents to benefit from the rich and powerful understanding of its social ecology system *and* from the amenities and efficiencies urban life affords remains one of the important challenges and opportunities for BES education in the future.

Acknowledgments   We gratefully acknowledge the generous support for BES education that has come from four NSF LTER grants (DEB-9714835, DEB-0423476, DEB-1027188, DEB-1637661) along with SLTER/EDEN, REU, RET and RAHSS supplements to those grants; an NSF Biocomplexity Grant (DRL-0628189); an NSF Communicating Climate Change Grant (CFDA 47.076); an NSF MSP grant (G-3062-5—CFDA 47.076); an NSF STEM+C grant for Comp Hydro (G96702-1); the NSF Responsive teaching study grant (ESI 0455711); and the NSF UMEB grant (DBI-9975463). Additional support has come from the U.S. Forest Service, small grants from the Chesapeake Bay Trust, and the Maryland Department of Natural Resources.

Many people have contributed to the success of BES education, and we would like to acknowledge them here. They include the BES scientists Steward Pickett, Morgan Grove, Peter Groffman, Emma Rosi, Jonathan Walsh, Shannon LaDeau, Janet Coffey, Richard Pouyat, Quintaniay Holifield, Gordon Heisler, Kenneth Belt, John Hom, Guy Hager, Claire Welty, Andy Miller,

Robert Shedlock, Ed Doheny, and Christopher Swan; the science education collaborators John Moore, Andy Anderson, Ray Tschillard, Ali Whitmer, Sylvia Parker, Julie Bianchini, Carol Rinke, Laurel Hartley, Jennifer Doherty, Hui Jin, Kristin Gunckel, Beth Covitt, Shandy Hauk, Gel Alvarado, Jennifer Forester, Sarah Haines, Kristin Holfelder, William Hoyt, Robert Mayes, Amanda Morrison, Kimberly Melville-Smith, Gregory Newman, Scott Simon, Andrew Warnock, Sara Syswerda, Tobias Irish, Eric Keeling, LaTisha Hammond, Robert Panoff, Garrett Love, Randy Boone, Greg Newman, Joshua Gabrielse, Keisha Matthews, Katya Denisova, George Newberry, Andrea Bowden, Susan Elias, and Jonathon Grooms; and numerous instrumental staff and students as well, including Vicki Fabiyi, Natalie Crabbs Mollett, Cornelia Harris, Terry Grant, Mary Hardcastle, Mary Washington, Mary Cox, Pearline Tyson, Tanaira Cullens, Sharon Schueler, Katie Jagger, Richard Foot, Molly Charnes, Tissa Thomas, Tammy Newcomer, Julie Baynard, Molly Van Appledorn, Trevor Shattuck, Tim Meyers, and Patrick Bond.

Finally, the students and teachers of Baltimore have been an endless source of inspiration, insight, and creativity. We acknowledge their contributions with heartfelt appreciation and respect.

Afterword

FROM A VIEW OF THE PAST
TO A VISION FOR THE FUTURE

THE ESTABLISHMENT OF THE Baltimore Ecosystem Study (BES) in
1997, supported by the National Science Foundation (NSF) and the USDA
Forest Service, heralded a new era in the science of urban ecology. At the
same time, NSF also began its support of the Central Arizona–Phoenix
LTER (CAP), which has been a partner in irrevocably altering the science of
urban ecology. Previously, urban ecology in the United States, and indeed
in most countries, involved the translation of the "biological" science of
ecology to familiar places in cities, suburbs, and towns that were like the
places ecologists usually studied. This comfortable transfer was labeled ecology *in* the city. Important, interesting, and useful knowledge came from
this approach. Indeed, such ecological research in urban places continues
to produce novel and useful knowledge.

But BES and CAP added something new to the toolkit of American
urban ecology. The novelty was at that time the most well developed expression of "the long march to cross-disciplinary research" (chapter 2). This
new approach took cities, suburbs, and towns to be ecosystems. To consider
an urban place as an ecosystem required not only that biological and physical components be recognized—as they are in the classical ecosystem concept in ecology—but also that humans, their institutions, and their artifacts
be investigated. In other words, urban areas are manifestly human ecosystems that inextricably link four kinds of components: biological organisms,

371

physical environments, humans in their various aggregations, and their constructed places. Each of these four components is itself complex and multifaceted. Consequently, the number and kinds of interactions in urban areas, as an instance of the human ecosystem concept, are myriad. The feedbacks among all components of urban ecosystems constitute a new level of complexity that urban ecology must address. Ecology has long struggled over how to deal with this new complexity.

The new approach exemplified by BES and CAP stimulated a new field: the ecology *of* the city. This required BES to embrace a disciplinarily diverse team of researchers. It also required that this BES team communicate deeply and openly about their theories, their assumptions and biases, and their methodologies as well as explore new ways to apply their findings. Indeed, the continuity of interactions and growth of this team over the twenty-plus years of the project opened significant new channels of cross-disciplinary communication and shared thinking. These channels generated new interpretations of social-ecological phenomena and led to the development of new frameworks for urban ecology. These frameworks have encompassed a richness of urban processes and phenomena that reach well beyond the history and situations in Baltimore. Therefore, the continuity of effort and interaction in BES as an ongoing project is as important in the conceptual and practical realms as the long-term runs of data are in the empirical realm.

This book has summarized many of the key empirical findings over the twenty-year span of BES so far. The contributors to this book are rightly proud of their scientific contributions, illustrated by their more than fourteen hundred papers, books, chapters, and published reports. But they are also humbled and pleased that many of their findings have been used by communities to better their situations, by managers to understand the outcomes of their decisions, and by policy makers to design new solutions to the pressing problems that confront the core city, inner-ring suburbs, the exurban fringe, and the entire metropolitan region as a whole. But BES findings have also pointed to opportunities to better link the quality of environment with the quality of life for all residents of the metropolis.

Twenty years is an important milestone for any scientific endeavor. Much of the productivity of BES to date has relied on its principal funders, the Division of Environmental Biology (DEB) at NSF and, via in-kind support and staff time, the USDA Forest Service. This funding has permitted

BES to establish an adaptive culture of inclusiveness, openness, and productivity. This support has facilitated the construction and maintenance of a project that, first, generates new and integrated interdisciplinary social-ecological research findings; second, effectively engages with people who need and can use better scientific knowledge for their neighborhoods, their city, and their metropolitan complex; third, conducts powerful educational programs in minority-serving school districts, coupled with research on teaching and learning about ecology; and, finally, works toward a more diverse scientific workforce. Notably, this culture and its effectiveness have always been leveraged by funding sources beyond DEB.

Twenty years also marks a milestone to compare with the growth of other disciplines. BES has grown in parallel with ongoing improvements in the "metascience" described by the history, philosophy, and sociology of science. Advances in the philosophy of science—apparently little known in ecology—counter the regressive message of positivism, which emphasizes a predictive science that can operate only under certain ideal conditions. Contemporary philosophy of science counters the hegemony of experimental approaches often suitable in population and community ecology, where some sort of equilibrium or ideal behavior can be articulated as a hypothetical reference in presumably uniform places. The philosophy of science and the growth of the science of complexity reinforce the need to address the multifaceted, nonlinear interactions among human agency, institutional responses, and the hybrid natural/social entities and phenomena of the rapidly changing urban realm. BES has increasingly found its science moving from *coupled* understandings of social and ecological systems to examining the simultaneity of *coproduced* social-ecological systems, and from a science of the academy to a transdisciplinary science that more fully engages diverse perspectives and embraces the need to produce actionable science in concert with decision makers. Cities, in the broadest sense, are substantially new things in the twenty-first century, requiring a science that can adapt to the changing reality of the urban world. This reality is complex, in both the figurative and the technical senses, and would seem ideally suited to long-term examination.

In the next twenty years of research in Baltimore, there is the opportunity to seek a broader portfolio of support for our radically interdisciplinary social-ecological research and application. The bold history of interdisciplinarity that promotes understanding of human-inflected systems (described in chapter 2) must be complemented by an equally bold future that builds on the advances of the past twenty years of urban ecological science in BES. Core institutional partners that have shaped and promoted BES are committed to this unique project and its contributions to the science required for the rapidly changing global urban reality. The people and institutions that have contributed to this volume envision continued research into, and deep thinking about, the "city yet to come," to use AbdouMalique Simone's cogent phrase, as a high priority. What form and functions this emerging city takes in the Baltimore region is a pressing question. The answer is relevant to shrinking cities, to expanding urban megaregions, to cities of consumption and of service, to sustainable cities, and many more. The dimensions of urban change in the Baltimore region echo those occurring in many other towns and cities worldwide that share some subset of the attributes that now exist or are set to appear in Baltimore. These attributes, changes, and novelties are the raw material for producing the sustainable, resilient approach to the urbanizing, globalizing world of the twenty-first century. Nothing could be a more compelling justification for continued investigation of humankind's most intimate and most dynamic habitat, already so well understood in Baltimore. The community represented in this book is committed to the idea that Baltimore continue as a home for urban ecological science, a beneficiary of the findings of that science, and a beacon for integrated, synthetic urban ecology throughout the world.

BIBLIOGRAPHIC ESSAY

To enhance the flow of the text, we have forgone the usual practice in scientific writing of referencing individual statements and facts with numerous citations. However, in almost all cases the research that is presented, synthesized, and discussed in this book has been published and is readily available in peer-reviewed scientific journals and books. Where relevant, the citations supporting data in tables and figures are supplied in the accompanying captions. In this essay, we point out for each chapter the major articles from the Baltimore Ecosystem Study (BES) and the broader literature that provide important background or specific support for the statements found in each chapter. In this way, we seek to provide a general overview to the subject as an entry point to the literature for further consultation. A complete list of all literature cited follows. The complete bibliography of BES research is found on the project website (http://beslter.org).

FOREWORD
Scott Collins refers to the Risser report, the first decadal review of the Long-Term Ecological Research program of the U.S. National Science Foundation. A diverse review committee was cochaired by Paul Risser and Jane Lubchenco. The report was developed in 1993 but bears a copyright date of 2003 and is available at https://lternet .edu/wp-content/uploads/2010/12/ten-year-review-of-LTER.pdf.

CHAPTER 1. GOALS OF THE BALTIMORE SCHOOL OF ECOLOGICAL SCIENCE
Urban ecological science in the United States has a long, fraught history. Sharon Kingsland's *The Evolution of American Plant Ecology, 1890–2000* (2005) puts the origin and interdisciplinary struggle of the BES Long-Term Ecological Research project in its largest context. Chapter 2 in this volume is an important complement to Kingsland's

375

book. Early attempts to create an American urban ecology are best exemplified by Stearns and Montag's *The Urban Ecosystem—A Holistic Approach* (1974), which brought intellectually adventuresome biophysical and social scientists together to provide a scope for a new interdiscipline. Boyden and colleagues, "The Ecology of a City and Its People: The Case of Hong Kong," from 1981 summarizes a pioneering interdisciplinary project that unfortunately did not continue beyond what was intended to be an establishment phase. The birth of modern American urban ecology can be traced to Mark McDonnell and Richard Pouyat, whose leadership is represented, respectively, by "Ecosystem Processes Along an Urban-to-Rural Gradient" from 1997 and "Investigative Approaches to Urban Biogeochemical Cycles: New York Metropolitan Area and Baltimore as Case Studies." How this work has evolved into the present-day BES is described by Cadenasso and Pickett in "Three Tides: The Development and State of the Art of Urban Ecological Science," from 2013, and as a mature school of thought by Grove and colleagues in *The Baltimore School of Urban Ecology: Space, Scale, and Time for the Study of Cities,* in 2015. The difficulty of overcoming the interdisciplinary misconceptions and the uses of scientific ideas as metaphors rather than technical terms associated with clear theoretical structures is explained by Christopher Eliot in "The Legend of Order and Chaos: Communities and Early Community Ecology" and by Jennifer Light in *The Nature of Cities: Ecological Visions and the American Urban Professions 1920–1960.*

Details of Baltimore as a research site and the structure of BES research can be found in several publications. Grace Brush's "Historical Land Use, Nitrogen, and Coastal Eutrophication: A Paleoecological Perspective" (2008) provides information on Baltimore's geomorphology, deep ecological history, and changes since European arrival. Zhou and colleagues examined more recent changes in forest cover and fragmentation in the Gwynns Falls watershed in their paper "90 Years of Forest Cover Change in an Urbanizing Watershed: Spatial and Temporal Dynamics." Study design, motivation, and approaches are provided by Pickett and others in "Urban Ecological Systems: Linking Terrestrial Ecological, Physical, and Socioeconomic Components of Metropolitan Areas" (2001); Cadenasso et al.'s "Integrative Approaches to Investigating Human-Natural Systems: The Baltimore Ecosystem Study" (2006); and Grove and colleagues' "Building an Urban LTSER: The Case of the Baltimore Ecosystem Study and the D.C./B.C. ULTRA-Ex Project" (2013). A more recent empirical update appears in Pickett et al., "Urban Ecological Systems: Scientific Foundations and a Decade of Progress" (2011).

The guiding questions for BES are motivated by theory from many disciplines. Background on the structure and change of ecological theory is presented by Pickett et al. in *Ecological Understanding: The Nature of Theory and the Theory of Nature,* 2nd edition, 2007. The deep ecological theory emphasizing spatial and temporal heterogeneity is described by Scheiner and Willig's *The Theory of Ecology* (2011). Ecological heterogeneity reflects BES's foundational patch dynamics approach, illustrated by Wu and Loucks, "From Balance of Nature to Hierarchical Patch Dynamics: A Paradigm Shift in Ecology" (1995). Hierarchical patch dynamics has been extended to an integrated social-ecological approach via Pickett and others in "Dynamic Heterogeneity: A

Framework to Promote Ecological Integration and Hypothesis Generation in Urban Systems" in 2017, and this is complemented by a 2017 conceptual model for applying disturbance as a mechanism within urban systems by Grimm and others in "Does the Ecological Concept of Disturbance Have Utility in Urban Social-Ecological-Technological Systems?" The focus of BES on the hybrid biophysical-social nature of urban ecosystems has led Cadenasso and colleagues to propose a more nuanced urban land classification in "Spatial Heterogeneity in Urban Ecosystems: Reconceptualizing Land Cover and a Framework for Classification." Social sciences motivations are exemplified by Eleanor Ostrom's *Understanding Institutional Diversity*, from 2005; and urban design and planning motivations by Anne Whiston Spirn's *The Granite Garden* and Grahame Shane's *Recombinant Urbanism*. Within ecology, the watershed approach, summarized, for example, by Gene Likens in *The Ecosystem Approach: Its Use and Abuse*, has been key to BES. The development of the BES conception of the human ecosystem was founded on the seminal 1997 paper by the social scientists Gary Machlis and colleagues, "The Human Ecosystem 1. The Human Ecosystem as an Organizing Concept in Ecosystem Management." The human ecosystem framework is the subject of a 2017 monograph by Burch and others who developed the original formulation, as *The Structure and Dynamics of Human Ecosystems: Toward a Model for Understanding and Action*. An important refinement of the frameworks employed by BES is written by Cadenasso et al. as "Dimensions of Ecosystem Complexity: Heterogeneity, Connectivity, and History," which identified the kinds of complexity necessary for understanding urban areas as social-ecological systems. Grimm et al. have specified that technology is an unavoidable facet of urban ecosystems in their 2016 paper "A Broader Framing of Ecosystem Services in Cities: Benefits and Challenges of Built, Natural, or Hybrid System Function." The first-generation BES approaches and key unexpected findings were summarized by Pickett and others in the 2008 "Beyond Urban Legends: An Emerging Framework of Urban Ecology, as Illustrated by the Baltimore Ecosystem Study," while the continued growth toward the current array of approaches and questions was presented in Pickett et al., "Urban Ecological Systems: Scientific Foundations and a Decade of Progress" (2011). Gaps and needs for further development of urban ecological science are proposed by McPhearson and colleagues in "Advancing Urban Ecology Towards a Science of Cities" (2016), and by Groffman and colleagues in "Moving Towards a New Urban Systems Science" (2017).

The traditional core areas of the LTER program appear at http://www.lternet.edu/research/core-areas. The new LTER core areas required of urban projects were reflected in the contrast between ecology in versus ecology of the city, first articulated in 1997 by Pickett et al. in "Integrated Urban Ecosystem Research," a viewpoint expanded by Grimm et al. in the 2000 publication "Integrated Approaches to Long-Term Studies of Urban Ecological Systems." The contrast between ecology in versus ecology of the city has been documented to have empirical as well as conceptual significance by Zhou et al. in "Shifting Concepts of Urban Spatial Heterogeneity and Their Implications for Sustainability" (2017). Childers and colleagues suggested that application of urban ecology to sustainability would require something more than the move from ecology in to ecology of the city: "An Ecology for Cities: A Transformational Nexus of Design and Ecology to Advance Climate Change Resilience and Urban Sustainability."

CHAPTER 2. URBAN ECOLOGICAL SCIENCE IN AMERICA

Discussions within the Ecological Society of America about the relationship of ecology to human problems and social sciences are in early presidential addresses by B. Moore, "The Scope of Ecology" and S. A. Forbes, "The Humanizing of Ecology" and in articles by C. C. Adams, "The Relation of General Ecology to Human Ecology" and "A Note for Social-Minded Ecologists and Geographers."

On the links between ecology, Darwinism, and sociology, see R. E. Park, "The City: Suggestions for the Investigation of Human Behavior in the City Environment" and his book *Human Ecology.* A. H. Hawley's critique of the Chicago School is in "Ecology and Human Ecology" and in *Human Ecology: A Theory of Community Structure.* On the history of human ecology at Chicago, see J. Light, *The Nature of Cities: Ecological Visions and the American Urban Professions, 1920–1960.* E. Cittadino offers a broad survey in "The Failed Promise of Human Ecology," which supports the assessment by H. W. Odum in *American Sociology: The Story of Sociology in the United States through 1950.* Odum linked ecology and regional sociology in *American Social Problems: An Introduction to the Study of the People and Their Dilemma* and in H. W. Odum and H. E. Moore, *American Regionalism: A Cultural-Historical Approach to National Integration.* Odum was influenced by the Indian sociologist R. Mukerjee, especially by Mukerjee's book *Regional Sociology* and articles such as "The Ecological Outlook in Sociology" and "The Concepts of Distribution and Succession in Social Ecology."

On postwar concerns about human population growth and emerging environmental problems, two bestsellers in 1948 were W. Vogt, *Road to Survival,* and F. Osborn, *Our Plundered Planet.* Discussion continued into the 1950s with such books as J. Boyd Orr, *The White Man's Dilemma: Food and the Future,* and H. Brown, *The Challenge of Man's Future.* Ecologists promoted human and urban ecology in light of these challenges, notably P. B. Sears in "Human Ecology: A Problem in Synthesis" and E. Anderson in "The City Watcher." The Princeton conference on the environmental challenges of the postwar world was published as W. L. Thomas, ed., *Man's Role in Changing the Face of the Earth,* a landmark in environmental thought and discussion of human ecology.

The plight of American cities was a central theme of the emerging environmental movement of the 1960s. A. Rome discusses suburbanization in *The Bulldozer in the Countryside: Suburban Sprawl and the Rise of American Environmentalism,* while the impact of race riots on Baltimore is assessed in J. L. Elfenbein, T. L. Hollowak, and E. M. Nix, eds., *Baltimore '68: Riots and Rebirth in an American City.* Many new approaches to urban ecology emerged in the 1960s. Discussions of the "metabolism of cities" are in A. Wolman, "The Metabolism of Cities," S. Boyden, "An Integrative Ecological Approach to the Study of Human Settlements," J. Tarr, "The Metabolism of the Industrial City: The Case of Pittsburgh," C. Kennedy, J. Cuddihy, and J. Engel-Yan, "The Changing Metabolism of Cities," and, for Baltimore, S. Olson, "Urban Metabo-

lism and Morphogenesis" and "Downwind, Downstream, Downtown: The Environmental Legacy in Baltimore and Montreal." A. Spilhaus's futuristic ideas were out-
lined in "The Experimental City" and were the subject of a doctoral dissertation by
T. A. Wildermuth, *Yesterday's City of Tomorrow: The Minnesota Experimental City and
Green Urbanism.* The operations and history of the Institute of Ecology are discussed
in R. F. Inger, "Inter-American Institute of Ecology: Operational Plan," J. M. Neuhold,
"The Institute of Ecology: TIE," and J. Doherty and A. W. Cooper, "The Short Life and
Early Death of the Institute of Ecology." TIE's project on urban ecology was published
as F. Stearns and T. Montag, eds., *The Urban Ecosystem: A Holistic Approach.* Ian
McHarg extended ecological ideas into urban planning and landscape architecture in
Design with Nature.

The ecosystem concept emerged as a central organizing idea within ecology in
conjunction with these environmental discussions. A. G. Tansley introduced the term
"ecosystem" in "The Use and Abuse of Vegetational Concepts and Terms," and
F. Evans recognized its relevance to urban ecology in "Biology and Urban Areal Research." E. P. Odum made the ecosystem central to ecology in his textbook *Fundamentals of Ecology,* the third edition of which drew attention to the urgent need to develop
human ecology. H. T. Odum also contributed to the development of the ecosystem
concept from the 1950s (H. T. Odum and R. C. Pinkerton, "Time's Speed Regulator:
The Optimum Efficiency for Maximum Power Output in Physical and Biological Systems") and offered a synthetic approach in his book *Environment, Power, and Society.*
For the history of the ecosystem concept, see Joel Hagen, *An Entangled Bank: The Origins of Ecosystem Ecology,* Frank B. Golley, *A History of the Ecosystem Concept in Ecology:
More Than the Sum of the Parts,* David C. Coleman, *Big Ecology: The Emergence of Ecosystem Science,* and Jean Craige, *Eugene Odum: Ecosystem Ecologist and Environmentalist.* On the concept of "ecosystem services" as a vehicle for linking natural and social
systems, see E. P. Odum and H. T. Odum, "Natural Areas as Necessary Components
of Man's Total Environment," W. E. Westman, "How Much Are Nature's Services
Worth?" and G. Daily, ed., *Nature's Services.* On the link to ecological economics and
the BES, see R. Costanza and H. E. Daly, *Ecological Economics,* and R. Costanza et al.,
"Integrated Ecological Economics of the Patuxent River Watershed, Maryland."

The evolution of thought about ecology's relation to social science can be seen in
an address by ESA President F. H. Bormann ("Unlimited Growth: Growing, Growing, Gone?") and in a report he commissioned by C. S. Holling and G. Orians, "Toward an Urban Ecology." Subsequent developments important for the BES included
the articulation of the concept of ecological "resilience," for example, in C. S. Holling,
"Resilience and Stability of Ecological Systems," and L. H. Gunderson and C. S.
Holling, *Panarchy: Understanding Transformations in Human and Natural Systems.* Critiques of equilibrium thinking paralleled these conceptual developments, for instance, by D. Botkin in *Discordant Harmonies: A New Ecology for the 21st Century,* and
J. Wu and O. Loucks, in "From Balance of Nature to Hierarchical Patch Dynamics:
A Paradigm Shift in Ecology."

Historical background on the BES is in S. E. Kingsland, *The Evolution of American Ecology, 1890–2000.* BES drew on work by faculty at the Yale School of Forestry (see

F. H. Bormann, "Ecology: A Personal History"). The small-watershed model of eco-system analysis was developed by F. H. Bormann and G. Likens, with results summa-rized in *Pattern and Process in a Forested Ecosystem: Disturbance, Development, and the Steady State Based on the Hubbard Brook Ecosystem Study*. W. R. Burch Jr. offered a balanced critique of environmental debates and helped develop the human ecosystem framework guiding BES's direction. See W. R. Burch Jr., *Daydreams and Nightmares: A Sociological Essay on the American Environment*, G. E. Machlis, J. E. Force, and W. R. Burch Jr., "The Human Ecosystem, Part 1: The Human Ecosystem as an Organizing Concept in Resource Management," J. M. Grove and W. R. Burch Jr., "A Social Ecology Approach and Applications of Urban Ecosystem and Landscape Analyses: A Case Study of Baltimore, Maryland," J. M. Grove et al., *The Baltimore School of Urban Ecol-ogy: Space, Scale, and Time for the Study of Cities*, and W. R. Burch Jr., G. E. Machlis, and J. E. Force, *The Structure and Dynamics of Human Ecosystems: Toward a Model for Un-derstanding and Action*.

CHAPTER 3. THE HISTORY AND LEGACIES OF MERCANTILE, INDUSTRIAL, AND SANITARY TRANSFORMATIONS

There are essential books to understanding the history of Baltimore that this chapter leans on heavily. Sherry Olson's *Baltimore* is a masterwork, one of the models for writing a history of an American city. John McGrain's work on the county side of the line, including both *Grist Mills in Baltimore County, Maryland* and *From Pig Iron to Cot-ton Duck: A History of Manufacturing Villages in Baltimore County*, provide a more spe-cialized but undoubtedly essential treasure trove of information. All of the subsequent work on the connections between the Chesapeake Bay and the upland—including Baltimore—builds on Louis Gottschalk, "Effects of Soil Erosion on Navigation in Upper Chesapeake Bay" (1948). Of that work, Grace Brush's is particularly influential on the perspective and scope of this chapter. Her 2017 monograph *Decoding the Deep Sediments* is an excellent introduction to her work. Ed Orser's *Blockbusting in Balti-more: The Edmondson Village Story* clearly and powerfully illustrates a whole package of ugliness that is blockbusting, something that continues to haunt Baltimore and most U.S. cities. The G.M. Hopkins Company Atlas of 1876 is a crucial window into a crucial period. The scans of this atlas that Jim Gillispie at the JHU Eisenhower Li-brary had the foresight to acquire were essential to this work.

There are specific concepts directly alluded to in the chapter that will confuse without context. Matt Luck et al.'s "Urban Funnel Model" (2001) is an interesting example of a "footprint" metric. The term "sanitary city" arises from Martin Melosi's book of the same name. The demarcation of "signal" and "structural" legacy effects is made in Bain et al.'s 2012 BioScience paper. Groffman et al.'s 2002 paper details the degradation of ecosystem services (e.g., nutrient assimilation) in buried floodplains. The Olmsted Brothers' *Report upon the Development of Public Grounds for Greater Balti-more* (1904) is a template for the modern big parks in Baltimore. Watershed 263 is laid out in Felson and Pickett in 2005 in "Designed Experiments: New Approaches to Studying Urban Ecosystems." Fred Besley's work on Maryland forests is accessible through the "Maryland's Forest Resources: A Preliminary Report" (1909) and *The*

Forests of Maryland (1916). Andrew Giguere's master's thesis is one of several sources to document the parks/expressway story. Cleaves et al.'s description of the serpentine barrens remains a definitive description of serpentine landscapes ("Chemical Weathering of Serpentinite in the Eastern Piedmont of Maryland," 1974). Red Run Lake permit data are used by Colosimo and Wilcock in "Alluvial Sedimentation and Erosion in an Urbanizing Watershed, Gwynns Falls, Maryland" (2007) to characterize the "before" condition of this watershed.

Like most urban system science, this work leans heavily on scholars that lie on a continuum between history and historical geography. Craig Colten's early work on historical hazards (for example, his paper in *The Professional Geographer*, 1990) recognized the signal legacies that arise during urbanization of an industrial fringe. Garret Power, "Parceling Out Land in the Vicinity of Baltimore: 1632–1796, Part 2" (1992) reconstructs the emergence of Baltimore's downtown literally from the Northwest Branch of the Patapsco. Terry Sharrer, "Flour Milling in the Growth of Baltimore, 1750–1830" (1976) clearly explained the connection between Baltimore flour and Caribbean monoculture. Keach Johnson, "The Genesis of the Baltimore Ironworks" (1953) gathered essential sources for examining the early role of iron refining, and Ronald Lewis explored the labor force in early Baltimore iron refining as an important case study in *Coal, Iron, and Slaves: Industrial Slavery in Maryland and Virginia, 1715–1865* (1979). Stephen Whitman *The Price of Freedom: Slavery and Manumission in Baltimore and Early National Maryland* (1999) laid out the demographics of slaves and freed slaves in pre–Civil War Baltimore. Groves and Muller, "The Evolution of Black Residential Areas in Late Nineteenth-Century Cities" (1975) traced the reverberations of this history through the turn of the twentieth century, recognizing the roots of persistent housing patterns. Geoffrey L. Buckley, *America's Conservation Impulse: A Century of Saving Trees in the Old Line State* (2010) details the forestry history in much greater detail.

CHAPTER 4. ENVIRONMENTAL JUSTICE AND ENVIRONMENTAL HISTORY

BES researchers are interested in identifying patterns and uncovering the processes that produce those patterns. Charles Lord and Keaton Norquist's 2010 article in *Environmental Law* proposes a method for understanding cities not as chaotic and mysterious but as complex and emergent systems. They argue that viewing cities through the lens of emergence theory can reveal and make sense of urban patterns. Their work builds on Jane Jacobs's classic *The Death and Life of Great American Cities* (1961) as well as on that of Michael Batty, whose article in *Science* in 2008 contends that cities are complex systems that mainly grow from the bottom up, their size and shape following well-defined scaling laws that result from intense competition for space, and Steven Johnson, whose 2001 book *Emergence: The Connected Lives of Ants, Brains, Cities, and Software* explores and explains the concept of bottom-up intelligence and self-organizing systems.

Breaking down disciplinary barriers is a priority of the BES. Several studies highlight the benefits of collaborative environmental justice research. In a 2013 chapter entitled "Ecology and Environmental Justice: Understanding Disturbance Using Ecological Theory," Steward Pickett, Christopher Boone, and Mary Cadenasso explore theoretical and practical linkages between ecology and environmental justice that can generate new insights into urban form and function, including the occurrence of environmental hazards and benefits. They maintain that ecological theory, specifically disturbance, can deepen our understanding of the persistence of environmental injustices and long-term processes. Boone, Cadenasso, Kirsten Schwartz, Morgan Grove, and Geoffrey Buckley, in an article published in *Urban Ecosystems* in 2010, argue that because landscape features develop over time we must take into account the social characteristics and priorities of past residents to truly understand present-day patterns. Using Rutherford Platt's book from 2006, *The Humane Metropolis: People and Nature in the 21st-Century City*, as a starting point, Pickett, Buckley, Sujay Kaushal, and Yvette Williams's 2011 article in *Urban Ecosystems* stresses the critical role science can play in charting a course for a more just and sustainable future. Whether it's a more equitable distribution of trees, access to fresh fruits and vegetables, or the removal of asphalt from schoolyards, Cynthia Merse ("Street Trees and Urban Renewal: A Baltimore Case Study," 2009), Michelle Corrigan ("Growing What You Eat: Developing Community Gardens in Baltimore, Maryland," 2011), and Buckley, Boone, and Grove ("The Greening of Baltimore's Asphalt Schoolyards," 2017) show that behind every successful sustainability effort is a collaborative community-based approach.

The potential for environmental injustice hinges on residential segregation. Garrett Power's 1983 "Apartheid Baltimore Style: The Residential Segregation Ordinances of 1910–1913," published in the *Maryland Law Review*, and Buckley and Boone's 2011 study of neighborhood improvement associations cast light on the policies and practices that established and sustained de jure and de facto segregation in Baltimore. Using this matrix, BES researchers have examined the distribution of disamenities and amenities using both distributive and procedural justice approaches. Boone used a historical perspective in his 2002 "An Assessment and Explanation of Environmental Inequity in Baltimore" and 2008 "Improving Resolution of Census Data in Metropolitan Areas Using a Dasymetric Approach" to explain how decades of disadvantage for African Americans in Baltimore, including segregation away from sought-after factory jobs, created a modern landscape that concentrates polluting industry in majority white neighborhoods. A 2014 article in *Cities* by Boone, Michail Frakias, Buckley, and Grove not only confirms Boone's 2002 and 2008 findings but also shows that the spatial distribution of environmental hazards has shifted over time and that low educational attainment is correlated with high density of polluting facilities.

With respect to the distribution of environmental amenities, Boone, Buckley, Grove, and Sister's "Parks and People: An Environmental Justice Inquiry in Baltimore, Maryland," published in the *Annals of the Association of American Geographers* in 2009, used both distributive and procedural methods to determine whether African Americans have equitable access to Baltimore's parks. Site-specific studies, such as Cassandra Korth's 2006 "Leakin Park: Frederick Law Olmsted, Jr.'s Critical Advice," James Wells et al.'s 2008 "Separate But Equal? Desegregating Baltimore's Golf Courses," and Michelle Chevalier-Flick's 2009 "Toxic Playground: A Retrospective Case Study of Environmental Justice in Baltimore, Maryland" remind us that the road to environmental equity is fraught with detours and delays and, in the case of Wells and Chevalier-Flick, that park quality—not just access—matters. Buckley's 2010 *America's Conservation Impulse: A Century of Saving Trees in the Old Line State* offers a detailed account of the early history of tree planting and management in Baltimore. Meanwhile, Michael Battaglia's 2014 "It's Not Easy Going Green: Obstacles to Tree-Planting Programs in East Baltimore" challenges the notion that there is consensus regarding what constitutes an amenity.

CHAPTER 5. STEWARDSHIP NETWORKS AND THE EVOLUTION OF ENVIRONMENTAL GOVERNANCE FOR THE SUSTAINABLE CITY

Elinor Ostrom's seminal work *Governing the Commons* and subsequent books and articles such as "A Diagnostic Approach for Going Beyond Panaceas" offer a foundational understanding of the complexity of the interactions among multiple groups, organizations, and institutions involved in the governance of natural resources. Durant et al.'s edited volume *Environmental Governance Reconsidered: Challenges, Choices, and Opportunities* provides an excellent understanding of how environmental policies, management, and governance have changed since the "environmental decade" of the 1970s.

Applications of social network analysis to examine organizations are relatively recent. S. Borgatti and P. Foster provide a solid state of the organizational research in "The Network Paradigm in Organizational Research: A Review and Typology," and M. Diani and D. McAdam's edited book "Social Movements and Networks: Relational Approaches to Collective Action" offers insight into how networks are used in social movements. K. Provan and others created a body of research examining social networks in public administration in such papers as "Do Networks Really Work? A Framework for Evaluating Public-Sector Organizational Networks" and "Modes of Network Governance: Structure, Management, and Effectiveness" and in C. Koliba et al.'s *Governance Networks in Public Administration and Public Policy*. Social network analysis of natural resources governance in particular have been reviewed in Ö. Bodin and B. Crona's "The Role of Social Networks in Natural Resource Governance: What Relational Patterns Make a Difference?" and are well described in Ö. Bodin and C. Prell's *Social Networks and Natural Resource Management*. Research focusing on network governance of natural resources in urban areas includes H. Ernstson's papers on Stockholm, e.g., "Scale-Crossing Brokers and Network Governance of Urban Ecosystem Services: The Case of Stockholm, Sweden" and "Social Movements and Ecosystem Services—The Role of Social Network Structure in Protecting and Managing Urban Green Areas in Stockholm," and the multi-city USDA Forest Service research project described in E. Svendsen et al.'s *Stewardship Mapping and Assessment Project: A Framework for Understanding Community-Based Environmental Stewardship*.

S. Dalton's dissertation, *The Gwynns Falls Watershed: A Case Study of Public and Non-Profit Sector Behavior in Natural Resource Management*, provides the full 1999 case study, and M. Romolini et al.'s "A Social-Ecological Framework for Urban Stewardship Network Research to Promote Sustainable and Resilient Cities"; "Towards an Understanding of Citywide Urban Environmental Governance: An Examination of

Stewardship Networks in Baltimore and Seattle"; and "Assessing and Comparing Relationships Between Urban Environmental Stewardship Networks and Land Cover in Baltimore and Seattle" provide the background information on the STEW-MAP social-ecological network analysis framework, data collection, and applications for the 2011 study described in the chapter. T. Allen and T. Hoekstra's *Toward a Unified Ecology* formed the basis for the social-ecological "lens" approach to studying urban stewardship networks. T. Muñoz-Erickson et al.'s paper "Demystifying Governance and Its Role for Transitions in Urban Social-Ecological Systems" synthesizes urban governance research in several U.S. cities and is intended for an audience of ecologists and other researchers seeking to better understand this field and its relevance to their work.

CHAPTER 6. EFFECTS OF DISAMENITIES AND AMENITIES ON HOUSING MARKETS AND LOCATIONAL CHOICES

Urban ecological features can increase surrounding housing values, but differences in their type and pattern can generate substantial differences in their impact on housing values (see E. Irwin "The Effects of Open Space on Residential Property Values"). Purposeful investments to enhance or restore ecological amenities can increase urban housing values but often depend on other spatially varying attributes of the location (see Towe et al.'s "A Valuation of Restored Streams Using Repeat Sales"). Some ecological features may be a disamenity for residents, such as the case in the work of Irwin et al.'s "Do Stormwater Basins Generate Co-Benefits? Evidence from Baltimore County, Maryland." The large variation in the value that households ascribe to urban ecological features is due to the heterogeneity of these attributes, e.g., as M. Livy and H. A. Klaiber, in "Maintaining Public Goods: The Capitalized Value of Local Park Renovations," illustrate in the case of local public parks, and differences across locations (see A. Troy et al., "Predicting Opportunities for Greening and Patterns of Vegetation on Private Urban Lands"). Differences in household preferences for open space also play an important role. H. A. Klaiber and D. Phaneuf (in "Valuing Open Space in a Residential Sorting Model of the Twin Cities") use this insight to estimate the value that heterogeneous households assign to local open space amenities. Open space amenities may also convey multiple environmental, economic, and social benefits. For example, several papers have examined the role of urban tree canopy in promoting or deterring neighborhood safety and social order (see A. Troy et al. in "The Relationship Between Tree Canopy and Crime Rates Across an Urban–Rural Gradient in the Greater Baltimore Region" and "The Relationship Between Residential Yard Management and Neighborhood Crime: An Analysis from Baltimore City and County" and Branas et al. in "A Difference-in-Differences Analysis of Health, Safety, and Greening Vacant Urban Space"). These papers provide empirical evidence of the classic thesis by Jane Jacobs contained in *The Death and Life of Great American Cities* that suggests that factors such as trees could increase "eyes on the street" and thereby improve safety and social order.

CHAPTER 7. THE ROLE OF REGULATIONS AND NORMS IN LAND USE CHANGE

Evolving land use patterns and land use regulations are interdependent processes, each influencing the other. To understand the historical context of land use regulation in the Baltimore region, a wide variety of sources and data sets are available. Sherry Olson's seminal work *Baltimore: The Building of an American City* is essential reading and is undoubtedly the best place to begin an investigation into the city's past. Christopher G. Boone's article in the 2002 volume of *Historical Geography,* "Obstacles to Infrastructure Provision: The Struggle to Build Comprehensive Sewer Works in Baltimore," and Geoffrey L. Buckley's 2010 book *America's Conservation Impulse: A Century of Saving Trees in the Old Line State* cast light on Baltimore's efforts to build a sewer system, design and construct a network of parks, and expand and care for the city's tree canopy. Numerous scholars have written about the power and influence of homeowners' associations, especially when it comes to protecting the interests of property owners (see Mark Purcell's "Neighborhood Activism Among Homeowners as a Politics of Space" and Marcia England's "When 'Good Neighbors' Go Bad: Territorial Geographies of Neighborhood Associations"). Specific to Baltimore, G. Buckley and C. Boone's 2011 chapter "'To Promote the Material and Moral Welfare of the Community': Neighborhood Improvement Associations in Baltimore, Maryland, 1900–1945" uses the meeting minutes and promotional materials of several neighborhood improvement associations, including those of the Mount Royal Improvement Association, Mount Washington Improvement Associations, and the Peabody Heights Improvement Association, to acquire a better understanding of the role these groups played in guiding the allocation of city resources. Erin Pierce's 2010 master's thesis, "The Historic Roots of Green Urban Policy in Baltimore County, Maryland," offers a detailed account of the greater Baltimore area's post–World War II planning history and the development of the urban growth boundary. Subdivsion maps were derived from the Maryland State Archives online resource of www.plats.net.

Empirical evidence shows that zoning and other land use regulations can have substantial impacts on urban land use patterns, including the type and density of development. In many cases the research shows that greater restrictions on land development leads to both intended and unintended consequences, e.g., as Gnagey finds with the federal wetland policy in his paper "Buried Streams: An Analysis of Regulations and Land Development." D. Wrenn and E. Irwin show evidence of the unintended effects of regulatory delay in their paper "Time Is Money: An Empirical Examination of the Dynamic Effects of Regulatory Delay on Residential Subdivision Development." Likewise, D. Newburn and J. Ferris, in "The Effect of Downzoning for Managing Residential Development and Density," find that the unintended effects of downzoning in Baltimore County have increased low-density development. Nonetheless, the combined effect of all downzoning policies across multiple counties appears to have reduced the overall amount of leapfrog development (see Zhang et al., "Spatial Heterogeneity, Accessibility, and Zoning: An Empirical Investigation of Leapfrog Development") over time. An ongoing debate is whether regulatory policies or incentives are more effective in shaping development and redevelopment patterns. In his work "Urban Revitalization and Targeted Demolitions: How Do Housing Markets Respond to City-Led Redevelopment?" N. Irwin shows that incentives can be very effective and may minimize unintended effects. In all this work, BES researchers have made use of long timeseries data that provide detailed historical accounts of land use and regulation changes over time, including Maryland's Property View GIS database, as well as administrative records, e.g., on zoning regulations and demolition and renovation permits.

CHAPTER 8. HUMAN INFLUENCES ON URBAN SOIL DEVELOPMENT

Soil science is an old, well-conceptualized discipline and is well represented by the comprehensive and foundational framework introduced by Hans Jenny in 1941. On the other hand, urban soil research is relatively new to soil science and has generated a significant number of publications only during the past two decades. The term "urban soil" was first used by Zemlyanitskiy in 1963 to describe the characteristics of highly disturbed soils in urban areas. Urban soil was later defined by Craul in 1992 as "a soil material having a non-agricultural, man-made surface layer more than 50-cm thick that has been produced by mixing, filling, or by contamination of land surface in urban and suburban areas." The definition was derived from, and thus similar to, earlier definitions by Bockheim in 1974 and Craul and Klein in 1980. Since these earlier characterizations, Evans et al. suggested the term "anthropogenic soil," which places urban soils in a broader context of human-altered soils rather than limiting the definition to urban landscapes alone (see also review on anthropogenic soils by Capra et al.). In the same vein and to better incorporate a wider set of observations, Pouyat and Effland, Lehmann and Stahr, and recently Morel et al. more broadly defined urban soils to include not only those soils that are physically disturbed, such as those of vacant lots and landfills, but also those that are undisturbed yet altered by urban environmental change, e.g., the urban heat island effect. Finally, Huggett suggested that soils "evolve" through continual creation and destruction and thus can both progress or retrogress depending on changes that may occur in factors that affect soil development—which for this chapter includes anthropogenic and urban factors.

CHAPTER 9. APPLYING THE WATERSHED APPROACH TO URBAN ECOSYSTEMS

The watershed approach in ecosystem ecology is put in its largest context by Gene Likens in *The Ecosystem Approach: Its Use and Abuse.* Its explicitly urban application was conceptualized by S. Kaushal and K. Belt, in "The Urban Watershed Continuum: Evolving Temporal and Spatial Dimensions." The watershed has been a key tool for evaluating whole ecosystem function and response to disturbance and environmental change because it allows ecologists to quantify inputs and outputs of water, energy, nutrients, and carbon to hydrologically defined drainage basins, as exemplified by S. R. Carpenter in "The Need for Large-Scale Experiments to Assess and Predict the Response of Ecosystems to Perturbation," and P. M. Vitousek and J. M. Mellilo in "Nitrate Losses from Disturbed Forests: Patterns and Mechanisms." The watershed approach also has been fundamental to the development of theory about patterns of ecosystem development in time and space and about the regulation and maintenance of ecosystem function via mechanisms of resistance and resilience, as explained by E. P. Odum in *Fundamentals of Ecology,* 3rd edition, C. S. Holling in "Resilience and Stability in Ecological Systems," and G. E. Likens and F. H. Bormann in *Biogeochemistry of a Forested Ecosystem,* 2nd edition. Watershed-based approaches have been applied to fundamental studies of forest, grassland, wetland, desert, agricultural, and other ecosystem types, as illustrated by J. Ruegg et al., "Baseflow Physical Characteristics Differ at Multiple Spatial Scales in Stream Networks Across Diverse Biomes."

The watershed approach is a central component of BES. We hypothesized that the watershed approach would allow us to compare the novel urban, suburban, and exurban ecosystems of Baltimore with the fluxes of water, energy, carbon, and nutrients in natural ecosystems such as those at the Hubbard Brook Experimental Forest in New Hampshire, which are summarized by F. H. Bormann and G. E. Likens in *Patterns and Processes in a Forested Ecosystem* and by Likens and Bormann in *Biogeochemistry of a Forested Ecosystem*.

As is common in watershed studies, long-term monitoring data provide a signal of watershed ecosystem behavior that drives an iterative process of question generation, hypothesis testing, experimental work, model development, and comparative studies leading to comprehensive understanding of a complex system, as discussed by S. Carpenter in "The Need for Large-Scale Experiments to Assess and Predict the Response of Ecosystems to Perturbations." Such signals have been investigated by S. S. Kaushal and K. T. Belt in "The Urban Watershed Continuum," Kaushal et al. in "Longitudinal Patterns in Carbon and Nitrogen Fluxes and Stream Metabolism along an Urban Watershed Continuum," S. Schwartz and B. Smith in "Slowflow Fingerprints of Urban Hyrdology," B. Smith et al. in "Spectrum of Storm Event Hydrologic Response in Urban Watersheds," and K. G. Hopkins et al. in "Assessment of Regional Variation in Streamflow Responses to Urbanization and the Persistence of Physiography." The long-term nitrate signals we have observed in forested and agricultural watersheds are as expected, exemplified by the work of P. Gundersen et al. in "Leaching of Nitrate from Temperate Forests—Effects of Air Pollution and Forest Management" and by Webster et al. in "Evidence for a Regime Shift in Nitrogen Export from a Forested Watershed."

The nitrate signal from the agricultural watershed is also not a surprise, fueled as it is by high nitrogen inputs from fertilizer, such as the patterns documented by D. Boesch et al. in "Chesapeake Bay Eutrophication: Scientific Understanding, Ecosystem Restoration, and Challenges for Agriculture." W. Broussard and R. E. Turner, in "A Century of Changing Land-Use and Water-Quality Relationships in the Continental US," document concentrations of nitrate in agricultural streams throughout the eastern United States that are similar to those discovered in our agricultural stream in Baltimore.

CHAPTER 10. URBAN INFLUENCES ON THE ATMOSPHERES OF BALTIMORE,
MARYLAND, AND PHOENIX, ARIZONA
A fundamental reference to urban climate knowledge is provided by T. R. Oke, G. Mills, A. Christen, and J. Mills in the book *Urban Climates*. G. Heisler and A. Brazel present a summary of the urban heat island phenomenon in "The Urban Physical Environment: Temperature and Urban Heat Islands."

In "Climatology at Urban Long-Term Ecological Research Sites: Baltimore Ecosystem Study and Central Arizona-Phoenix," A. Brazel and G. Heisler described the major climate research goal of both BES and CAP, which was to develop and test methods to quantify the patterns, in both time and space, of urban influences on energy balances, air and surface temperatures, and solar irradiance. This was an important goal because, given the patterns and magnitude of the influences, decisions can be made on where efforts and resources should be applied in remediating adverse urban influences on climate in cities, suburbs, and exurbs.

Urban energy balances, and consequently air and surface temperatures, are strongly influenced by general climate, especially by precipitation. Quantification of the significantly different general climates of Baltimore and Phoenix is provided by a variety of sources, such as data that appear in *Climate Variability and Ecosystem Response at Long-Term Ecological Research Sites,* edited by D. Greenland, D. Goodin, and R. C. Smith. The Greenland et al. book includes a chapter by A. J. Brazel and A. W. Ellis, "The Climate of Central Arizona and Phoenix Long-Term Ecological Research Site (CAP LTER) and Links to ENSO." We also used an Internet source described by S. C. Sheridan in "The Redevelopment of Weather-Type Classification Scheme for North America" to picture the different weather types throughout an average year at the two cities. P. M. Groffman and his colleagues described, in "Ecological Homogenization of Urban USA," how despite large differences in climate and natural vegetation biomes most large cities have developed to be in many respects similar in structure, and this is true of Baltimore and Phoenix as well. However, though landscape form is more similar in these cities now than predevelopment, urban energy balances and heat island effects remain different, in large part owing to the differing moisture regimes as made evident by the S. C. Sheridan data, the book chapter by Brazel and Ellis, and by urban energy balance measurements such as those by W. T. L. Chow and colleagues in "Seasonal Dynamics of a Suburban Energy Balance in Phoenix, Arizona." The importance of acclimatization to general climates on health of people in Baltimore and Phoenix is apparent in "An Evaluation of Climate/Mortality Relationships in Large U.S. Cities and the Possible Impacts of a Climate Change" by L. S. Kalkstein and J. S. Green (1997).

A much-referenced paper by A. Brazel, N. Selover, R. Vose, and G. Heisler, "The Tale of Two Climates—Baltimore and Phoenix Urban LTER Sites," analyzed trends of urban heat island (UHI) intensity in the two cities during the twentieth century. By the end of the century Phoenix had a larger UHI at night than Baltimore, which is in part a function of the rural landscapes, as found in a Phoenix-area experiment by T. W. Hawkins and colleagues in "The Role of Rural Variability in Urban Heat Island Determination for Phoenix, Arizona."

In both CAP and BES remote sensing was used to analyze spatial patterns of landscape elements, including for Phoenix by S. Myint and colleagues, "Does the Spatial Arrangement of Urban Landscapes Matter? Examples of Urban Warming and Cooling in Phoenix and Las Vegas," and by W. Zhou, G. Huang, and M. L. Cadenasso for Baltimore: "Does Spatial Configuration Matter? Understanding the Effects of Land Cover Pattern on Land Surface Temperature in Urban Landscapes." These works illustrate complex patterns of urban land surface temperatures (LST) that are related to surface land cover ecology, composition, and configuration compared to rural environments. Methods of analyzing the pattern of urban surface cover are provided by

W. Zhou, M. L. Cadenasso, K. Schwarz, and S. Pickett in "Quantifying Spatial Hetero-geneity in Urban Landscapes: Integrating Visual Interpretation and Object-Based Classification" and by M. L. Cadenasso, S. T. A. Pickett, and K. Schwarz in "Spatial Heterogeneity in Urban Ecosystems: Reconceptualizing Land Cover and a Framework for Classification." G. M. Heisler, A. Ellis, D. J. Nowak, and I. Yesilonis used remotely sensed data to study air temperature patterns at high resolution in "Modeling and Imaging Land-Cover Influences on Air Temperature In and Near Baltimore, MD."

Most research on urban climate has concentrated on warm-season conditions. An exception is a CAP study of fifty-year temperature trends in Phoenix by L. A. Baker and colleagues, "Urbanization and Warming of Phoenix (Arizona, USA): Impacts, Feedbacks and Mitigation," which examined trends in energy used for heating and cooling of buildings as affected by urban climate changes.

A series of studies have been produced by CAP climate scientists, sometimes in collaboration with City of Phoenix personnel that were aimed at developing manage-ment decisions to alleviate extreme temperatures and their effects. These include "Urban Heat Island Research in Phoenix, Arizona: Theoretical Contributions and Policy Applications" by W. T. L. Chow, D. Brennan, and A. J. Brazel; "Using Watered Landscapes to Manipulate Urban Heat Island Effects: How Much Water Will It Take to Cool Phoenix?" by P. Gober and colleagues; and "Impact of Urban Form and De-sign on Mid-Afternoon Microclimate in Phoenix Local Climate Zones" by A. Middel, A. K. Häbba, A. J. Brazel, C. A. Martin, and S. Guhathakurta. In "Microclimates in a Desert City Were Related to Land Use and Vegetation Index," L. Stabler, C. A. Martin, and A. Brazel found that effective management for urban temperature reduction varies with existing land use.

Both BES and CAP research involved social scientists in developing methods to prioritize locations for urban heat island remediation for especially vulnerable populations. Valuable examples of the socioeconomic nexus with climate processes in Phoenix include S. H. Harlan and colleagues' "Neighborhood Microclimates and Vul-nerability to Heat Stress" and G. D. Jenerette and colleagues' "Regional Relationships Between Surface Temperature, Vegetation, and Human Settlement in a Rapidly Ur-banizing Ecosystem." Similarly, G. Huang, W. Zhou, and M. L. Cadenasso dealt with the question "Is Everyone Hot in the City? Spatial Pattern of Land Surface Tempera-tures, Land Cover and Neighborhood Socioeconomic Characteristics in Baltimore City, MD." Another health-related issue is exposure to ultraviolet radiation. G. M. Heisler, R. H. Grant, W. Gao, and J. R. Slusser reported on the effect of the Baltimore atmo-sphere on incidence of ultraviolet radiation in "Solar Ultraviolet-B Radiation in Urban Environments: The Case of Baltimore, MD." The effect of tree cover on exposure of pedestrians in different land use types in Baltimore is explored in "Effect of Cloud Cover on UVB Exposure Under Tree Canopies: Will Climate Change Affect UVB Ex-posure?" by R. H. Grant and G. M. Heisler.

The potential effect of urban areas on global-scale climate change depends in part on their effect on the net emission of CO_2. B. Crawford, C. S. B. Grimmond, and A. Christen reported these emissions in Baltimore as measured at a tall tower in "Five Years of Carbon Dioxide Flux Measurements in a Highly Vegetated Suburban Area."

In 2009 A. J. Brazel and G. H. Heisler, in "Climatology at Urban Long-Term Ecological Research Sites: Baltimore Ecosystem Study and Central Arizona-Phoenix," concluded that the BES and CAP LTER urban sites should remain excellent laboratories in which to refine concepts in urban ecology and climatology and links to social, political, economic, and ecological processes over a long time period.

CHAPTER 11. RESPONSES OF BIODIVERSITY TO FRAGMENTATION AND MANAGEMENT
Leibold et al., in "The Metacommunity Concept: A Framework for Multi-Scale Community Ecology," describe biodiversity as the balance between local environmental constraints and regional dispersal patterns, which helps ecologists understand how patterns in species associations occur locally, regionally, and between habitats. Whittaker's classic paper "Vegetation of the Siskiyou Mountains, Oregon and California" conceptualized biodiversity as having a regional species pool, or gamma (γ) diversity, composed of local, or alpha (α), diversity within specific habitats and compositional turnover, or beta (β) diversity, between different habitats. In urban ecosystems such local and regional effects become exacerbated due to the substantial spatial heterogeneity driven by human management and interventions, as pointed out by Cadenasso et al. in "Spatial Heterogeneity in Urban Ecosystems: Reconceptualizing Land Cover and a Framework for Classification." Three explanations for how biodiversity is maintained in nonurban systems have been invoked for the urban setting. Gaston, in "Biodiversity and Extinction: Species and People," suggests that since cities are very productive they should attract more species. The intermediate disturbance hypothesis, described in Connell's seminal "Diversity in Tropical Rain Forests and Coral Reefs," suggests species diversity should peak at intermediate levels of disturbance, as both weakly competitive but highly mobile species should coexist with strong competitors that lack the ability to withstand disturbance. Lastly, Rapport's work "Ecosystem Behavior Under Stress" points out that ecosystem stress is hypothesized to result in a decline in richness as humans impose a harsh geophysical environment. Together, though, these phenomena strive to describe patterns at only the local scale, not regionally or between habitats.

To move the field forward, studies of community assembly in urban ecosystems are embracing theories applied in nonurban ecosystems. Approaches proposed in Webb's "Phylogenies and Community Ecology" and in Knapp et al.'s "Challenging Urban Species Diversity: Contrasting Phylogenetic Patterns Across Plant Functional Groups in Germany" suggest investigating biological structure beyond the taxonomic level, extending to the ecological trait and phylogenetic levels. Such approaches may aid in our understanding of how shifts in the regional species pool, as Pyšek offers in "Alien and Native Species in Central European Urban Floras: A Quantitative Comparison," which is what often happens in urban areas with species introductions. Colautti et al. provide, in "Propagule Pressure: A Null Model for Biological Invasions," a relevant example of how human preferences for specific floral displays of nonnative ornamental species can shift the urban regional species pool composition. Nassauer et al. goes on, in "What Will the Neighbors Think? Cultural Norms and Ecological Design," to describe at what spatial scale such preferences should be managed.

Biotic homogenization, as defined by McKinney and Lockwood in "Biotic Homogenization: A Few Winners Replacing Many Losers in the Next Mass Extinction," is the reduction in biodiversity due to an increase in the similarity of species composition as a result of both the introduction of nonnative species and the extinction of endemic species. McKinney further explains in "Urbanization as a Major Cause of Biotic Homogenization" that urbanization is the primary driver of the pattern. As an example, Pautasso's work in "Scale Dependence of the Correlation between Human Population Presence and Vertebrate and Plant Species Richness" shows this for plants. Knapp et al.'s work, mentioned above, attempts to provide a mechanism to explain homogenization by demonstrating shifts in functional trait distributions between urban and rural plant communities. Further work by Johnson et al., in "Human Legacies Differentially Organize Functional and Phylogenetic Diversity of Urban Herbaceous Plant Communities at Multiple Spatial Scales," studied how variation in human legacy effects in vacant lots drove shifts in herbaceous plant community structure.

CHAPTER 12. LAWNS AS COMMON GROUND FOR SOCIETY AND THE FLUX OF WATER AND NUTRIENTS

Lawns and residential lands are two of the most obvious components of urban ecosystems. While Cadenasso et al., in "Spatial Heterogeneity in Urban Ecosystems: Reconceptualizing Land Cover and a Framework for Classification," and Byrne, in "Habitat Structure: A Fundamental Concept and Framework for Urban Soil Ecology," approach urban ecosystems as heterogeneous and variable mixes of paved surfaces, trees, shrubs, and grass, Blanco-Montero et al., in "Potential Environmental and Economic Impacts of Turfgrass in Albuquerque, New Mexico (USA)," and Robbins and Birkenholtz, in "Turfgrass Revolution: Measuring the Expansion of the American Lawn," quantify grass itself as a dominant land cover, representing 20–30 percent of typical residential parcels. There are over 150,000 km² of lawns in the United States, according to Milesi et al.'s "Mapping and Modeling the Biogeochemical Cycling of Turf Grasses in the United States."

Early ideas about lawns in BES were rooted in theory and data relevant to highly disturbed, heavily managed ecosystems with significant inputs and outputs to surrounding environments, as articulated in Odum's classic textbook *Fundamentals of Ecology*, and, specifically relevant to agricultural ecosystems, by Robertson et al.'s "Long-Term Ecological Research in a Human-Dominated World." Early ecological analyses of lawns focused on concerns about their environmental performance. Work by Falk in "Energetics of a Suburban Lawn Ecosystem" and "The Primary Productivity of Lawns in a Temperate Environment" as well as Bormann et al.'s *Redesigning the American Lawn* and two works by Robbins and colleagues, "Lawns and Toxins" and "Turfgrass Revolution" focused especially on outputs of nutrients and pesticides and intensive use of energy and water. Bormann et al.'s *Redesigning the American Lawn* and Robbins' *Lawn People* also addressed some of the social-ecological aspects of lawns and residential lands with ideas about the benefits (ecosystem services) that people derive from lawns and some of the philosophical, emotional, social, and political motivations behind the establishment and maintenance of lawns.

More recently, BES has attempted to use theories and concepts from grasslands and rangelands in our studies of lawns, developing the term "urban grasslands," which is defined in Groffman et al., "Nitrate Leaching and Nitrous Oxide Flux in Urban Forests and Grasslands" and in Thompson and Kao-Kniffien's "Diversity Enhances NPP, N Retention, and Soil Microbial Diversity in Experimental Urban Grassland Assemblages."

While ecologists have paid some attention to lawns over the past fifty years, other disciplines have paid much greater academic and commercial attention to lawns. This body of work is represented by Morton and colleagues in "Influence of Overwatering and Fertilization on Nitrogen Losses from Home Lawns," Gold and colleagues in "Nitrate-Nitrogen Losses to Groundwater from Rural and Suburban Land Uses," and Petrovic in "The Fate of Nitrogenous Fertilizers Applied to Turfgrass." Thus two very different threads of theory and practice, one from basic ecology and one from applied turfgrass science, have provided a foundation for lawn research in BES.

Novel methods have been required in BES to examine the social-ecological dynamics of lawns and residential lands because of the fine-scale spatial heterogeneity of urban areas as conceptualized in Cadenasso et al., "Spatial Heterogeneity in Urban Ecosystems." These methods have included a combination of intensive surveys of household perceptions, attitudes, and behaviors; improving the characterization of urban heterogeneity patterns using remote sensing and geographic information systems as described and quantified in papers by Zhou et al., including "Modeling Residential Lawn Fertilization Practices," "Object-Based Land Cover Classification in Shaded Areas," "Mapping Urban Landscape Heterogeneity," "90 Years of Forest Cover Change," and "Quantifying Spatial Heterogeneity in Urban Landscapes," and by Zhou and Cadenasso, "Effects of Patch Characteristics and Within Patch Heterogeneity"; and integration with existing administrative and marketing data (see Grove et al., *The Baltimore School of Urban Ecology*).

CHAPTER 13. LONG-TERM TRENDS IN URBAN FOREST SUCCESSION

This chapter is influenced by many excellent volumes that describe the interactions among changes in forests, water, and humans over time. Hugh Raup's 1966 "The View from John Sanderson's Farm" detailed the effects of economic pressures on forest succession and how those pressures dictated human activities. William Cronon's *Changes in the Land: Indians, Colonists, and the Ecology of New England* documents the influence of two cultures on the land, and his *Nature's Metropolis* explains the dependence and influence of an urban economy on rural land. In *The Big Muddy* Christopher Morris chronicles the effects of the conversion of Mississippi marshes to agricultural land and the introduction of exotic species on the Gulf of Mexico. The many papers in Foster and Aber's marvelous book *Forests in Time* describe changes primarily in the Harvard Forest brought about by generations of humans. Gordon Whitney's *From Coastal Wilderness to Fruited Plain: A History of Environmental Change in Temperate North America from 1500 to the Present* contains lists of species for different parts of the eastern United States before and after European colonization. Going back further in history, Charles Mann wrote about connections between people and the land in the 1400s.

The chapter leans more specifically on changes in land and water recorded in sediments deposited over time periods ranging from a century to millennia. Local forest histories, including Besley's *The Forests of Maryland*, Shreve's *The Plant Life of Maryland*, and Wayne Tyndall's papers on the serpentine barrens, are compared with records of plant populations during climate shifts throughout the Holocene. The paleo-ecological studies, based on pollen and seeds of plants preserved in sediments deposited in lakes, bogs, and estuaries, were pioneered in North America by Ed Deevey, Margaret Davis, and Herb Wright. Their studies, among others, revolutionized theories of plant stability from one of little or no change over long periods of time to a dynamic system with the capability of rapid response to change. William Mayre's descriptions of the original drainage of the eastern Piedmont Barrens and Lois Green Carr's historical documents describing how a wetland was drained to plant crops explained the rapid shift in the 1700s from pollen of wet to dry herbs in Chesapeake sediment cores. The chapter also draws on studies by Stoermer and others, who showed, using the fossil record of organisms and chemicals in sediment cores from the Great Lakes, how rapidly algal systems respond to human-caused changes such as adding excess nutrients in the form of fertilizers and sewage into lake waters.

John Goodlett, "Vegetation Adjacent to the Border of the Wisconsin Drift in Potter County Pennsylvania," as well as his unpublished maps of the distribution of oak species in New England, influenced the design of Brush et al.'s study "The Natural Forests of Maryland," showing a close relationship with geologic substrate and species distributions. Forrest Shreve in *The Plant Life in Maryland*, cited above, recorded species growing in the late 1800s on abandoned fields underlain by different soils. Brush et al., in their 1980 Ecological Monographs paper, showed that species similar to those recorded by Shreve were occupying similar substrates three-quarters of a century later in mid-Atlantic Maryland.

Studies on the effect of land use on the reduction of forest cover and forest succession drew on earlier publications by Pickett and White, where the concept of patch dynamics was introduced. And, finally, much motivation for these retrospective studies based both on history and paleoecology has been inspired by readings from George Perkins Marsh's *Man and Nature*.

CHAPTER 14. PRINCIPLES OF URBAN ECOLOGICAL SCIENCE

Principles are used to achieve two goals in this chapter. First, principles are conceptual components of theory, and we use them here to contribute to the formulation of a general theory of urban ecology. Second, because principles are abstractions, idealizations, and summarization we use them here to synthesize empirical research conducted as part of BES. The role of principles in theory development is informed by S. Pickett et al., *Ecological Understanding: The Nature of Theory and the Theory of Nature*, and we draw on earlier attempts to summarize urban ecological principles from BES research, including M. Cadenasso and S. Pickett's "Urban Principles for Ecological

Landscape Design and Management: Scientific Fundamentals" and S. Pickett and M. Cadenasso's "How Many Principles of Urban Ecology Are There?" The chapter is organized into thirteen principles, each of which is described in general terms invoking fundamental literature sources followed by examples from specific empirical research conducted as part of BES. We organize this bibliographic essay similarly.

Principle 1 states that cities are human ecosystems in which social and ecological processes feed back into one another. An ecosystem was defined as an ecological unit of study by A. Tansley in "The Use and Abuse of Vegetation Concepts and Terms" as the interaction between biotic and abiotic complexes. To apply this concept to urban systems, this definition was expanded to include two additional complexes—built and social—as described by S. Pickett and J. M. Grove in "Urban Ecosystems: What Would Tansley Do?" We used two examples of empirical work from BES to illustrate Principle 1. The first, by A. Johnson et al., "Human Legacies Differentially Organize Functional and Phylogenetic Diversity of Urban Herbaceous Plant Communities at Multiple Spatial Scales," investigated the differences in plant communities inside and outside the footprints of demolished buildings. The second example quantified the feedback between land surface temperature and patterns of land cover (W. Zhou et al., "Does Spatial Configuration Matter? Understanding the Effects of Land Cover Pattern on Land Surface Temperature in Urban Landscapes") and between land surface temperature and a suite of variables describing the human population (G. Huang et al., "Is Everyone Hot in the City? Spatial Pattern of Land Surface Temperatures, Land Cover and Neighborhood Socioeconomic Characteristics in Baltimore, MD").

Principle 2 recognizes the ecological functioning of remnant and newly emergent patches within urban areas. This principle was illustrated by research from P. Groffman et al., in "Down by the Riverside: Urban Riparian Ecology," which focused on remnant riparian zones, and S. Raciti et al., in "Nitrate Production and Availability in Residential Soils," which investigated the functioning of residential lawns, an example of a newly emergent patch type. Both remnant and newly emergent patches contain a diverse assemblage of flora and fauna (Principle 3). The work by A. Johnson et al., referred to under Principle 1, focused on plant diversity and two additional examples that focused on birds were also used. C. Nilon, in "Urban Biodiversity and the Importance of Management and Conservation," determined that common bird species occupied distinct urban habitats, and C. Carlson, as part of her thesis research ("The Relationship Between Breeding Bird Community Structure in Urban Forest Patches and the Human-Mediated Resources in the Surrounding Residential Matrix"), concluded that the species richness of birds occupying remnant patches of woodlands in the City of Baltimore was influenced by the vegetation structure of adjacent residential parcels.

Human values, perceptions, and differential exposures to ecosystem benefits and burdens are captured in Principles 4 and 5. Because the definition of ecosystem services is the benefits humans derive from ecological systems, it necessarily considers the values and perceptions people have about environmental structures and processes. M. Battaglia et al., in "It's Not Easy Going Green: Obstacles to Tree-Planting Programs in East Baltimore," highlight that not all residents view a tree canopy as an unmitigated good. The perception, however, that trees may obscure visibility in neighborhoods and thereby make the neighborhoods less safe was refuted by A. Troy et al., in "The Relationship Between Tree Canopy and Crime Rates Across an Urban–Rural Gradient in the Greater Baltimore Region." The differential exposure to environmental benefits and burdens across the urban landscape necessarily invokes concern for environmental justice. Research by C. Boone focused on the relationship between the distribution of toxic release inventory sites and neighborhood racial demographics, as reported in "An Assessment and Explanation of Environmental Inequity in Baltimore," and differential access to parks in Baltimore, as quantified in C. Boone et al., "Parks and People: An Environmental Justice Inquiry in Baltimore, Maryland." How the perceptions of parks as amenities or disamenities influence the value of homes in proximity to parks was investigated by A. Troy and J. M. Grove, "Property Values, Parks, and Crime: A Hedonic Analysis in Baltimore, MD."

Principles 6, 7, and 8 address three aspects of the urban form: that it varies in space, reflects a multitude of decisions, and changes through time. From its beginning BES has been concerned with quantifying the structure of the urban system and understanding how that structure came to be, how it changes, and how it influences ecological and social processes. Spatial heterogeneity is of fundamental importance in the field of ecology, as evidenced by several important works such as that by J. Kolasa and S. Pickett, *Ecological Heterogeneity*, M. Turner, *Landscape Heterogeneity and Disturbance*, and "Landscape Ecology: The Effect of Pattern on Process," S. Pickett and M. Cadenasso, "Landscape Ecology: Spatial Heterogeneity in Ecological Systems," and J. Wiens, "Ecological Heterogeneity: An Ontogeny of Concepts and Approaches." BES has characterized the heterogeneity of Baltimore by considering social differentiation using the PRIZM classification provided by Claritas, *PRIZM Cluster Snapshots; Getting to Know the 62 Clusters*. In addition, a new land cover classification named HERCULES was developed to describe the biophysical heterogeneity of Baltimore, and the conceptual underpinnings of HERCULES are described in M. Cadenasso et al., "Spatial Heterogeneity in Urban Ecosystems: Reconceptualizing Land Cover and a Framework for Classification," and W. Zhou et al. presents the rule set used to generate the HERCULES classification in "Quantifying Spatial Heterogeneity in Urban Landscapes: Integrating Visual Interpretation and Object-Based Classification." A new conceptual and visual explanation of HERCULES appears in Marhsall et al., *Patch Atlas*.

We draw on two primary sources to describe the history of the built form of Baltimore: M. Hayward's *Baltimore's Alley Houses: Homes for Working People Since the 1780s* and M. Hayward and C. Belfoure's *The Baltimore Rowhouse*. C. Boone's research on the impact of the cholera epidemic on the dispersion of people from low-lying areas in Baltimore to higher ground is reported in "An Assessment and Explanation of Environmental Inequity in Baltimore." The relationship between social and biophysical heterogeneity is explored by J. M. Grove et al. in "Data and Methods Comparing Social Structure and Vegetation Structure of Urban neighborhoods in Baltimore, Maryland," and work by Boone et al., "Landscape, Vegetation Characteristics, and Group Identity in an Urban and Suburban Watershed: Why the 60s Matter," demonstrates the often temporally lagged nature of social and biophysical heterogeneities such that at any one time period the vegetation structure may be more closely related to the social characteristics of time past rather than to current characteristics.

Urban form is created and modified by urban designers, and the ecological and social outcomes of those designs can be measured (Principle 9). The use of design as a mechanism of experimentation was introduced to ecologists by A. Felson and S. Pickett in "Designed Experiments: New Approaches to Studying Urban Ecosystems" and to design practitioners by A. Felson in "The Design Process as a Framework for Collaboration between Ecologists and Designers." In "Promoting Earth Stewardship through Urban Design Experiments," A. Felson et al. demonstrate the utility of the approach for research. BES has employed this strategy by quantifying the ecological impact of a series of interventions designed to affect stormwater management in a small residential watershed (see G. Hager et al., "Socioecological Revitalization of an Urban Watershed").

Principles 10 and 11 address different methodological issues of conducting research in urban systems. Because these systems are defined as ecosystems, boundaries to determine what is internal and external to the system must be defined, but boundaries are also based on specific research questions. The small-watershed approach pioneered at the Hubbard Brook Long-Term Ecological Research site (see G. Likens et al., *Biogeochemistry of a Forested Ecosystem*) was adopted by the BES, and the Gwynns Falls watershed was selected as the primary research watershed for the program, as described by S. Pickett et al. in "Watersheds in Baltimore, Maryland: Understanding and Application of Integrated Ecological and Social Processes." A second methodological precursor to BES is the concept of the urban-rural gradient, first introduced by M. McDonnell and S. Pickett in "Ecosystem Structure and Function along Urban-Rural Gradients: An Unexploited Opportunity for Ecology" and updated by M. McDonnell and A. Hahs in "The Use of Gradient Analysis Studies in Advancing Our Understanding of the Ecology of Urbanizing Landscapes: Current Status and Future Directions." The gradient approach has its roots in the foundational work of R. Whittaker, "Vegetation of the Great Smoky Mountains," and is considered a key concept in the field of ecology, as demonstrated by its inclusion as a chapter by G. Fox et al., "Ecological Gradient Theory: A Framework for Aligning Data and Models," in the edited volume by S. Scheiner and M. Willig, *The Theory of Ecology*.

The final two principles look outward, connecting the urban system with other systems and regions that may or may not be immediately adjacent. Principle 12 focuses on the expansion of urban land into wildlands, which has been quantified by the work of V. Radeloff and colleagues in "The Wildland Urban Interface in the United States." An influential book by J. Garreau, *Edge City: Life on the New Frontier*, focuses on transportation infrastructure as nodes of development and eventual expansion of development. Policy plays an important role in regulating, or not, outward expansion, and, in the Baltimore region, research by E. Irwin and colleagues aims to model such expansion and its drivers. For example, see D. Wren and E. Irwin, "Time Is Money: An Empirical Examination of the Effects of Regulatory Delay on Residential Subdivision Development." The flux of water also connects urban areas with surrounding systems, which is the topic of Principle 13. As discussed under Principle 10, the watershed concept is a key theoretical construct for BES that guides much empirical research on water quality and quantity in the Gwynns Falls watershed as an integrated measure of both terrestrial and aquatic dynamics. BES research has focused on riparian function as cited under Principle 10 but also on pollution from personal care products (E. Rosi-Marshall et al., "Pharmaceuticals Suppress Algal Growth and Microbial Respiration and Alter Bacterial Communities of Stream Biofilms") and road salt (S. Kaushal et al., "Increased Salinization of Fresh Water in the Northeastern United States") as two examples.

CHAPTER 15. LESSONS LEARNED FROM THE ORIGIN AND DESIGN OF LONG-TERM SOCIAL-ECOLOGICAL RESEARCH

The origins for the *Baltimore School of Urban Ecology* (Grove et al. 2015) came from several sources. Early work in social forestry, captured in Khon Kaen University's *Proceedings of the 1985 International Conference on Rapid Rural Appraisal* and Burch and Parker's (1992) book on social science for agroforestry, provided examples of frameworks, concepts, and approaches that Burch and Grove (1993, 1996) transferred to urban areas. Grove and Burch wrote an urban community forestry publication for the Forest Service (Grove et al. 1993) and a more research-oriented publication (1997), which gave both a practical and research orientation for the work in Baltimore. Later, Grimm and others (2000) would discuss the urban LTER approach, soon after Baltimore and Phoenix were established through the NSF's LTER program.

The ecological roots for the Baltimore School can be found both in rural, watershed approaches described by Bormann and Likens (1979) and Black (1991) and a concentrated focus on urban ecology and urban ecosystems, which dates to Stearns and Montag's (1974) book, *The Urban Ecosystem: A Holistic Approach*. Later, the Cary Institute's URGE team developed an urban ecology program in the New York City metropolitan region (McDonnell et al., 1993) using an urban-rural gradient approach. Pickett et al. (1997), who was a member of the URGE team, expanded on the urban-rural gradient approach to advance a mosaic perspective, which conceptually extended urban ecology beyond an ecology *in* cities to an ecology *of* cities. The ecology *of* cities perspective built on the patch dynamics approach from ecology (Pickett and White, 1985) and the social areas approach (Shevky and Bell, 1955) from sociology. The political scientist Donald E. Stokes identified use-inspired basic research as a major scientific goal in his 1997 book *Pasteur's Quadrant: Basic Science and Technological Innovation*.

CHAPTER 16. ECOLOGICAL URBAN DESIGN

The literature cited in this chapter presents a critical view of traditional urban design practice as it intersects with theories of social justice and ecosystem science. Baltimore was an important locus for theorizing the spatial injustice of North American urban policy, especially in the post–World War II period. The basis for a social justice critique of normative urban design practice lies in two important urban geographers at Johns Hopkins University in the 1970s and 1980s: David Harvey and Sherry Olson. Harvey's *Social Justice and the City* (1973) is a seminal theoretical text in the field of urban justice, based on empirical rent data for inner-city Baltimore (Harvey 1974a and 1974b). His critique is based on the theory of the metabolic rift between urban and rural as well as the worker and the land as elaborated by J. B. Foster (1999). While Harvey's texts may have had a wider impact, Olson, at that time also on the faculty at Johns Hopkins, provides more expansive evidence of the uneven social impacts of sex, race, class, and ethnic segregation in the city (1976).

The development of normative urban design theory as a challenge to modernist planning orthodoxy is presented here as articulated by Kevin Lynch (1960), Jane Jacobs (1961), Jonathan Barnett (1982), Ian McHarg (1969), and David Grahame Shane (2005, 2011). Sydney Wong (2013) has related the struggle as urban designers attempted to "fix" massive urban renewal projects in postwar Baltimore City, while D. C. Outen (2007) described the land use controls set in place in suburban Baltimore County in the late 1960s. William Burch provided a new model for social-ecological planning for parks in Baltimore (1991) and the human ecosystem approach, as elaborated with his coauthors G. E. Machlis, and J. E. Force (1997).

The ecological theory of patch dynamics served as the theoretical framework to expand urban design theory through ecological and social research in academic urban design studios (Pickett et al. 1999 and 2004; Pickett and Cadenasso 2007b; McGrath et al. 2007, Shane 2013). The studios theorized adaptive cycles in urban design through new digital tools for mapping cities (B. McGrath 1994, 2000, 2008). Theories of social ecological models (Grove et al 2006a, 2006b, and 2007) were used in the studios in order to design for implementing change. Integrative approaches to urban ecology and design using boundaries as landscape components that manage flows of organisms, nutrients, and information in cities were introduced by Cadenasso et al. (2006, 2007b, and 2008), while watershed frameworks were central to reframing urban design theory (Pickett et al. 2007a, Hager et al. 2013). Ecological urban design was theorized as a form of resilience, as articulated by J. Wu and T. Wu (2013), S. T. A. Pickett and Cadenasso (2008), and Pickett et al. (2013, 2014).

Background for this chapter is a decade of research on new forms of ecological urban design praxis, including an integrated land cover classification of urban spatial heterogeneity by Cadenasso et al. (2007a), Pickett and Cadenasso (2008), and Cadenasso (2013a). Literature on linking spatial heterogeneity to the social-ecological functioning and design of urban systems includes Cadenasso and Pickett (2007b), Cadenasso (2013a), and Cadenasso et al (2013b); in particular, the working of feedback loops of ecological and social cooperation, as articulated by Pickett et al. (2005), Pickett and Cadenasso (2008), and Cadenasso et al. (2008). Guida (2012) and Torre (1981) have presented more open and flexible ideas for architectural and urban space as matrix rather than rooms and corridors. Mitrasinovic (2015) has articulated new methods of inclusive urban design praxis further developed as the metacity framework by McGrath (2013), McGrath and Pickett (2011 and 2016), and Pickett et al. (2013). An action-based ecology for the city has been developed by Childers et al. (2015) and an ecological urban design, choice-oriented process by Pickett et al. (2013).

CHAPTER 17. TEACHING AND LEARNING THE BALTIMORE ECOSYSTEM

The BES approach to teaching the urban ecosystem leans heavily on that of G. E. Likens, *The Ecosystem Approach: Its Use and Abuse,* and the writings of Pickett and coworkers. The early work on urban–rural gradients by M. McDonnell et al., "Ecosystem Processes Along an Urban-to-Rural Gradient," and, more recently, on a framework of ecology IN, OF, and FOR cities by S. T. A. Pickett et al., "Evolution and Future of Urban Ecological Science: Ecology In, Of, and For the City," are particularly noteworthy. BES emphasis on going beyond negative stereotypes of cities, as presented eloquently by S. T. A. Pickett et al., "Beyond Urban Legends: An Emerging Framework of Urban Ecology, as Illustrated by the Baltimore Ecosystem Study," helps frame BES education's approach, just as the idea of management as experiment of A. Felson et al., "Promoting Earth Stewardship Through Urban Design Experiments," inspires our partnership work with schools and communities. Many chapters in the 2003 book based on the 1999 Cary Conference, A. R. Berkowitz et al., *Understanding Urban Ecosystems,* provided early guidance and inspiration. Seminal research papers from BES, including those on urban soils by R. Pouyat et al., "Chemical, Physical, and Biological Characteristics of Urban Soils," energy use by A. Troy, *The Very Hungry City: Urban Energy Efficiency and the Economic Fate of Cities,* watershed and stream biogeochemistry by P. Groffman et al., "Nitrogen Fluxes and Retention in Urban Watershed Ecosystems," and S. Kaushal and K. T. Belt, "The Urban Watershed Continuum: Evolving Spatial and Temporal Dimensions," and environmental justice by Boone et al., "A Long View of Polluting Industry and Environmental Justice in Baltimore," have been invaluable.

A rich diversity of oft-disparate visions for ecology and environmental literacy provide the foundation of BES education. The term "environmental literacy" was coined by Charles "Chuck" Roth fifty years ago in his essay "On the Road to Conservation," published in the magazine of the Massachusetts Audubon Society. The first substantive discussion of ecological literacy was in the presidential address of Paul Risser, "Ecological Literacy," presented to the members of the Ecological Society of America (ESA) in 1986. Since then, David Orr's *Ecological Literacy* and Fritjof Capra's *The Web of Life: A New Scientific Understanding of Living Systems,* in which the term "ecoliteracy" was first used, continued the already long-standing tradition of diverse approaches to framing the purpose, mission, and methods of the environment-related enterprise. The history of the evolution of ideas in this arena and substantive similarities and differences are summarized by B. B. McBride et al. in "Environmental Literacy, Ecological Literacy, Ecoliteracy: What Do We Mean and How Did We Get Here?"

Closer to home, members of the ESA built on Risser's inspiration and model with a series of efforts to define ecological literacy for the nascent field of ecology education, including K. Klemow's "Basic Ecological Literacy: A First Cut," A. R. Berkowitz's "Defining Environmental Literacy: A Call for Action," A. R. Berkowitz et al., "A Framework for Integrating Ecological Literacy, Civics Literacy and Environmental Citizenship in Environmental Education," and R. Jordan et al., "What Should Every Citizen Know About Ecology?" ESA educators also provided valuable scientific information to the North American Association of Environmental Education (NAAEE), *Excellence in Environmental Education: Guidelines for Learning (K-12).* The results of a survey on ecological literacy to which over one thousand ESA members responded, which emphasized the importance of human dependency and impacts on the environment, summarized in B. B. McBride's dissertation "Essential Elements of Ecological Literacy and Pathways to Achieve it: Perspectives of Ecologists," provide added impetus to the shift from "natural ecosystems" to "social-ecological systems" as a focus of twenty-first-century ecology and environmental education.

A. K. Poole et al., "A Call for Ethics Literacy in Environmental Education," presents a strong case for making the understanding of ethical thinking a target of instruction, an important enhancement of our earlier view of environmental citizenship. B. Covitt et al.'s "The Role of Practices in Scientific Literacy" makes a strong case for interweaving practices into our view of science literacy, while a strong argument is made for the importance of computational thinking by S. Grover and R. Pea, "Computational Thinking: A Competency Whose Time Has Come," J. Wofford, "K–16 Computationally Rich Science Education: A Ten-Year Review of the Journal of Science Education and Technology (1998–2008)," and P. Sengupta et al., "Integrating Computational Thinking with K-12 Science Education Using Agent-Based Computation: A Theoretical Framework."

A number of approaches have been taken to building on and blending science and environmental literacy for a distinctively urban ecosystem literacy. Several chapters in the book from the 1999 Cary Conference, A. R. Berkowitz et al., *Understanding Urban Ecosystems*, stand out in this regard. These include R. Bybee, "Integrating Urban Ecosystem Education into Educational Reform," B. Simmons, "An Interdisciplinary Approach to Urban Ecosystems," S. Keiny, "'Ecological Thinking' as a Tool for Understanding Urban Ecosystems: A Model from Israel," and C. Fialkowski, "Approaches to Urban Ecosystem Education in Chicago: Perspectives and Processes from an Environmental Educator." Our vision for the future of urban ecosystem education was laid out in a chapter by A. R. Berkowitz et al., "Urban Ecosystem Education in the Coming Decade: What Is Possible and How Can We Get There?" Marianne Krasny and colleagues at the Civic Ecology Group at Cornell (https://civicecology.org/outreach/urbanee/) have a rich collection of publications, including the recent edited volume K. G. Tidball and M. E. Krasny, *Urban Environmental Education Review,* and K. G. Tidball and M. E. Krasny, "Urban Environmental Education from a Social-Ecological Perspective: Conceptual Framework for Civic Ecology Education."

Both the goals of our work for student learning and achievement and our vision for effective pedagogy have been shaped largely by national and state standards. The 1990s produced both the National Research Council's *National Science Education Standards* and *Benchmarks for Science Literacy* from the American Association for the Advancement of Science's Project 2061. Each identified important big ideas in ecology, placed a strong emphasis on inquiry and learning-by-doing, and argued for the development of students' deeper understanding over time. The AAAS's *Atlas for Science Literacy* provides invaluable guidance on expected pathways of increasing sophistication and for linking across topics. More recently, the National Research Council produced *A Framework for K-12 Science Education: Practices, Crosscutting Concepts, and Core Ideas,* representing a significant advancement in promoting depth over breadth of coverage. The *Framework* and NGSS Lead States' *Next Generation Science Standards,* published the following year, together provide very compelling and nearly nationwide guidance and inspiration for curriculum and instruction. These identify a smaller set of big ideas for students to learn, while adding new topics like engineering and placing new emphasis on old ones, such as Earth and space science, and widen the scope of science practices to embrace quantitative and computational thinking, data sensemaking, and modeling. Finally, Maryland State's *Maryland Standards State Science Curriculum* and *Maryland Environmental Literacy Standards* are important resources and inspirations for environmental science education in the Baltimore and the region.

The constructivist theory of learning first developed by L. S. Vygotsky in two seminal books, *Thought and Language* and *Mind in Society: The Development of Higher Psychological Processes,* serves as the foundation for much current education theory and practice. G. J. Posner's "Accommodation of a Scientific Conception: Toward a Theory of Conceptual Change," G. D. Hendry's "Constructivism and Educational Practice," and M. G. and J. G. Brooks's "The Courage to be Constructivist" are but a few of the large body of useful work applying this theory to practice. An excellent example of constructivism applied to ecology learning can be found in the curriculum developed by the Cary Institute of Ecosystem Studies educator K. Hogan, who also made important research contributions influential in BES education. Important publications by Hogan include her *Eco-Inquiry* and "Assessing Students' Systems Reasoning in Ecology," K. Hogan and J. Fisherkeller, "Representing Students' Thinking about Nutrient Cycling in Ecosystems: Bidimensional Coding of Complex Topics," and K. Hogan and D. Thomas, "Cognitive Comparisons of Students' Systems Modeling in Ecology." The high school curriculum to which we contributed employs may of the educational strategies used in BES and has been published as "Environmental Science and Complexity in 2016," written by G. Puttick et al.

Of more recent significance in shaping BES education has been the development of learning progressions, perhaps best articulated for our work by K. L. Gunckel et al., "Addressing Challenges in Developing Learning Progressions for Environmental Science Literacy." Further elaboration of learning progressions has been accompanied by fascinating work exploring student thinking and learning about carbon cycling, e.g., H. Jin et al., "Developing a Fine-Grained Learning Progression Framework for Carbon-Transforming Processes," and J. W. Schramm et al., "Improved Student Reasoning About Carbon-Transforming Processes Through Inquiry-Based Learning Activities Derived from an Empirically Validated Learning Progression," about water, e.g., K. L. Gunckel, "A Learning Progression for Water in Socio-Ecological Systems," and about biodiversity, e.g., C. Harris et al., "Exploring Biodiversity's Big Ideas in Your School Yard."

BES education's work with teachers builds on insights from the Eco-Inquiry curriculum mentioned earlier and the Schoolyard Ecology for Elementary School Teachers (SYEFEST) project, summarized in part by K. Hogan and A. R. Berkowitz, "Teachers as Inquiry Learners." Important guidance has come from the work of L. Darling-Hammond et al., *Professional Learning in the Learning Profession: A Status Report on Teacher Development in the United States and Abroad.* E. A. Davis and J. Krajcik's "Designing Educative Curriculum Materials to Promote Teacher Learning" provides insights into the role that curriculum designed to teach both teachers and students can play. Our approach to fostering professional learning communities is featured in R. C. Jordan et al., "A Collaborative Model of Science Teacher Professional Development?" and wisdom from the field summarized by S. Mundry and K. E. Stiles in *Professional Learning Communities for Science Teaching: Lessons from Research and Practice*. The approach to working with teachers, including in the Data Jam competitions, was inspired in large part by the education efforts of colleagues at other Long Term Ecological Research sites, as featured in S. V. Bestelmeyer's "Collaboration, Interdisciplinary Thinking, and Communication: New Approaches to K–12 Ecology Education." Finally, S. A. Yoon et al.'s "Teaching About Complex Systems Is No Simple Matter: Building Effective Professional Development for Computer-Supported Complex Systems Instruction" provides useful guidance for our recent work to incorporate computational thinking into our work with teachers.

BES and collaborating education researchers have contributed to important fields of teacher professional development through a number of publications. This includes contributions to the understanding of pedagogical content knowledge by K. L. Gunckel, "Learning Progressions as Supports for Teacher Content Knowledge and Pedagogical Content Knowledge About Water in Environmental Systems." S. Hauk et al.'s "Multiple Perspectives on Teacher Implementation of Learning Progression Teaching Strategies in Environmental Science" summarizes insights into how learning progressions can be put to work in classrooms. The Responsive Teaching study produced useful insights into when and how attention to student thinking can be used in science teaching, including J. E. Coffey et al., "The Missing Disciplinary Substance of Formative Assessment," and D. T. Levin et al., *Becoming a Responsive Science Teacher: Focusing on Student Thinking in Secondary Science.*" More recent work by B. Covitt et al., "Teachers' Use of Learning Progression-Based Formative Assessment in Water Instruction," shows how formative assessments can help teachers, while also finding that many continue to look for right and wrong answers rather than delving deeper into student thinking. Finally, H. Jin et al., "A Learning Progression Approach to Incorporate Climate Sustainability into Teacher Education" and "Promoting Students' Progressions in Science Classrooms: A Video Study," described a positive connection between both teacher knowledge and key teaching practices and student learning.

Our work with urban youth in building their interest in environmental career paths is grounded in the socio-cognitive "theory of interest" of A. Bandura in *Social Foundations of Thought and Action: A Social Cognitive Theory* and "Human Agency in Social Cognitive Theory" and applied to vocational interest by Lent et al., "Toward a Unifying Social Cognitive Theory of Career and Academic Interest, Choice, and Performance." This theory has been well applied to understanding students in the SEEDS program of ESA in the work of M. J. Armstrong, "Understanding Why Underrepresented Students Pursue Ecology Careers: A Preliminary Case Study," and with students in Baltimore in J. L. Quimby et al., "Social Cognitive Predictors of Interest in Environmental Science: Recommendations for Environmental Educators" and "Social Cognitive Predictors of African American Adolescents' Career Interests." Among the many other important contributions in this area are the research of J. Rahm and J. C. Moore, "A Case Study of Long-Term Engagement and Identity-in-Practice: Insights into the STEM Pathways of Four Underrepresented Youths," S. J. Basu and A. Calabrese-Barton, "Developing a Sustained Interest in Science among Urban Minority Youth," M. Barnett et al., "Using the Urban Environment to Engage Youths in Urban Ecology Field Studies," E. A. Hashimoto-Martell et al., "Connecting Urban Youth with Their Environment: The Impact of an Urban Ecology Course on Student Content Knowledge, Environmental Attitudes and Responsible Behaviors," and A. Kudryavtsev et al., "Sense of Place in Environmental Education."

AFTERWORD

The afterword draws on the insights expressed throughout the seventeen chapters of this book. In particular, by pointing to a future that extends key elements of this work, it builds on the deep ecological history provided by chapter 2. The author of chapter 2, Sharon Kingsland, is a doyen of the history of American ecology, and her other works can profitably be consulted, especially *History of American Ecology, 1890–2000* (2005). Important points from that chapter that inform the afterword are the human motivation of the idea of resilience that has come to be associated with C. S. Holling, for example, in "Engineering Resilience Versus Ecological Resilience" (1996). Importantly, Kingsland in chapter 2 notes that Holling's turn toward resilience was initiated during his work with Gordon Orians for the Ecological Society of America in a report that pointed out that the study of urban systems would require conceptual devices that were different from the equilibrium ecology dominant at the time. Shifts toward a nonequilibrium paradigm in ecology are signaled by Simberloff in "Succession of Paradigms in Ecology—Essentialism to Materialism and Probabilism" (1980) and, more recently, discussed in "Balance of Nature—Evolution of a Panchreston" (2014). The shifting paradigm in ecology has also been summarized by Pickett and colleagues' "The new paradigm in ecology: Implications for conservation biology above the species level" (1992). The relevance of contemporary philosophy of science to ecology and pointers to the demise of positivism as a guide for contemporary scientific practice are discussed at length in Pickett and colleagues' *Ecological Understanding: The Nature of Theory and the Theory of Nature,* 2nd edition (2007). Rademacher, Cadenasso, and Pickett have pointed to the need to conceive of the simultaneity of social-ecological systems and their coproduction in their 2018 article, "From feedbacks to coproduction: toward an integrated conceptual framework for urban ecosystems." Some parts of NSF see the need for transdisciplinary approaches in the study of urban ecological systems; notably described in a 2018 report to NSF, "Sustainable Urban Systems: Articulating a Long-Term Convergence Research Agenda: A Report from the NSF Advisory Committee for Environmental Research and Education." AbdouMalique Simone's *For the City Yet to Come: Changing African Life in Four Cities,* published in 2004 exemplifies the great flexibility of trajectories of urban transformations. McHale et al.'s 2015 exemplifies the wide range of urban transitions particularly relevant to ecology in "The New Global Urban Realm: Complex, Connected, Diffuse, and Diverse Social-ecological Systems."

BIBLIOGRAPHY

Adams, C. C. 1935. "The relation of general ecology to human ecology." *Ecology* 16:316–35.

———. 1938. "A note for social-minded ecologists and geographers." *Ecology* 19:500–502.

Advisory Committee for Environmental Research and Education. 2018. "Sustainable Urban Systems: Articulating a Long-Term Convergence Research Agenda. A Report from the NSF Advisory Committee for Environmental Research and Education." Prepared by the Sustainable Urban Systems Subcommittee.

Allen, T. F. H., and T. W. Hoekstra. 1992. *Toward a Unified Ecology.* New York: Columbia University Press.

American Association for the Advancement of Science. 1993. *Benchmarks for Science Literacy.* New York: Oxford University Press.

———. 2001. *Atlas of Science Literacy.* Washington, D.C.

Anderson, E. 1958. "The city watcher." *Landscape* 8:7–8.

Anderson, J. R., E. E. Hardy, J. T. Roach, and R. E. Witmer. 1976. "Land use and land cover classification systems for use with remote sensor data." *U.S. Geological Survey Professional Paper* 964:1–28.

Armstrong, M. J., A. R. Berkowitz, L. A. Dyer, and J. Taylor. 2007. "Understanding why underrepresented students pursue ecology careers: A preliminary case study." *Frontiers in Ecology and the Environment* 5:415–20.

Bain, D. J., M. B. Green, J. L. Campbell, J. F. Chamblee, S. Chaoka, J. M. Fraterrigo, S. S. Kaushal, S. L. Martin, T. E. Jordan, A. J. Parolari, W. V. Sobczak, D. E. Weller, W. M. Wollheim, E. R. Boose, J. M. Duncan, G. M. Gettel, B. R. Hall, P. Kumar, J. R. Thompson, J. M. Vose, E. M. Elliott, and D. S. Leigh. 2012.

"Legacy effects in material flux: Structural catchment changes predate long-term studies." *BioScience* 62:575–84.

Baker, L. A., A. J. Brazel, N. Selover, C. Martin, N. McIntyre, F. R. Steiner, A. Nelson, and L. Musacchio. 2002. "Urbanization and warming of Phoenix (Arizona USA): Impacts, feedbacks and mitigation." *Urban Ecosystems* 6:183–203.

Bandura, A. 1986. *Social Foundations of Thought and Action: A Social Cognitive Theory.* Englewood Cliffs: Prentice Hall.

———. 1989. "Human agency in social cognitive theory." *American Psychologist* 44:1175–84.

Barnett, J. 1982. *An Introduction to Urban Design.* New York: Harper and Row.

Barnett, M., C. Lord, E. Strauss, C. Rosca, H. Langford, D. Chavez, and L. Deni. 2006. "Using the urban environment to engage youths in urban ecology field studies." *Journal of Environmental Education* 37:3–11.

Basu, S. J., and A. C. Barton. 2007. "Developing a sustained interest in science among urban minority youth." *Journal of Research in Science Teaching* 44:466–89.

Battaglia, M., G. Buckley, M. Galvin, and M. Grove. 2014. "It's not easy going green: Obstacles to tree-planting programs in East Baltimore." *Cities and the Environment (CATE)* 7.

Batty, M. 2008. "The size, scale, and shape of cities." *Science* 319:769–71.

Berkowitz, A. R., M. Archie, and D. Simmons. 1997. "Defining environmental literacy: A call for action." *Bulletin of the Ecological Society of America* 78:170–72.

Berkowitz, A. R., M. E. Ford, and C. A. Brewer. 2005. "A framework for integrating ecological literacy, civics literacy and environmental citizenship in environmental education." Pages 227–66 in E. A. Johnson and M. J. Mappin, editors, *Environmental Education and Advocacy: Changing Perspectives of Ecology and Education.* Cambridge: Cambridge University Press.

Berkowitz, A. R., K. S. Hollweg, and C. H. Nilon. 2003a. "Urban ecosystem education in the coming decade: What is possible and how might we get there?" Pages 476–501 in A. R. Berkowitz, C. H. Nilon, and K. S. Hollweg, editors, *Understanding Urban Ecosystems: A New Frontier for Science and Education.* New York: Springer-Verlag.

Berkowitz, A. R., C. H. Nilon, and K. S. Hollweg, editors. 2003b. *Understanding Urban Ecosystems: A New Frontier for Science and Education.* New York: Springer-Verlag.

Besley, F. W. 1909. "Maryland's forest resources: A preliminary report." *Forestry Leaflet* 7.

———. 1916. *"The Forests of Maryland."* Baltimore: Press of the Advertiser-Republican.

Bestelmeyer, S. V., M. M. Elser, K. V. Spellman, E. B. Sparrow, S. S. Haan-Amato, and A. Keener. 2015. "Collaboration, interdisciplinary thinking, and communication: New approaches to K–12 ecology education." *Frontiers in Ecology and the Environment* 13:37–43.

Bettez, N. D., J. M. Duncan, P. M. Groffman, L. E. Band, J. O'Neil-Dunne,

S. S. Kaushal, K. T. Belt, and N. Law. 2015. "Climate variation overwhelms efforts to reduce nitrogen delivery to coastal waters." *Ecosystems* 18:1319–31.

Bhaskar, A. S., and C. Welty. 2012. "Water balances along an urban-to-rural gradient of metropolitan Baltimore, 2001–2009." *Environmental & Engineering Geoscience* 18:37–50.

Black, P. E. 1991. *Watershed Hydrology*. Englewood Cliffs: Prentice Hall.

Blanco-Montero, C. A., T. B. Bennett, P. Neville, C. S. Crawford, B. T. Milne, and C. R. Ward. 1995. "Potential environmental and economic impacts of turfgrass in Albuquerque, New Mexico (USA)." *Landscape Ecology* 10:121–28.

Bockheim, J. G. 1974. "Nature and properties of highly-disturbed urban soils." Annual Meeting of the Soil Science Society of America, Chicago.

Bodin, O., and B. Crona. 2009. "The role of social networks in natural resource governance: What relational patterns make a difference?" *Global Environmental Change* 19:366–74.

Bodin, O., and C. Prell. 2011. *Social Networks and Natural Resource Management*. Cambridge: Cambridge University Press.

Boesch, D. F., R. B. Brinsfield, and R. E. Magnien. 2001. "Chesapeake Bay eutrophication: Scientific understanding, ecosystem restoration, and challenges for agriculture." *Journal of Environmental Quality* 30:303–20.

Boone, C. 2008. "Improving resolution of census data in metropolitan areas using a dasymetric approach: Applications for the Baltimore Ecosystem Study." *Cities and the Environment (CATE)* 1.

Boone, C., M. L. Cadenasso, J. M. Grove, K. Schwartz, and G. L. Buckley. 2010. "Landscape, vegetation characteristics, and group identity in an urban and suburban watershed: Why the 60s matter." *Urban Ecosystems* 13:255–71.

Boone, C., M. Fragkias, G. Buckley, and J. Grove. 2014. "A long view of polluting industry and environmental justice in Baltimore." *Cities* 36:41–49.

Boone, C. G. 2002. "An assessment and explanation of environmental inequity in Baltimore." *Urban Geography* 23:581–95.

———. 2003. "Obstacles to infrastructure provision: The struggle to build comprehensive sewer works in Baltimore." *Historical Geography* 31:151–68.

Boone, C. G., G. B. Buckley, J. M. Grove, and C. Sister. 2009. "Parks and people: An environmental justice inquiry in Baltimore, Maryland." *Annals of the Association of American Geographers* 99:1–21.

Borgatti, S. P. 2003. "The network paradigm in organizational research: A review and typology." *Journal of Management* 29:991–1013.

Bormann, F. H. 1972. "Unlimited growth: Growing, growing, gone?" *BioScience* 22:706–9.

———. 1996. "Ecology: A personal history." *Annual Review of Energy and the Environment* 21:1–29.

Bormann, F. H., D. Balmori, and G. T. Geballe. 2001. *Redesigning the American Lawn: A Search for Environmental Harmony*. New Haven: Yale University Press.

Bormann, F. H., and G. E. Likens. 1979. *Pattern and Processes in a Forested Ecosystem:*

Disturbance, Development, and the Steady State Based on the Hubbard Brook Ecosystem Study. New York: Wiley and Sons.

Botkin, D. B. 1990. *Discordant Harmonies: A New Ecology for the Twenty-First Century.* New York: Oxford University Press.

Boyden, S. 1979. "An integrative ecological approach to the study of human settlements." *Man and the Biosphere technical notes,* volume 12. Paris: UNESCO.

Boyden, S., S. Millar, K. Newcombe, and B. O'Neill. 1981. *The Ecology of a City and Its People: The Case of Hong Kong.* Canberra: Australian National University Press.

Branas, C. C., R. A. Cheney, J. M. MacDonald, V. W. Tam, T. D. Jackson, and T. R. Ten Have. 2011. "A difference-in-differences analysis of health, safety, and greening vacant urban space." *Journal of Epidemiology* 174:1296–1306.

Brazel, A. J., and K. Crewe. 2002. "Preliminary test of a surface heat island model (SHIM) and implications for a desert urban environment, Phoenix, Arizona." *Journal of the Arizona-Nevada Academy of Science* 34:98–105.

Brazel, A. J., and A. W. Ellis. 2003. "The climate of central Arizona and Phoenix Long-Term Ecological Research site (CAP LTER) and links to ENSO." Pages 117–40 in D. Greenland, D. Goodwin, and R. Smith, editors, *Climate Variability and Ecosystem Response in Long-Term Ecological Research Sites.* New York: Oxford University Press.

Brazel, A. J., and G. M. Heisler. 2009. "Climatology at urban long-term ecological research sites: Baltimore Ecosystem Study and Central Arizona-Phoenix." *Geography Compass* 3:22–44.

Brazel, A., N. Selover, R. Voce, and G. Heisler. 2000. "The tale of two climates—Baltimore and Phoenix LTER sites." *Climate Research* 15:123–35.

Brooks, M. G., and J. G. Brooks. 1999. "The courage to be constructivist." *Educational Leadership* 57:18–24.

Broussard, W., and R. E. Turner. 2009. "A century of changing land-use and water-quality relationships in the continental US." *Frontiers in Ecology and the Environment* 7:302–7.

Brown, H. 1954. *The Challenge of Man's Future: An Inquiry Concerning the Condition of Man During the Years That Lie Ahead.* New York: Viking Press.

Brush, G. S. 2009. "Historical land use, nitrogen, and coastal eutrophication: A paleoecological perspective." *Estuaries and Coasts* 32:18–28.

———. 2017. *Decoding the Deep Sediments: The Ecological History of the Chesapeake Bay.* College Park: Sea Grant.

Brush, G. S., C. Lenk, and J. Smith. 1980. "The natural forests of Maryland: An explanation of the vegetation map of Maryland." *Ecological Monographs* 50:77–92.

Buckley, G. L. 2010. *America's Conservation Impulse: A Century of Saving Trees in the Old Line State.* Chicago: Center for American Places at Columbia College Chicago.

Buckley, G. L., and C. G. Boone. 2011. "'To promote the material and moral welfare of the community': Neighborhood Improvement Associations in Baltimore,

Maryland, 1900–1945." Pages 43–65 in G. Massard-Guilbaud and R. Rodger, editors, *Environmental and Social Justice in the City: Historical Perspectives*. Cambridge: White Horse Press.

Buckley, G. L., C. G. Boone, and J. M. Grove. 2017. "The greening of Baltimore's asphalt schoolyards." *Geographical Review* 107:516–35.

Burch, W. 1991. *The Strategic Plan for Action*. Baltimore: Baltimore City Department of Recreation and Parks.

Burch, W. R. 1971. *Daydreams and Nightmares: A Sociological Essay on the American Environment*. New York: Harper and Row.

Burch, W. R., Jr. 1992. *Social Science Applications in Asian Agroforestry*. New Delhi: IBH Publishing.

Burch, W. R., Jr., and J. M. Grove. 1993. "People, trees, and participation on the urban frontier." *Unasylva* 44:19–27.

———. 1996. "Life on the city streets: Some lessons from Baltimore for reaching out to grow trees, kids, and communities." *Journal of Public Service & Outreach* 1:46–55.

Burch, W. R., Jr., G. E. Machlis, and J. E. Force. 2017. *The Structure and Dynamics of Human Ecosystems: Toward a Model for Understanding and Action*. New Haven: Yale University Press.

Bybee, R. W. 2003. "Integrating urban ecosystem education into educational reform." Pages 430–449 in A. R. Berkowitz, C. H. Nilon, and K. S. Hollweg, editors, *Understanding Urban Ecosystems: A New Frontier for Science and Education*. New York: Springer.

Byrne, L. B. 2007. "Habitat structure: A fundamental concept and framework for urban soil ecology." *Urban Ecosystems* 10:255–74.

Cadenasso, M. L. 2013. "Designing ecological heterogeneity." Pages 271–81 in *Urban Design Ecologies: AD Reader*. Hoboken: John Wiley.

Cadenasso, M. L., and S. T. A. Pickett. 2007. "Boundaries as structural and functional entities in landscapes: Understanding flows in ecology and urban design." Pages 116–31 in B. McGrath, V. Marshall, M. L. Cadenasso, J. M. Grove, S. T. A. Pickett, R. Plunz, and J. Towers, editors, *Designing Patch Dynamics*. New York: Columbia, Graduate School of Architecture, Planning and Preservation.

———. 2008. "Urban principles for ecological landscape design and management: Scientific fundamentals." *Cities and the Environment* 1:article 4.

———. 2013. "Three tides: The development and state of the art of urban ecological science." Pages 29–46 in S. T. A. Pickett, M. L. Cadenasso, and B. McGrath, editors, *Resilience in Ecology and Urban Design: Linking Theory and Practice for Sustainable Cities*. New York: Springer.

Cadenasso, M. L., S. T. A. Pickett, P. Groffman, L. E. Band, G. S. Brush, M. F. Galvin, J. M. Grove, G. W. Hager, V. Marshall, B. McGrath, J. P. M. O'Neil-Dunne, W. P. Stack, and A. R. Troy. 2008. "Exchanges across land-water-scape boundaries in urban systems: Strategies for reducing nitrate pollution." *Annals of the New York Academy of Sciences* 1134:213–32.

Cadenasso, M. L., S. T. A. Pickett, and J. M. Grove. 2006a. "Dimensions of ecosystem complexity: Heterogeneity, connectivity, and history." *Ecological Complexity* 3:1–12.

———. 2006b. "Integrative approaches to investigating human-natural systems: The Baltimore ecosystem study." *Natures Sciences Societies* 14:1–14.

Cadenasso, M. L., S. T. A. Pickett, B. McGrath, and V. Marshall. 2013. "Ecological heterogeneity in urban ecosystems: Reconceptualized land cover models as a bridge to urban design." Pages 107–29 in S. T. A. Pickett, M. L. Cadenasso, and B. McGrath, editors, *Resilience in Ecology and Urban Design: Linking Theory and Practice for Sustainable Cities*. New York: Springer.

Cadenasso, M. L., S. T. A. Pickett, and K. Schwarz. 2007. "Spatial heterogeneity in urban ecosystems: Reconceptualizing land cover and a framework for classification." *Frontiers in Ecology and the Environment* 5:80–88.

Cadotte, M. W. 2006. "Dispersal and species diversity: A meta-analysis." *American Naturalist* 167:913–24.

Capra, C. F., A. Ganga, E. Grilla, S. Vacca, and A. Buondonno. 2015. "A review on anthropogenic soils from a worldwide perspective." *Journal of Soils and Sediments* 15:1602–18.

Capra, F. 1997. *The Web of Life: A New Scientific Understanding of Living Systems.* New York: Anchor Books.

Carlson, C. E. 2006. "The relationship between breeding bird community structure in urban forest patches and the human-mediated resources in the surrounding residential matrix." M.S. diss., University of Georgia, Athens.

Carpenter, S. R. 1998. "The need for large-scale experiments to assess and predict the response of ecosystems to perturbation." Pages 287–312 in M. L. Pace and P. M. Groffman, editors, *Successes, limitations, and frontiers in ecosystem science.* New York: Springer.

Carr, L. G. 1992. *The Chesapeake and Beyond—A Celebration.* Crownsville: Maryland Historical and Cultural Publications.

Chevalier-Flick, M. M. 2009. "Toxic playground: A retrospective case study of environmental justice in Baltimore, Maryland." M.S. diss., Ohio University, Athens.

Childers, D. L., M. L. Cadenasso, J. M. Grove, V. Marshall, B. McGrath, and S. T. A. Pickett. 2015. "An ecology for cities: A transformational nexus of design and ecology to advance climate change resilience and urban sustainability." *Sustainability* 7:3774–91.

Chow, W. T. L., D. Brennan, and A. J. Brazel. 2012. "Urban heat island research in Phoenix, Arizona: Theoretical contributions and policy applications." *Bulletin of the American Meteorological Society* 93:517–30.

Chow, W. T. L., T. J. Volo, E. R. Vivoni, G. D. Jenerette, and B. L. Ruddell. 2014. "Seasonal dynamics of a suburban energy balance in Phoenix, Arizona." *International Journal of Climatology* 34:3863–80.

Cittadino, E. 1993. "The failed promise of human ecology." Pages 251–83 in M. Shortland, editor, *Science and Nature: Essays in the History of the Environmental Sciences.* Stanford-in-the-Vale, UK: British Society for the History of Science.

Claritas. 1999. "PRIZM cluster snapshots: Getting to know the 62 clusters." Ithaca: Claritas Corporation.

Cleaves, E. T., D. W. Fisher, and O. P. Bricker. 1974. "Chemical weathering of serpentinite in the eastern piedmont of Maryland." *Geological Society of America Bulletin* 85:437–44.

Coffey, J. E., D. Hammer, D. M. Levin, and T. Grant. 2011. "The missing disciplinary substance of formative assessment." *Journal of Research in Science Teaching* 48:1109–36.

Colautti, R., I. A. Grigorovich, and H. J. MacIsaac. 2006. "Propagule pressure: A null model for biological invasions." *Biological Invasions* 8:1023–37.

Coleman, D. C. 2010. *Big Ecology: The Emergence of Ecosystem Science*. Berkeley: University of California Press.

Colosimo, M. F., and P. R. Wilcock. 2007. "Alluvial sedimentation and erosion in an urbanizing watershed, Gwynns Falls, Maryland." *Journal of the American Water Resources Association* 43:499–521.

Colten, C. E. 1990. "Historical hazards: The geography of relict industrial wastes." *Professional Geographer* 42:143–56.

Connell, J. H. 1978. "Diversity in tropical rain forests and coral reefs." *Science* 199:1302–10.

Corrigan, M. 2011. "Growing what you eat: Developing community gardens in Baltimore, Maryland." *Applied Geography* 31:1232–41.

Costa, K. H., and P. M. Groffman. 2013. "Factors regulating net methane flux in urban forests and grasslands." *Soil Science Society of America Journal* 77:850–55.

Costanza, R., and H. E. Daly, editors. 1987. *Ecological Economics*. Amsterdam: Elsevier.

Costanza, R., A. Voinov, R. Boumans, T. Maxwell, F. Villa, L. Wainger, and H. Voinov. 2002. "Integrated ecological economic modeling of the Patuxent River watershed, Maryland." *Ecological Monographs* 72:203–31.

Covitt, B. A., K. L. Gunckel, B. Caplan, and S. Syswerda. 2018. "Teachers' use of learning progression-based formative assessment in water instruction." *Applied Measurement in Education* 31:128–42.

Covitt, B., J. Dauer, and C. Anderson. 2017. "The role of practices in scientific literacy." Pages 59–84 in C. Schwarz, C. Passmore, and B. Reiser, editors, Supporting *Next Generation Scientific and Engineering Practices in K-12 Classrooms*. Arlington: NSTA Press.

Craige, B. J. 2001. *Eugene Odum: Ecosystem Ecologist and Environmentalist*. Athens: University of Georgia Press.

Craul, P. J. 1992. *Urban Soil in Landscape Design*. New York: John Wiley.

Craul, P. J., and C. J. Klein. 1980. "Characterization of streetside soils of Syracuse, New York." *METRIA* 3:88–101.

Crawford, B., C. S. B. Grimmond, and A. Christen. 2011. "Five years of carbon dioxide fluxes measurements in a highly vegetated suburban area." *Atmospheric Environment* 45:896–905.

Cronon, W. 1983. *Changes in the Land: Indians, Colonists, and the Ecology of New England*. New York: Hill and Wang.

———. 1991. *Nature's Metropolis: Chicago and the Great West*. New York: Norton.

Daily, G. C., editor. 1997. *Nature's Services: Societal Dependence on Natural Ecosystems*. Washington, D.C.: Island Press.

Dalton, S. E. 2001. "The Gwynns Falls watershed: A case study of public and non-profit sector behavior in natural resource management." Ph.D. diss., Johns Hopkins University.

Darling-Hammond, L., R. C. Wei, A. Andree, N. Richardson, and S. Orphanos. 2009. *Professional Learning in the Learning Profession: A Status Report on Teacher Development in the United States and Abroad*. Dallas: National Staff Development Council.

Davis, E. A., and J. S. Krajcik. 2005. "Designing educative curriculum materials to promote teacher learning." *Educational Researcher* 34:3–14.

Davis, M. B. 1986. "Climatic instability, time lags, and community disequilibrium." Pages 269–284 in J. Diamond and T. J. Case, editors, *Community Ecology*. New York: Harper and Row.

Deevey, E. S. 1949. "Pleistocene research." *Bulletin of the Geological Society of America* 60:1315–1416.

Diani, M., and D. McAdam, editors. 2003. *Social Movements and Networks: Relational Approaches to Collective Action*. Oxford: Oxford University Press.

Doherty, J., and A. W. Cooper. 1990. "The short life and early death of the Institute of Ecology: A case study in institution building." *Bulletin of the Ecological Society of America* 71:6–17.

Duncan, J. M., L. E. Band, P. M. Groffman, and E. S. Bernhardt. 2015. "Mechanisms driving the seasonality of catchment scale nitrate export: Evidence for riparian ecohydrologic controls." *Water Resources Research* 51:3982–97.

Duncan, J. M., C. Welty, J. T. Kemper, P. M. Groffman, and L. E. Band. 2017. "Dynamics of nitrate concentration-discharge patterns in an urban watershed." *Water Resources Research* 53:7349–65.

Durant, R. F., D. J. Fiorino, and R. O'Leary, editors. 2004. *Environmental Governance Reconsidered: Challenges, Choices, and Opportunities*. Cambridge: MIT Press.

Elfenbein, J. I., T. L. Hollowak, and E. M. Nix, editors. 2011. *Baltimore '68: Riots and Rebirth in an American City*. Philadelphia: Temple University Press.

Eliot, C. 2010. "The legend of order and chaos: Communities and early community ecology." Pages 61–122 in K. deLaplante, B. Brown, and K. A. Peacock, editors, *Philosophy of Ecology*. Amsterdam: Elsevier.

Ellis, E. C., H. Q. Wang, H. S. Xiao, K. Peng, X. P. Liu, S. C. Li, H. Ouyang, X. Cheng, and L. Z. Yang. 2006. "Measuring long-term ecological changes in densely populated landscapes using current and historical high resolution imagery." *Remote Sensing of Environment* 100:457–73.

England, M. 2008. "When 'good neighbors' go bad: Territorial geographies of neighborhood associations." *Environment and Planning A* 40:2879–94.

Ernstson, H., S. Barthel, E. Andersson, and S. Borgstrom. 2010. "Scale-crossing brokers and network governance of urban ecosystem services: The case of Stockholm, Sweden." *Ecology & Society* 15:1–25.

Ernstson, H., S. Sorlin, and T. Elmqvist. 2008. "Social movements and ecosystem services: The role of social network structure in protecting and managing urban green areas in Stockholm." *Ecology & Society* 13.

Evans, C. V., D. S. Fanning, and J. R. Short. 2000. "Human-influenced soils." *Agronomy Monographs* 39:33–67.

Evans, F. 1951. "Biology and urban areal research." *Scientific Monthly* 73:37–38.

Falk, J. H. 1976. "Energetics of an urban lawn ecosystem." *Ecology* 57:141–50.

———. 1980. "The primary productivity of lawns in a temperate environment." *Journal of Applied Ecology* 17:689–95.

Felson, A. J. 2013. "The design process as a framework for collaboration between ecologists and designers." Pages 365–82 in S. T. A. Pickett, M. L. Cadenasso, and B. McGrath, editors, *Resilience in Ecology and Urban Design: Linking Theory and Practice for Sustainable Cities*. New York: Springer.

Felson, A. J., M. A. Bradford, and T. M. Terway. 2013. "Promoting Earth Stewardship through urban design experiments." *Frontiers in Ecology and the Environment* 11:362–67.

Felson, A. J., and S. T. A. Pickett. 2005. "Designed experiments: New approaches to studying urban ecosystems." *Frontiers in Ecology and the Environment* 3:549–56.

Fialkowski, C. 2003. "Approaches to urban ecosystem education in Chicago: Perspectives and processes from an environmental educator." Pages 343–54 in A. R. Berkowitz, C. H. Nilon, and K. S. Hollweg, editors, *Understanding Urban Ecosystems*. New York: Springer.

Forbes, S. A. 1922. "The humanizing of ecology." *Ecology* 2:89–92.

Foster, D. R., and J. D. Aber, editors. 2004. *Forests in Time: The Environmental Consequences of 1,000 Years of Change in New England*. New Haven: Yale University Press.

Foster, J. B. 1999. "Marx's theory of metabolic rift: Classical foundations for environmental sociology." *Americal Journal of Sociology* 105:366–405.

Fox, G. A., S. M. Scheiner, and M. R. Willig. 2011. "Ecological gradient theory: A framework for aligning data and models." Pages 283–307 in S. M. Scheiner and M. R. Willig, editors, *The Theory of Ecology*. Chicago: University of Chicago Press.

Fraser, J. C., J. T. Bazuin, L. E. Band, and J. M. Grove. 2013. "Covenants, cohesion, and community: The effects of neighborhood governance on lawn fertilization." *Landscape and Urban Planning* 115:30–38.

Garreau, J. 1991. *Edge City: Life on the New Frontier*. New York: Doubleday.

Gaston, K. J. 2005. "Biodiversity and extinction: Species and people." *Progress in Physical Geography* 29:239–47.

Giguere, A. M. 2009. "'. . . and never the twain shall meet': Baltimore's east-west expressway and the construction of the 'Highway to Nowhere.'" M.A. diss., Ohio University, Athens.

Gnagey, M. K. 2016. "Buried streams: An analysis of regulations and land development." Working paper, Weber State University.

Gober, P., A. Brazel, R. Quay, S. Myint, S. Grossman-Clarke, A. Miller, and S. Rossi. 2010. "Using watered landscapes to manipulate urban heat island effects: How much water will it take to cool Phoenix?" *Journal of American Planning Association* 76:109–21.

Gold, A. J., W. R. Deragon, W. M. Sullivan, and J. L. Lemunyon. 1990. "Nitrate-nitrogen losses to groundwater from rural and suburban land uses." *Journal of Soil and Water Conservation* 45:305–10.

Golley, F. B. 1993. *A History of the Ecosystem Concept in Ecology: More Than the Sum of the Parts.* New Haven: Yale University Press.

Goodlett, J. C. 1954. "Vegetation adjacent to the border of the Wisconsin drift in Potter County, Pennsylvania." *Harvard Forest Bulletin* 25.

Gottschalk, L. C. 1945. "Effects of soil erosion on navigation in upper Chesapeake Bay." *Geographical Review* 35:219–38.

Grant, R. H., and G. M. Heisler. 2006. "Effect of cloud cover on UVB exposure under tree canopies: Will climate change affect UVB exposure?" *Photochemistry and Photobiology* 82:487–94.

Greenland, D., D. Goodwin, and R. C. Smith, editors. 2003. *Climate Variability and Ecosystem Response at Long-Term Ecological Research Sites.* New York: Oxford University Press.

Grimm, N. B., E. M. Cook, R. L. Hale, and D. M. Iwaniec. 2015. "A broader framing of ecosystem services in cities: Benefits and challenges of built, natural, or hybrid system function." Pages 203–12 in K. C. Seto and W. D. Solecki, editors, *The Routledge Handbook of Urbanization and Global Environmental Change.* New York: Routledge.

Grimm, N. B., S. T. A. Pickett, R. L. Hale, and M. L. Cadenasso. 2017. "Does the ecological concept of disturbance have utility in urban social-ecological-technological systems?" *Ecosystem Health and Sustainability* 3:e01255.10.1002/ehs2.1255.

Grimm, N., J. M. Grove, S. T. A. Pickett, and C. Redman. 2000. "Integrated approaches to long-term studies of urban ecological systems." *BioScience* 50:571–84.

Grimmond, C. S. B., and T. R. Oke. 2002. "Turbulent heat fluxes in urban areas: Observations and a local-scale urban meteorological parameterization scheme (LUMPS)." *Journal of Applied Meteorology* 41:792–810.

Groffman, P. M., M. Avolio, J. Cavender-Bares, N. D. Bettez, J. M. Grove, S. Hall, S. E. Hobbie, K. L. Larson, S. B. Lerman, D. H. Locke, J. B. Heffernan, J. L. Morse, C. Neill, K. C. Nelson, J. O'Neil-Dunne, D. E. Pataki, C. Polsky, R. V. Pouyat, R. Roy Chowdhury, M. Steele, and T. L. E. Trammel. 2017a. "Ecological homogenization of residential macrosystems." *Nature Ecology & Evolution* article 0191.

Groffman, P. M., D. J. Bain, L. E. Band, K. T. Belt, G. S. Brush, J. M. Grove, R. V. Pouyat, I. C. Yesilonis, and W. C. Zipperer. 2003. "Down by the riverside: Urban riparian ecology." *Frontiers in Ecology and the Environment* 1:315–21.

Groffman, P. M., N. J. Boulware, W. C. Zipperer, R. V. Pouyat, L. E. Band, and M. F. Colosimo. 2002. "Soil nitrogen cycling processes in urban riparian zones." *Environmental Science & Technology* 36:4547–52.

Groffman, P. M., M. L. Cadenasso, J. Cavender-Bares, D. L. Childers, N. B. Grimm, J. M. Grove, S. E. Hobbie, L. R. Hutyra, G. D. Jenerette, T. McPhearson, D. E. Pataki, S. T. A. Pickett, R. V. Pouyat, E. J. Rosi-Marshall, and B. L. Ruddell. 2017b. "Moving towards a new urban systems science." *Ecosystems* 20:38–43.

Groffman, P. M., J. Cavender-Bares, N. D. Bettez, J. M. Grove, S. J. Hall, J. B. Heffernan, S. E. Hobbie, K. L. Larson, J. L. Morse, C. Neill, K. Nelson, J. O'Neil-Dunne, L. Ogden, D. E. Pataki, C. Polsky, R. R. Chowdhury, and M. K. Steele. 2014. "Ecological homogenization of urban USA." *Frontiers in Ecology and the Environment* 12:74–81.

Groffman, P. M., J. M. Grove, C. Polsky, N. D. Bettez, J. L. Morse, J. Cavender-Bares, S. J. Hall, J. B. Heffernan, S. E. Hobbie, K. L. Larson, C. Neill, K. Nelson, L. A. Ogden, J. O'Neil-Dunne, D. E. Pataki, R. Roy Chowdhury, and D. H. Locke. 2016. "Satisfaction, water and fertilizer use in the American residential macrosystem." *Environmental Research Letters* 11:034004.

Groffman, P. M., N. L. Law, K. T. Belt, L. E. Band, and G. T. Fisher. 2004. "Nitrogen fluxes and retention in urban watershed ecosystems." *Ecosystems* 7:393–403.

Groffman, P. M., and R. V. Pouyat. 2009. "Methane uptake in urban forests and lawns." *Environmental Science & Technology* 43:5229–35.

Groffman, P. M., R. V. Pouyat, M. L. Cadenasso, W. C. Zipperer, K. Szlavecz, I. D. Yesilonis, L. E. Band, and G. S. Brush. 2006. "Land use context and natural soil controls on plant community composition and soil nitrogen and carbon dynamics in urban and rural forests." *Forest Ecology and Management* 236:177–92.

Groffman, P. M., C. O. Williams, R. V. Pouyat, L. E. Band, and I. Yesilonis. 2009. "Nitrate leaching and nitrous oxide flux in urban forests and grasslands." *Journal of Enviornmental Quality* 38:1848–60.

Grove, J. M., and W. R. Burch. 1997. "A social ecology approach and applications of urban ecosystem and landscape analyses: A case study of Baltimore, Maryland." *Urban Ecosystems* 1:259–75.

Grove, J. M., W. R. Burch, M. Wilson, and A. W. Vemuri. 2007. "The mutual dependence of social meanings, social capital, and the design of urban green infrastructure." Pages 66–77 in B. McGrath, V. Marshall, M. L. Cadenasso, J. M. Grove, S. T. A. Pickett, R. Plunz, and J. Towers, editors, *Designing Patch Dynamics*. New York: Columbia, Graduate School of Architecture, Planning, and Preservation.

Grove, J. M., M. L. Cadenasso, W. R. Burch, S. T. A. Pickett, J. P. M. O'Neil-Dunne, K. Schwarz, and M. Wilson. 2006a. "Data and methods comparing social structure and vegetation structure of urban neighborhoods in Baltimore, Maryland." *Society & Natural Resources* 19:117–36.

Grove, J. M., S. T. A. Pickett, A. Whitmer, and M. L. Cadenasso. 2013. "Building an urban LTSER: The case of the Baltimore Ecosystem Study and the D.C./B.C.

ULTRA-Ex Project." Pages 369–408 in J. S. Singh, H. Haberl, M. Chertow, M. Mirtl, and M. Schmid, editors, *Long Term Socio-Ecological Research: Studies in Society: Nature Interactions Across Spatial and Temporal Scales*. New York: Springer.

Grove, J. M., A. R. Troy, J. P. M. O'Neil-Dunne, W. R. Burch, M. L. Cadenasso, and S. T. A. Pickett. 2006b. "Characterization of households and its implications for the vegetation of urban ecosystems." *Ecosystems* 9:578–97.

Grove, J. M., K. Vachta, M. McDonough, and W. R. Burch Jr. 1993. "The urban resources initiative: Community benefits from forestry." In P. H. Gobster, editor, *Managing Urban and High-Use Recreation Settings*. General Technical Report, NC-163, USDA Forest Service, St. Paul.

Grove, M., M. L. Cadenasso, S. T. A. Pickett, G. Machlis, and W. R. Burch Jr. 2015. *The Baltimore School of Urban Ecology*. New Haven: Yale University Press.

Grover, S., and R. Pea. 2018. "Computational thinking: A competency whose time has come." Pages 20–38 in S. Sentence, S. Carsten, and E. Barendsen, editors, *Computer Science Education: Perspectives on Teaching and Learning*. London: Bloomsbury.

Groves, P. A., and E. K. Muller. 1975. "The evolution of black residential areas in late nineteenth-century cities." *Journal of Historical Geography* 1:169–91.

Guida, I. 2012. "Corridors." Ph.D. diss., IUAV University of Venice, Venice.

Gunckel, K. L., B. A. Covitt, and I. Salinas. 2018. "Learning progressions as tools for supporting teacher content knowledge and pedagogical content knowledge about water in environmental systems." *Journal of Research in Science Teaching* 55 DOI: 10.1002/tea.21454.

Gunckel, K. L., B. A. Covitt, I. Salinas, and C. W. Anderson. 2012a. "A learning progression for water in socio-ecological systems." *Journal of Research in Science Teaching* 49:843–68.

Gunckel, K. L., L. Mohan, B. A. Covitt, and C. W. Anderson. 2012b. "Addressing challenges in developing learning progressions for environmental science literacy." Pages 39–76 in A. Alonzo and A. W. Gotwals, editors, *Learning Progressions in Science*. Boston: Sense Publishers.

Gundersen, P., I. K. Schmidt, and K. Raulund-Rasmussen. 2006. "Leaching of nitrate from temperate forests—effects of air pollution and forest management." *Environmental Reviews* 14:1–57.

Gunderson, L. H., and C. S. Holling, editors. 2002. *Panarchy: Understanding Transformations in Human and Natural Systems*. Washington, D.C.: Island Press.

Hagen, J. B. 1992. *An Entangled Bank: The Origins of Ecosystem Ecology*. New Brunswick: Rutgers University Press.

Hager, G. W., K. T. Belt, W. Stack, K. Burgess, J. M. Grove, B. Caplan, M. Hardcastle, D. Shelley, S. T. A. Pickett, and P. M. Groffman. 2013. "Socioecological revitalization of an urban watershed." *Frontiers in Ecology and the Environment* 11:28–36.

Harlan, S. L., A. Brazel, L. Prashad, W. L. Stefanov, and L. Larsen. 2006. "Neighborhood microclimates and vulnerability to heat stress." *Social Science & Medicine* 63:2847–63.

Harris, C., A. R. Berkowitz, J. H. Doherty, and L. M. Hartley. 2013. "Exploring biodiversity's big ideas in your school yard." *Science Scope* 36:20–27.

Harvey, D. "Class-monopoly rent, finance capital and the urban revolution." *Regional Studies* 197:239–55.

Harvey, D., and L. Chatterjee. 1974. "Absolute rent and the structuring of space by government and financial institutions." *Antipode* 6:22–36.

Hashimoto-Martell, E. A., K. L. McNeill, and E. M. Hoffman. 2012. "Connecting urban youth with their environment: The impact of an urban ecology course on student content knowledge, environmental attitudes and responsible behaviors." *Research in Science Education* 42:1007–26.

Hauk, S., N. Yestness, K. Roach, A. R. Berkowitz, and G. Alvarado. 2014. *Multiple Perspectives on Teacher Implementation of Learning Progression Teaching Strategies in Environmental Science.* Philadelphia: American Educational Research.

Hawkins, T. W., A. J. Brazel, W. L. Stefanov, W. Bigler, and E. M. Saffell. 2004. "The role of rural variability in urban heat island determination for Phoenix, Arizona." *Journal of Applied Meteorology* 43:476–86.

Hawley, A. H. 1944. "Ecology and human ecology." *Social Forces* 22:398–405.

———. 1950. *Human Ecology: A Theory of Community Structure.* New York: Ronald Press.

Hayward, M. E. 2008. *Baltimore's Alley Houses: Homes for Working People Since the 1780s.* Baltimore: Johns Hopkins University Press.

Hayward, M. E., and C. Belfoure. 1999. *The Baltimore Rowhouse.* New York: Princeton Architectural Press.

Heisler, G. M., and A. J. Brazel. 2010. "The urban physical environment: Temperature and urban heat islands." Pages 29–56 in J. Aitkenhead-Peterson and A. Volder, editors, *Urban Ecosystem Ecology.* Madison: American Society of Agronomy, Crop Science Society of America, and Soil Science Society of America (ASA-CSSA-SSSA).

Heisler, G. M., A. Ellis, D. J. Nowak, and I. Yesilonis. 2016. "Modeling and imaging land-cover influences on air temperature in and near Baltimore, MD." *Theoretical and Applied Climatology* 124:497–515.

Heisler, G. M., R. H. Grant, W. Gao, and J. R. Slusser. 2004. "Solar ultraviolet-B radiation in urban environments: The case of Baltimore, MD." *Photochemistry and Photobiology* 80:422–28.

Heisler, G. M., and Y. Wang. 2002. "Applications of a human thermal comfort model." Pages 70–71, *Fourth Symposium on the Urban Environment.* Norfolk: American Meteorological Society.

Hendry, G. 1996. "Constructivism and educational practice." *Australian Journal of Education* 40:19–45.

Hogan, K. 1994. *Eco-Inquiry: A Guide to Ecological Learning Experiences for the Upper Elementary/Middle Grades.* Dubuque: Kendall/Hunt.

————. 2000. "Assessing students' systems reasoning in ecology." *Journal of Biological Education* 35:22–28.

Hogan, K., and A. R. Berkowitz. 2000. "Teachers as inquiry learners." *Journal of Science Teacher Education* 11:1–25.

Hogan, K., and J. Fisherkeller. 1996. "Representing students' thinking about nutrient cycling in ecosystems: Bidimensional coding of complex topic." *Journal of Research in Science Teaching* 33:941–70.

Hogan, K., and D. Thomas. 2001. "Cognitive comparisons of students' systems modeling in ecology." *Journal of Science Education and Technology* 10:319–45.

Holling, C. S. 1973. "Resilience and stability of ecological systems." *Annual Review of Ecology and Systematics* 4:1–23.

————. 1996. "Engineering resilience versus ecological resilience." Pages 31–44 in P. C. Schulze, editor, *Engineering within Ecological Constraints*. Washington, D.C.: National Academies of Engineering.

Hopkins, G. M. 1877. *City Atlas of Baltimore, Maryland, and Environs: From Official Records, Private Plans and Actual Surveys, Based upon Plans Deposited in the Department of Surveys*. Philadelphia: F. Bourquin.

Hopkins, K. G., N. B. Morse, D. J. Bain, N. D. Bettez, N. B. Grimm, J. L. Morse, M. M. Palta, W. D. Schuster, A. R. Bratt, and A. K. Suchy. 2015. "Assessment of regional variation in streamflow responses to urbanization and the persistence of physiography." *Environmental Science & Technology* 49:2724–32.

Huang, G. L., W. Q. Zhou, and M. L. Cadenasso. 2011. "Is everyone hot in the city? Spatial pattern of land surface temperatures, land cover and neighborhood socioeconomic characteristics in Baltimore, MD." *Journal of Environmental Management* 92:1753–59.

Huggert, R. J. 1998. "Soil chronosequences, soil development, and soil evolution: A critical review." *Catena* 32:155–72.

Inger, R. F. 1971. "Inter-American institute of ecology: Operational plan." *Bulletin of the Ecological Society of America* 52:2–4.

Irwin, E. G. 2002. "The effects of open space on residential property values." *Land Economics* 78:465–80.

Irwin, N. 2017. "Urban revitalization and targeted demolitions: How do housing markets respond to city-led redevelopment?" Working paper, Ohio State University.

Irwin, N., H. A. Klaiber, and E. G. Irwin. 2017. "Do stormwater basins generate co-benefits? Evidence from Baltimore County, Maryland." *Ecological Economics* 141:202–12.

Jacobs, J. 1961. *The Death and Life of Great American Cities: The Failure of Town Planning*. New York: Random House.

Jenerette, G. D., S. L. Harlan, A. Brazel, N. Jones, L. Larsen, and W. L. Stefanov. 2007. "Regional relationships between surface temperature, vegetation, and human settlement in a rapidly urbanizing ecosystem." *Landscape Ecology* 22:353–65.

Jenny, H. 1941. *Factors of Soil Formation: A System of Quantitative Pedology.* New York: McGraw-Hill.

Jin, H., M. E. Johnson, H. J. Shin, and C. W. Anderson. 2017. "Promoting students' progressions in science classrooms: A video study." *Journal of Research in Science Teaching* 54:852–83.

Jin, H., M. E. Johnson, and R. N. Yestness. 2015. "A learning progression approach to incorporate climate sustainability into teacher education." Pages 121–39 in S. Stratton, R. Hagevik, A. Feldman, and M. Bloom, editors, *Educating Science Teachers for Sustainability.* Berlin: Springer.

Jin, H., L. Zhan, and C. W. Anderson. 2013. "Developing a fine-grained learning progression framework for carbon-transforming processes." *International Journal of Science Education* 35:1663–97.

Johnson, A. L., E. C. Tauzer, and C. M. Swan. 2014. "Human legacies differentially organize functional and phylogenetic diversity of urban herbaceous plant communities at multiple spatial scales." *Applied Vegetation Science* 18:513–27.

Johnson, K. 1953. "The genesis of the Baltimore ironworks." *Journal of Southern History* 19:157–79.

Johnson, S. 2001. "Emergence: *The Connected Lives of Ants, Brains, Cities, and Software.* New York: Simon and Schuster.

Jordan, R. C., J. R. DeLisi, W. R. Brooks, S. A. Gray, A. Alvarado, and A. R. Berkowitz. 2013. "A collaborative model of science teacher professional development." *International Journal of Modern Education Forum* 2:31–41.

Jordan, R., F. Singer, J. Vaughan, and A. R. Berkowitz. 2009. "What should every citizen know about ecology?" *Frontiers in Ecology and the Environment* 7:495–500.

Kalkstein, L. S., and J. S. Greene. 1997. "An evaluation of climate/mortality relationships in large U.S. cities and the possible impacts of a climate change." *Environmental Health Perspectives* 105:84–93.

Kaushal, S. S., and K. T. Belt. 2012. "The urban watershed continuum: Evolving spatial and temporal dimensions." *Urban Ecosystems* 15:409–35.

Kaushal, S. S., K. Delaney-Newcomb, S. E. G. Findlay, T. A. Newcomer, S. Duan, M. J. Pennino, G. M. Sivirichi, A. M. Sides-Raley, M. R. Walbridge, and K. T. Belt. 2014. "Longitudinal patterns in carbon and nitrogen fluxes and stream metabolism along an urban watershed continuum." *Biogeochemistry* 121:23–44.

Kaushal, S. S., S. S. Duan, T. R. Doody, S. Haq, R. M. Smith, T. A. Newcomer Johnson, K. D. Newcomb, J. Gorman, N. Bowman, P. M. Mayer, K. L. Wood, K. T. Belt, and W. P. Stack. 2017. "Human-accelerated weathering increases salinization, major ions, and alkalinization in fresh water across land use." *Applied Geochemistry* 83:121–35.

Kaushal, S. S., P. M. Groffman, G. E. Likens, K. T. Belt, W. P. Stack, V. R. Kelly, L. E. Band, and G. T. Fisher. 2005. "Increased salinization of fresh water in the northeastern United States." *Proceedings of the National Academy of Sciences* 102:13517–20.

Keiny, S., M. Shachak, and N. Avriel-Avni. 2003. "Ecological thinking: As a tool for understanding urban ecosystems." Pages 315–27 in A. R. Berkowitz, C. H. Nilon, and K. S. Hollweg, editors, *Understanding Urban Ecosystems*. New York: Springer.

Kennedy, C., J. Cuddihy, and J. Engel-Yan. 2007. "The changing metabolism of cities." *Journal of Industrial Ecology* 11:43–59.

Khon Kaen University. 1987. Proceedings of the 1985 International Conference on Rapid Rural Appraisal, Rural Systems Research and Farming Systems Research Projects, Khon Kaen, Thailand.

Kingsland, S. E. 2005. *The Evolution of American Ecology, 1890–2000*. Baltimore: Johns Hopkins University Press.

Klaiber, H. A., and D. J. Phaneuf. 2010. "Valuing open space in a residential sorting model of the Twin Cities." *Journal of Environmental Economics and Management* 60:57–77.

Klemow, K. 1991. "Basic ecological literacy: A first cut." *Ecological Society of America Education Section Newsletter* 2:4–5.

Knapp, S., I. Kuhn, O. Schweiger, and S. Klotz. 2008. "Challenging urban species diversity: Contrasting phylogenetic patterns across plant functional groups in Germany." *Ecology Letters* 11:1054–64.

Kolasa, J., and S. T. A. Pickett, editors. 1991. *Ecological Heterogeneity*. New York: Springer-Verlag.

Koliba, C., J. Meek, and A. Zia. 2010. *Governance Networks in Public Administration and Public Policy*. Boca Raton: CRC Press/Taylor & Francis.

Korth, C. A., and G. L. Buckley. 2006. "Leakin Park: Frederick Law Olmsted, Jr.'s critical advice." *The Olmstedian* 16.

Kudryavtsev, A., R. C. Stedman, and M. E. Krasny. 2012. "Sense of place in environmental education." *Environmental Education Research* 18:229–50.

Larson, K. L., K. C. Nelson, S. R. Samples, S. J. Hall, N. D. Bettez, J. Cavender-Bares, P. M. Groffman, J. M. Grove, J. B. Heffernan, S. E. Hobbie, J. Learned, J. L. Morse, C. Neill, L. A. Ogden, J. O'Neil-Dunne, D. E. Pataki, C. Polsky, R. Roy Chowdhury, M. Steele, and T. L. E. Trammel. 2016. "Ecosystem services in managing residential landscapes: Priorities, value dimensions, and cross-regional patterns." *Urban Ecosystems* 19:95–113.

Law, N. L., L. E. Band, and J. M. Grove. 2004. "Nitrogen input from residential lawn care practices in suburban watersheds in Baltimore County, MD." *Journal of Environmental Planning and Management* 47:737–55.

Lehman, A., and K. Stahr. 2007. "Nature and significance of anthropogenic urban soils." *Journal of Soils and Sediments* 7:247–60.

Leibold, M. A., M. Holyoak, N. Mouquet, P. Amarasekare, J. M. Chase, M. F. Hoopes, R. D. Holt, J. B. Shurin, R. Law, D. Tilman, M. Loreau, and A. Gonzalez. 2004. "The metacommunity concept: A framework for multi-scale community ecology." *Ecology Letters* 7:601–13.

Lent, R. W., S. D. Brown, and G. Hackett. 1994. "Toward a unifying social cognitive theory of career and academic interest, choice, and performance." *Journal of Vocational Behavior* 45:79–122.

Levin, D. T., D. Hammer, A. Elby, and J. Coffee. 2013. *Becoming a Responsive Science Teacher: Focusing on Student Thinking in Secondary Science*. Arkington: NSTA Press.

Lewis, R. L. 1979. *Coal, Iron, and Slaves: Industrial Slavery in Maryland and Virginia, 1715–1865*. Westport: Greenwood Press.

Light, J. S. 2009. *The Nature of Cities: Ecological Visions and the American Urban Professions, 1920–1960*. Baltimore: Johns Hopkins University Press.

Likens, G. E. 1992. *The Ecosystem Approach: Its Use and Abuse*. Oldendorf/Luhe, Germany: Ecology Institute.

Likens, G. E., and F. H. Bormann. 1995. *Biogeochemistry of a Forested Ecosystem*. 2nd edition. New York: Springer-Verlag.

Likens, G. E., F. H. Bormann, R. S. Pierce, J. S. Eaton, and N. M. Johnson. 1977. *Biogeochemistry of a Forested Ecosystem*. New York: Springer-Verlag.

Livy, M. R., and H. A. Klaiber. 2016. "Maintaining public goods: The capitalized value of local park renovations." *Land Economics* 92:96–116.

Locke, D. H., M. Avolio, T. L. E. Trammel, R. Roy Chowdhury, J. M. Grove, J. Rogan, D. G. Martin, N. Bettez, J. Cavender-Bares, P. M. Groffman, S. J. Hall, J. B. Heffernan, S. E. Hobbie, K. L. Larson, J. L. Morse, C. Neill, K. C. Nelson, L. A. Ogden, J. O'Neil-Dunne, D. E. Pataki, W. D. Pearse, C. Polsky, and M. M. Wheeler. 2018. "The relationships among public visibility, social norms, biodiversity, and ecosystem processes: Testing the landscape mullets concept on urban residential lands." *Landscape and Urban Planning* 178:102–11, doi:10.1016/j.landurbplan.2018.05.030.

Locke, D. H., C. Polsky, J. M. Grove, P. M. Groffman, K. C. Nelson, K. L. Larson, J. Cavender-Bares, J. B. Heffernan, R. Roy Chowdhury, S. E. Hobbie, N. D. Bettez, C. Neill, L. A. Ogden, and J. O'Neil-Dunne. In Revision. "Irrigation, fertilization, and pesticide application by households, among neighborhoods along an urban–rural gradient, and across climatically-diverse regions." *PLOS One*.

Lord, C., and K. Norquist. 2010. "Cities as emergent systems: Race as a rule in organized complexity." *Environmental Law* 40:551–96.

Luck, M., G. D. Jenrette, and J. G. N. B. Wu. 2001. "The urban funnel model and spatially explicit ecological footprint." *Ecosystems* 4:782–96.

Lynch, K. 1960. *The image of the city*. Cambridge: MIT Press.

Machlis, G. E., J. E. Force, and W. R. Burch. 1997. "The human ecosystem 1. The human ecosystem as an organizing concept in ecosystem management." *Society & Natural Resources* 10:347–67.

Mann, C. C. 2011. 1491. *New Revelations of the Americas Before Columbus*. New York: Vintage Books.

Marsh, G. P. 1864. *Man and Nature; or, Physical Geography as Modified by Human Action*. Cambridge: Harvard University Press.

Marshall, V., M. L. Cadenasso, B. McGrath, and S. T. A. Pickett. 2019. *Patch Atlas: Integrating Design Principles and Ecological Knowledge for Cities as Complex Systems*. New Haven: Yale University Press.

Martinez, N. G., N. D. Bettez, and P. M. Groffman. 2014. "Sources of variation in home lawn soil nitrogen dynamics." *Journal of Environmental Quality* 43:2146–51.

Marye, W. B. 1920. "The old Indian road." *Maryland Historical Magazine* 15:107–24.

Maryland State Department of Education. 2011. *Maryland Environmental Literacy Standards*. http://marylandpublicschools.org/programs/Documents/Environ mental/MDEnvironmentalLitStandards.pdf.

———. *Maryland Standards State Science Curriculum*. http://marylandpublicschools .org/about/Pages/DCAA/Science/index.aspx.

McBride, B. B. 2011. "Essential elements of ecological literacy and pathways to achieve it: Perspectives of ecologists." Ph.D. diss., University of Montana.

McBride, B. B., C. A. Brewer, A. R. Berkowitz, and W. T. Borrie. 2013. "Environmental literacy, ecological literacy, ecoliteracy: What do we mean and how did we get here?" *Ecosphere* 4:1–20.

McConnaughay, K. D. M., and F. A. Bazzaz. 1990. "Interactions among colonizing annuals: Is there an effect of gap size?" *Ecology* 71:1941–51.

McDonnell, M. J., and A. K. Hahs. 2008. "The use of gradient analysis studies in advancing our understanding of the ecology of urbanizing landscapes: Current status and future directions." *Landscape Ecology* 23:1143–55.

McDonnell, M. J., and S. T. A. Pickett. 1990. "Ecosystem structure and function along urban–rural gradients: An unexploited opportunity for ecology." *Ecology* 71:1232–37.

McDonnell, M. J., S. T. A. Pickett, and R. V. Pouyat. 1993. "The application of the ecological gradient paradigm to the study of urban effects." Pages 175–89 in M. J. McDonnell and S. T. A. Pickett, editors, *Humans as Components of Ecosystems: The Ecology of Subtle Human Effects and Populated Areas*. New York: Springer-Verlag.

McDonnell, M. J., S. T. A. Pickett, R. V. Pouyat, R. W. Parmelee, and M. M. Carreiro. 1997. "Ecosystem processes along an urban-to-rural gradient." *Urban Ecosystems* 1:21–36.

McGrain, J. W. 1980. *Grist Mills in Baltimore County, Maryland*. Towson, Md.: Baltimore County Public Library.

———. 1985. *From Pig Iron to Cotton Duck: A History of Manufacturing Villages in Baltimore County*. Towson, Md.: Baltimore County Public Library.

McGrath, B. 1994. *Transparent Cities*. New York: SITES Books.

———. 2000. *Manhattan Timeformations*. New York: Skyscraper Museum.

———. 2008. *Digital Modeling for Urban Design*. London: John Wiley.

———, editor. 2013. *Urban Design Ecologies*. London: John Wiley.

McGrath, B. P., V. Marshall, M. L. Cadenasso, J. M. Grove, S. T. A. Pickett, R. Plunz, and J. Towers, editors. 2007. *Designing Patch Dynamics*. New York: Columbia University School of Architecture, Planning and Preservation.

McGrath, B. P., and S. T. A. Pickett. 2016. "Archaeology of the metacity." Pages 104–34 in W. Ding, A. Graafland, and A. Lu, editors, *Cities in transition II: Power, environment, society*. Rotterdam: NAi 010.

———. 2011. "The metacity: A conceptual framework for integrating ecology and urban design." *Challenges* 2011:55–72.

McHale, M. R., S. T. A. Pickett, O. Barbosa, D. N. Bunn, M. L. Cadenasso, D. L. Childers, M. Gartin, G. R. Hess, D. M. Iwaniec, T. McPhearson, M. N. Peterson, A. K. Poole, L. Rivers, S. T. Shutters, and W. Zhou. 2015. "The new global urban realm: Complex, connected, diffuse, and diverse social-ecological systems." *Sustainability* 7:5211–40.

McHarg, I. 1969. *Design with Nature*. Garden City, N.J.: Doubleday/Natural History Press.

McKinney, M. L. 2006. "Urbanization as a major cause of biotic homogenization." *Biological Conservation* 127:247–60.

McKinney, M. L., and J. L. Lockwood. 1999. "Biotic homogenization: A few winners replacing many losers in the next mass extinction." *Trends in Ecology & Evolution* 14:450–53.

McPhearson, T., S. T. A. Pickett, N. B. Grimm, J. Niemelä, M. Alberti, T. Elmqvist, C. Weber, J. Breuste, D. Haase, and S. Qureshi. 2016. "Advancing urban ecology towards a science of cities." *BioScience* 66:198–212.

Melosi, M. V. 2000. *The Sanitary City: Urban Infrastructure in America from Colonial Times to the Present*. Baltimore: Johns Hopkins University Press.

Merse, C. L., G. L. Buckley, and C. G. Boone. 2009. "Street trees and urban renewal: A Baltimore case study." *Geographical Bulletin* 50.

Middel, A., K. Habba, A. J. Brazel, C. A. Martin, and S. Guhathakurta. 2014. "Impact of urban form and design on mid-afternoon microclimate in Phoenix Local Climate Zones." *Landscape and Urban Planning* 122:16–28.

Milesi, C., S. W. Running, C. D. Elvidge, J. B. Dietz, B. T. Tuttle, and R. R. Nemani. 2005. "Mapping and modeling the biogeochemical cycling of turf grasses in the United States." *Environmental Management* 36:426–38.

Mitrasinovic, M. 2015. *Concurrent Urbanities: Designing Infrastructures of Inclusion*. New York: Routledge.

Moore, B. 1920. "The scope of ecology." *Ecology* 1:3–5.

Morel, J. L., W. Burghardt, and K.-H. J. Kim. 2017. "The challenges for soils in the urban environment." Pages 1–6 in M. J. Levin, K.-H. J. Kim, J. L. Morel, W. Burghardt, P. Charzynski, R. K. Shaw, and IUSS Working Group SUITMA, editors, *Soils Within Cities*. Stuttgart: Schweizerbart.

Morris, C. 2012. *The Big Muddy: An Environmental History of the Mississippi and Its Peoples*. New York: Oxford University Press.

Morton, T. G., A. J. Gold, and W. M. Sullivan. 1998. "Influence of overwatering and fertilization on nitrogen losses from home lawns." *Journal of Environmental Quality* 17:124–30.

Mouquet, N., M. Loreau, and A. E. P. J. Morin. 2003. "Community patterns in source-sink metacommunities." *American Naturalist* 162:544–57.

Mount Royal Improvement Association. 1930. *The Mount Royal District: Baltimore's Best Urban Section*. Baltimore: Maryland Historical Society.

Mount Washington Improvement Association. 1900. *Picturesque Mount Washington*.
    Baltimore: Maryland Historical Society.
Mukerjee, R. 1926. *Regional Sociology*. New York: Century.
———. 1932a. "The concepts of distribution and succession in social ecology." *Social
    Forces* 11:1–7.
———. 1932b. "The ecological outlook in sociology." *American Journal of Sociology*
    38:349–55.
Mundry, S., and K. E. Stiles. 2009. "Professional learning communities for science
    teaching: Lessons from research and practice." Arlington: NSTA Press.
Muñoz-Erickson, T., L. K. Campbell, D. L. Childers, J. M. Grove, D. M. Iwaniec,
    S. T. A. Pickett, M. Romolini, and E. S. Svendsen. 2016. "Demystifying gover-
    nance and its role for transitions in urban social-ecological systems." *Ecosphere* 7.
Myint, S., E. Talen, B. Zheng, C. Fan, S. Kaplan, A. Middel, M. Smith, H.-P. Huang,
    and A. Brazel. 2015. "Does the spatial arrangement of urban landscape matter?
    Examples of urban warming and cooling in Phoenix and Las Vegas." *Ecosystem
    Health and Sustainability* 1:1–15.
Nassauer, J. I., Z. Wang, and E. Dayrell. 2009. "What will the neighbors think?
    Cultural norms and ecological design." *Landscape and Urban Planning*
    92:282–92.
National Research Council. 1996. *National Science Education Standards*. Washington,
    D.C.: National Academies Press.
———. 2012. *A Framework for K-12 Science Education: Practices, Crosscutting Concepts,
    and Core Ideas*. Washington, D.C.: National Academies Press.
Neuhold, J. M. 1975. "The Institute of Ecology: TIE." *Bulletin of the Ecological Society
    of America* 56:9–12.
Newburn, D. A., and J. S. Ferris. 2016. "The effect of downzoning for managing
    residential development and density." *Land Economics* 92:220–36.
Next Generation Science Standards Lead States. 2013. *Next Generation Science
    Standards: For States, by States*. Washington, D.C.: National Academies Press.
Nilon, C. H. 2011. "Urban biodiversity and the importance of management and
    conservation." *Landscape and Ecological Engineering* 7:45–52.
North American Association for Environmental Education. 2000. *Excellence in
    Environmental Education: Guidelines for Learning (K-12)*. Washington, D.C.:
    North American Association of Environmental Education.
Odum, E. P. 1971. *Fundamentals of Ecology*. 3rd edition. Philadelphia: Saunders.
Odum, E. P., and H. T. Odum. 1972. "Natural areas as necessary components of
    man's total environment." *Transactions of the North American Wildlife and
    Natural Resources Conferences* 37:178–79.
Odum, H. T. 1971. *Environment, Power, and Society*. New York: John Wiley.
Odum, H. T., and R. C. Pinkerton. 1955. "Time's speed regulator: The optimum
    efficiency for maximum power output in physical and biological systems."
    American Scientist 43:331–43.
Odum, H. W. 1939. *American Social Problems: An Introduction to the Study of the
    People and Their Dilemma*. New York: Henry Holt.

———. 1951. *American Sociology: The Story of Sociology in the United States Through 1950*. New York: Longmans, Green.

Odum, H. W., and H. E. Moore. 1938. *American Regionalism: A Cultural-Historical Approach to National Integration*. New York: Henry Holt.

Oke, T., G. Mills, A. Christen, and J. Voogt. 2017. *Urban Climates*. Cambridge: Cambridge University Press.

Oke, T. R. 1987. *Boundary Layer Climates*. New York: Methuen.

Olmsted Brothers. 1904. *Upon the Development of Public Grounds for Greater Baltimore*. Baltimore: Baltimore Municipal Arts Society.

Olson, S. 1976. *Baltimore*. Cambridge: Ballinger.

———. 1982. "Urban metabolism and morphogenesis." *Urban Geography* 3:87–109.

———. 2007. "Downwind, downstream, downtown: The environmental legacy in Baltimore and Montreal." *Environmental History* 12:845–66.

Olson, S. H. 1997. *Baltimore: The Building of an American City*. Baltimore: Johns Hopkins University Press.

Orr, D. 1992. *Ecological Literacy: Education and the Transition to a Postmodern World*. Albany: SUNY Press.

Orr, J. B. 1953. *The White Man's Dilemma: Food and the Future*. London: George Allen and Unwin.

Orser, W. E. 1994. *Blockbusting in Baltimore: The Edmondson Village Story*. Lexington: University Press of Kentucky.

Osborn, F. 1948. *Our Plundered Planet*. Boston: Little, Brown.

Ostrom, E. 1990. *Governing the Commons: The Evolution of Institutions for Collective Action*. New York: Cambridge University Press.

———. 2005. *Understanding Institutional Diversity*. Princeton: Princeton University Press.

———. 2007. "A diagnostic approach for going beyond panaceas." *Proceedings of the Natural Academy of Sciences* 104:15181–87.

Outen, D. C. 2007. "Pioneer on the frontier of smart growth: The Baltimore County, MD experience." 49 pages, Conference on Smart Growth @ 10, National Center for Smart Growth Research and Education, Maryland: College Park.

Park, R. E. 1915. "The city: Suggestions for the investigation of human behavior in the city environment." *American Journal of Sociology* 20:577–612.

———. 1936. "Human ecology." *American Journal of Sociology* 42:1–15.

Pautasso, M. 2007. "Scale dependence of the correlation between human population presence and vertebrate and plant species richness." *Ecology Letters* 10:16–24.

Peabody Heights Improvement Association. *Meeting Minutes 1909–1933*. Baltimore: Maryland Historical Society.

Peabody Heights Improvement Association of Baltimore City. 1908. *Constitution*. Baltimore: Maryland Historical Society.

Petrovic, A. M. 1990. "The fate of nitrogenous fertilizers applied to turfgrass." *Journal of Environmental Quality* 19:1–14.

Pickett, S. T. A., K. T. Belt, M. F. Galvin, P. Groffman, J. M. Grove, D. C. Outen, R. Pouyat, W. P. Stack, and M. L. Cadenasso. 2007a. "Watersheds in Baltimore,

Maryland: Understanding and application of integrated ecological and social processes." *Journal of Contemporary Water Research and Education* 136:44–55.

Pickett, S. T. A., C. G. Boone, and M. L. Cadenasso. 2013a. "Ecology and environmental justice: Understanding disturbance using ecological theory." Pages 27–47 in C. G. Boone and M. Fragkias, editors, *Urbanization and Sustainability—Linking Ecology, Environmental Justice, and Global Environmental Change.* Dordrecht: Springer.

Pickett, S. T. A., G. L. Buckley, S. S. Kaushal, and Y. Williams. 2011a. "Ecological science in the humane metropolis." *Urban Ecosystems* 14:319–39.

Pickett, S. T. A., and M. L. Cadenasso. 2007. "Patch dynamics as a conceptual tool to link ecology and design." Pages 16–29 in B. McGrath, M. L. Cadenasso, J. M. Grove, V. Marshall, S. Pickett, and J. Towers, editors, *Designing Urban Patch Dynamics.* New York: Columbia University Graudate School of Architecture, Planning and Preservation.

———. 1995. "Landscape ecology: Spatial heterogeneity in ecological systems." *Science* 269:331–34.

———. 2008. "Linking ecological and built components of urban mosaics: An open cycle of ecological design." *Journal of Ecology* 96:8–12.

———. 2017. "How many principles of urban ecology are there?" *Landscape Ecology* 32:699–705.

Pickett, S. T. A., M. L. Cadenasso, D. L. Childers, M. J. McDonnell, and W. Zhou. 2016. "Evolution and future of urban ecological science: Ecology in, of and for the city." *Ecosystem Health and Sustainability* 2:e01229.

Pickett, S. T. A., M. L. Cadenasso, and J. M. Grove. 2004. "Resilient cities: Meaning, models, and metaphor for integrating the ecological, socio-economic, and planning realms." *Landscape and Urban Planning* 69:369–84.

———. 2005. "Biocomplexity in coupled natural–human systems: A multidimensional framework." *Ecosystems* 8:225–32.

Pickett, S. T. A., M. L. Cadenasso, J. M. Grove, C. G. Boone, P. M. Groffman, E. Irwin, S. S. Kaushal, V. Marshall, B. P. McGrath, C. H. Nilon, R. V. Pouyat, K. Szlavecz, A. Troy, and P. Warren. 2011b. "Urban ecological systems: Scientific foundations and a decade of progress." *Journal of Environmental Management* 92:331–62.

Pickett, S. T. A., M. L. Cadenasso, J. M. Grove, P. M. Groffman, L. E. Band, C. G. Boone, W. R. Burch, C. S. B. Grimmond, J. Hom, J. C. Jenkins, N. L. Law, C. H. Nilon, R. V. Pouyat, K. Szlavecz, P. S. Warren, and M. A. Wilson. 2008. "Beyond urban legends: An emerging framework of urban ecology, as illustrated by the Baltimore Ecosystem Study." *BioScience* 58:139–50.

Pickett, S. T. A., M. L. Cadenasso, J. M. Grove, C. H. Nilon, R. V. Pouyat, W. C. Zipperer, and R. Costanza. 2001. "Urban ecological systems: Linking terrestrial ecological, physical, and socioeconomic components of metropolitan areas." *Annual Review of Ecology and Systematics* 32:127–57.

Pickett, S. T. A., M. L. Cadenasso, and B. McGrath, editors. 2013b. *Resilience in Ecology and Urban Design: Linking Thoery and Practice for Sustainable Cities.* New York: Springer.

Pickett, S. T. A., M. L. Cadenasso, E. J. Rosi-Marshall, K. T. Belt, P. M. Groffman, J. M. Grove, E. G. Irwin, S. S. Kaushal, S. L. LaDeau, C. H. Nilon, C. M. Swan, and P. S. Warren. 2017. "Dynamic heterogeneity: A framework to promote ecological integration and hypothesis generation in urban systems." *Urban Ecosystems* 20:1–14.

Pickett, S. T. A., and J. M. Grove. 2009. "Urban ecosystems: What would Tansley do?" *Urban Ecosystems* 12:1–8.

Pickett, S. T. A., W. R. Burch Jr., S. E. Dalton, and T. W. Foresman. 1997. "Integrated urban ecosystem research." *Urban Ecosystems* 1:183–84.

Pickett, S. T. A., J. Kolasa, and C. G. Jones. 2007b. *Ecological Understanding: The Nature of Theory and the Theory of Nature.* 2nd edition. Boston: Academic Press.

Pickett, S. T. A., B. McGrath, M. L. Cadenasso, and A. J. Felson. 2014. "Ecological resilience and resilient cities." *Building Research & Information* 42:143–57.

Pickett, S. T. A., V. T. Parker, and P. Fiedler. 1992. "The new paradigm in ecology: Implications for conservation biology above the species level." Pages 65–88 in P. Fiedler and S. Jain, editors, *Conservation Biology: The Theory and Practice of Nature Conservation, Preservation and Management.* New York: Chapman and Hall.

Pickett, S. T. A., and P. S. White, editors. 1985. *The Ecology of Natural Disturbance and Patch Dynamics.* Orlando: Academic Press.

Pickett, S. T. A., J. Wu, and M. L. Cadenasso. 1999. "Patch dynamics and the ecology of disturbed ground: A framework for synthesis." Pages 707–22 in L. R. Walker, editor, *Ecosystems of the World: Ecosystems of Disturbed Ground.* Amsterdam: Elsevier Science.

Pierce, E. 2010. "The historic roots of green urban policy in Baltimore County, Maryland." M.A. diss., Ohio University, Athens.

Platt, R. H., editor. 2006. *The Humane Metropolis: People and Nature in the 21st-Century City.* Amherst: University of Massachusetts Press.

Polsky, C., J. M. Grove, C. Knudson, P. M. Groffman, N. Bettez, J. Cavender-Bares, S. J. Hall, J. B. Heffernan, S. E. Hobbie, K. L. Larson, J. L. Morse, C. Neill, K. C. Nelson, L. A. Ogden, J. O'Neil-Dunne, D. E. Pataki, R. R. Chowdhury, and M. K. Steele. 2014. "Assessing the homogenization of urban land management with an application to US residential lawn care." *Proceedings of the National Academy of Sciences* 111:4432–37.

Poole, A. K., E. C. Hargrove, P. Day, W. Forbes, A. R. Berkowitz, P. Feinsinger, and R. Rozzi. 2013. "A call for ethics literacy in environmental education." Pages 349–71 in R. Rozzi, S. T. A. Pickett, C. Palmer, J. Armesto, and J. Callicott, editors, *Linking Ecology and Ethics for a Changing World.* Dordrecht: Springer.

Posner, G. J., K. A. Strike, P. W. Hewson, and W. A. Gertzog. 1982. "Accommodation of a scientific conception: Toward a theory of conceptual change." *Science Education* 66:211–27.

Pouyat, R. V., M. M. Carreiro, P. M. Groffman, and M. A. Pavao-Zuckerman. 2009. "Investigative approaches to urban biogeochemical cycles: New York metropolitan area and Baltimore as case studies." Pages 329–51 in M. J. McDonnell,

A. Hahs, and J. Breuste, editors, *Ecology of Cities and Towns: A Comparative Approach*. New York: Cambridge University Press.

Pouyat, R. V., and W. R. Effland. 1999. "The investigation and classification of humanly modified soils in the Baltimore Ecosystem Study." Pages 141–54 in J. M. Kimble, R. J. Ahrens, and J. P. Bryant, editors, *Classification, Correlation, and Management of Anthropogenic Soils, Proceedings—Nevada and California, September 21–October 2, 1998*. Lincoln, Neb.: USDA Natural Resource Conservation Service, National Survey Center.

Pouyat, R. V., K. Szlavecz, Y. D. Yesilonis, P. M. Groffman, and K. Schwarz. 2010. "Chemical, physical, and biological characteristics of urban soils." Pages 119–52 in J. Aitkenhead-Peterson and A. Volder, editors, *Urban Ecosystem Ecology*. Madison: American Society of Agronomy and Crop Science Society of America.

Power, G. 1983. "Apartheid Baltimore style: The residential segregation ordinances of 1910–1913." *Maryland Law Review* 42:289–328.

———. 1992. "Parceling out land in the vicinity of Baltimore: 1632–1796, Part 2." *Maryland Historical Magazine* 87:453–66.

Provan, K. G., and H. B. Milward. 2001. "Do networks really work? A framework for evaluating public-sector organizational networks." *Public Administration Review* 61:414–23.

Provan, K., and P. Kenis. 2008. "Modes of network governance: Structure, management, and effectiveness." *Journal of Public Administration Research and Theory* 18:229–52.

Purcell, M. 2011. "Neighborhood activism among homeowners as a politics of space." *Professional Geographer* 53:178–94.

Puttick, G., B. Drayton, J. Lockwood, M. Cole, M. Donovan and A. Berkowitz. 2016. Environmental Science and Biocomplexity. Activate Learning. Greenwich, Conn. url: https://activatelearning.com/high-school/other-curricula/environmental-science-and-biocomplexity.

Pyšek, P. 1998. "Alien and native species in Central European urban floras: A quantitative comparison." *Journal of Biogeography* 25:155–63.

Quimby, J. L., N. D. Seyala, and J. L. Wolfson. 2007a. Social cognitive predictors of interest in environmental science: Recommendations for environmental educators." *Journal of Environmental Education* 38:43–52.

Quimby, J. L., J. L. Wolfson, and N. D. Seyala. 2007b. "Social cognitive predictors of African American adolescents' career interests." *Journal of Career Development* 33:376–94.

Raciti, S. M., A. J. Burgin, P. M. Groffman, D. N. Lewis, and T. J. Fahey. 2011a. "Denitrification in suburban lawn soils." *Journal of Environmental Quality* 40:1392–1940.

Raciti, S. M., P. M. Groffman, and T. J. Fahey. 2008. "Nitrogen retention in urban lawns and forests." *Ecological Applications* 18:1615–26.

Raciti, S. M., P. M. Groffman, J. C. Jenkins, R. V. Pouyat, T. J. Fahey, M. L. Cadenasso, and S. T. A. Pickett. 2011b. "Accumulation of carbon and nitrogen in residential soils with different land use histories." *Ecosystems* 14:287–97.

Raciti, S. M., P. M. Groffman, J. C. Jenkins, R. V. Pouyat, T. J. Fahey, S. T. A. Pickett, and M. L. Cadenasso. 2011c. "Nitrate production and availability in residential soils." *Ecological Applications* 21:2357–66.

Radeloff, V. C., R. B. Hammer, S. I. Stewart, J. S. Fried, S. S. Holcomb, and J. F. McKeefry. 2005. "The wildland urban interface in the United States." *Ecological Applications* 15:799–805.

Rademacher, A., M. L. Cadenasso, and S. T. A. Pickett. 2018. "From feedbacks to coproduction: Toward an integrated conceptual framework for urban ecosystems." *Urban Ecosystems*, https://doi.org/10.1007/s11125.

Rahm, J., and J. C. Moore. 2016. "A case study of long-term engagement and identity-in-practice: Insights into the STEM pathways of four underrepresented youths." *Journal of Research in Science Teaching* 53:768–801.

Rapport, D. J., H. A. Regier, and T. C. Hutchinson. 1985. "Ecosystem behavior under stress." *American Naturalist* 125:617–40.

Raup, H. M. 1966. "The View From John Sanderson's Farm: A Perspective for the Use of the Land." *Forest & Conservation History* 10:2–11.

Reisinger, A. J., E. Woytowitz, E. Majcher, E. J. Rosi, K. T. Belt, J. M. Duncan, S. S. Kaushal, and P. M. Groffman. 2018. "Changes in long-term water quality of Baltimore streams are associated with both gray and green-infrastructure." *Limnology and Oceanography.* DOI: 10.1002/lno.10947

Richter, D., A. R. Bacon, L. M. Megan, C. J. Richardson, S. S. Andrews, L. West, S. Wills, S. Billings, C. A. Cambardella, N. Cavallaro, J. E. DeMeester, A. J. Franzluebbers, A. S. Grandy, S. Grunwald, J. Gruver, A. S. Hartshorn, H. Janzen, M. G. Kramer, J. K. Ladha, K. Lajtha, G. C. Liles, D. Markewitz, P. J. Megonigal, A. R. Mermut, C. Rasmussen, D. A. Robinson, P. Smith, C. A. Stiles, R. L. Tate, A. Thompson, A. J. Tugel, H. van Es, D. Yaalon, and T. M. Zobeck. 2011. "Human–soil relations are changing rapidly: Proposals from SSSA's cross-divisional soil change working group." *Soil Science Society of America Journal* 75:2079–84.

Risser, P. G. 1986. "Ecological literacy." *Bulletin of the Ecological Society of America* 66:455–60.

Risser, Paul G., Jane Lubchenco, Norman L. Christensen, Philip L. Johnson, Peter J. Dillon, Pamela Matson, Luis Diego Gomez, Nancy A. Moran, Daniel J. Jacob, Thomas Rosswall, and Michael Wright. 2003. Ten-year Review of the National Science Foundation's Longterm Ecological Research Program. https://lternet .edu/wp-content/uploads/2010/12/ten-year-review-of-LTER.pdf.

Robbins, P. 2007. *Lawn People: How Grasses, Weeds, and Chemicals Make Us Who We Are*. Philadelphia: Temple University Press.

Robbins, P., and T. Birkenholtz. 2003. "Turfgrass revolution: Measuring the expansion of the American lawn." *Land Use Policy* 20:181–94.

Robbins, P., A. Polderman, and T. Birkenholtz. 2001. "Lawns and toxins: An ecology of the city." *Cities* 18:369–80.

Robertson, G. P., S. L. Collins, D. R. Foster, N. Brokaw, H. W. Ducklow, T. L. Gragson, C. Gries, S. K. Hamilton, A. D. McGuire, J. C. Moore, E. H. Stanley,

R. B. Waide, and M. W. Williams. 2012. "Long-term ecological research in a human-dominated world." *BioScience* 62:342–53.

Rome, A. 2001. *The Bulldozer in the Countryside: Suburban Sprawl and the Rise of American Environmentalism*. Cambridge: Cambridge University Press.

Romolini, M., R. P. Bixler, and J. M. Grove. 2016a. "A social-ecological framework for urban stewardship network research to promote sustainable and resilient cities." *Sustainability* 8:956.

Romolini, M., and J. M. Grove. 2013. "Assessing and comparing relationships between urban environmental stewardship networks and land cover in Baltimore and Seattle." *Landscape and Urban Planning* 120:190–207.

Romolini, M., J. M. Grove, C. Koliba, D. Krymkowski, and C. Ventriss. 2016b. "Towards an understanding of citywide urban environmental governance: An examination of stewardship networks in Baltimore and Seattle." *Environmental Management* 58:254–67.

Rosi-Marshall, E. J., D. Kincaid, H. A. Bechtold, T. V. Royer, M. Rojas, and J. J. Kelly. 2013. "Pharmaceuticals suppress algal growth and microbial respiration and alter bacterial communities of stream biofilms." *Ecological Applications* 23:583–93.

Roth, C. 1968. "On the road to conservation." *Audubon* 38–41.

Ruddell, D. M., S. L. Harlan, S. Grossman-Clarke, and A. Buyantuyev. 2010. "Risk and exposure to extreme heat in microclimates of Phoenix, AZ." Pages 179–202 in P. S. Showalter and Y. Lu, editors, *Geospatial Techniques in Urban Hazard and Disaster Analysis*. New York: Springer.

Ruegg, J., W. K. Dodds, M. D. Daniels, K. R. Sheehan, C. L. Baker, W. B. Bowden, K. J. Farrell, M. B. Flinn, T. K. Harms, J. B. Jones, L. E. Koenig, J. S. Kominoski, W. H. McDowell, S. P. Parker, A. D. Rosemond, M. T. Trentman, M. Whiles, and W. M. Wollheim. 2016. "Baseflow physical characteristics differ at multiple spatial scales in stream networks across diverse biomes." *Landscape Ecology* 31:119–36.

Scheiner, S. M., and M. R. Willig, editors. 2011. *The Theory of Ecology*. Chicago: University of Chicago Press.

Schramm, J. W., H. Jin, E. G. Keeling, M. Johnson, and H. J. Shin. 2017. "Improved student reasoning about carbon-transforming processes through inquiry-based learning activities derived from an empirically validated learning progression." Research in Science Education 1–25 https://doi.org/10.1007/s11165-016-9584-0.

Schwartz, S. S., and B. Smith. 2014. "Slowflow fingerprints of urban hydrology." *Journal of Hydrology* 515:116–28.

Sears, P. B. 1954. "Human ecology: A problem in synthesis." *Science* 120:959–63.

Sengupta, P., J. S. Kinnebrew, S. Basu, G. Biswas, and D. Clark. 2013. "Integrating computational thinking with K–12 science education using agent-based computation: A theoretical framework. *Education and Information Technologies* 18:351–80.

Shane, D. G. 2005. *Recombinant Urbanism: Conceptual Modeling in Architecture, Urban Design, and City Theory*. Hoboken: John Wiley.

———. 2011. *Urban Design Since 1945: A Global Perspective*. Chichester, UK: Wiley London.

———. 2013. "Urban patch dynamics and resilience: Three London urban design ecologies." Pages 131–61 in S. T. A. Pickett, M. L. Cadenasso, and B. McGrath, editors, *Resilience in Ecology and Urban Design: Linking Theory and Practice for Sustainable Cities*. New York: Springer.

Sharrer, G. T. 1976. "Flour milling in the growth of Baltimore, 1750–1830." *Maryland Historical Magazine* 71:323.

Sheridan, S. C. 2002. "The redevelopment of weather-type classification scheme for North America." *International Journal of Climatology* 22:51–68.

Shevky, E., and W. Bell. 1955. *Social Area Analysis: Theory, Illustrative Application and Computational Procedures*. Stanford: Stanford University Press.

Shreve, F., M. A. Chrysler, F. H. Blodgett, and F. W. Besley. 1910. *The Plant Life of Maryland*. Baltimore: Johns Hopkins University Press.

Simberloff, D. 1980. "Succession of paradigms in ecology—Essentialism to materialism and probabilism." *Synthese* 43:3–39.

Simberloff, D. 2014. "The 'balance of nature': Evolution of a panchreston." *PLOS Biology* 12:e1001963.

Simmons, B. 2003. "An interdisciplinary approach to urban ecosystems." Pages 282–93 in A. R. Berkowitz, C. H. Nilon, and K. S. Hollweg, editors, *Understanding Urban Ecosystems*. New York: Springer-Verlag.

Smith, B. K., J. A. Smith, M. L. Baeck, G. Villarini, and D. B. Wright. 2013. "Spectrum of storm event hydrologic response in urban watersheds." *Water Resources Research* 49:2649–63.

Spilhaus, A. 1967. "The experimental city." *Daedalus* 96:1129–41.

Spirn, A. W. 1984. *The Granite Garden: Urban Nature and Human Design*. New York: Basic Books.

Stabler, L., C. A. Martin, and A. Brazel. 2005. "Microclimates in a desert city were related to land use and vegetation index." *Urban Forestry & Urban Greening* 3:137–47.

Stearns, F. W., and T. Montag. 1974. *The Urban Ecosystem: A Holistic Approach*. Stroudsburg, Pa.: Dowden, Hutchinson and Ross.

Stoermer, E. F., J. A. Wolin, C. L. Schelske, and D. L. Conley. 1985. "An assessment of ecological changes during the recent history of Lake Ontario based on siliceous microfossils preserved in the sediments." *Journal of Phycology* 21:257–76.

Stokes, D. E. 1997. *Pasteur's Quadrant: Basic Science and Technological Innovation*. Washington, D.C.: Brookings Institution Press.

Svendsen, E. S., L. K. Campbell, D. R. Fisher, J. J. Connolly, M. L. Johnson, N. F. Sonti, D. H. Locke, L. M. Westphal, C. L. Fisher, J. M. Grove, M. Romolini, D. J. Blahna, and K. L. Wolf. 2016. "Stewardship mapping and assessment project: A framework for understanding community-based environmental stewardship. General Technical Report 156, U.S. Department of Agriculture, Forest Service, Northeastern Research Station, Newtown Square, Pa.

Swan, C. M., A. E. Johnson, and D. Nowak. 2017. "Differential organization of taxonomic and functional diversity in an urban woody plant metacommunity." *Applied Vegetation Science* 20:7–17.

Swan, C. M., S. T. A. Pickett, K. Szlavecz, P. Warren, and K. T. Willey. 2011. "Biodiversity and community composition in urban ecosystems: Coupled human, spatial, and metacommunity processes." Pages 179–86 in J. Niemelä, editor, *Handbook of Urban Ecology*. New York: Oxford University Press.

Tansley, A. G. 1935. "The use and abuse of vegetational concepts and terms." *Ecology* 16:284–307.

Tarr, J. A. 2002. "The metabolism of the industrial city: The case of Pittsburgh." *Journal of Urban History* 28:511–45.

Thomas, W. L., editor. 1956. *Man's Role in Changing the Face of the Earth*. Chicago: University of Chicago Press.

Thompson, G. L., and J. Kao-Kniffien. 2016. "Diversity enhances NPP, N retention, and soil microbial diversity in experimental urban grassland assemblages." *PLOS One* 11(5):e0155986. https://doi.org/10.1371/journal.pone.0155986.

Tidball, K. G., and M. E. Krasny. 2010. "Urban environmental education from a social-ecological perspective: Conceptual framework for civic ecology education." *Cities and the Environment* 3 (1):A 11.

———. 2017. *Urban Education Review*. Ithaca: Cornell University Press.

Torre, S. S. 1981. "Space as matrix. Making room: Women in architecture." *Heresies* 11:51–52.

Towe, C., H. A. Klaiber, and J. Maher. 2015. "A valuation of restored streams using repeat sales." Working paper, University of Connecticut.

Troy, A. 2012. *The Very Hungry City: Urban Energy Efficiency and the Economic Fate of Cities*. New Haven: Yale University Press.

Troy, A., J. M. Grove, and J. O'Neil-Dunne. 2012. "The relationship between tree canopy and crime rates across an urban–rural gradient in the greater Baltimore region." *Landscape and Urban Planning* 106:262–70.

Troy, A., A. Nunery, and J. M. Grove. 2016. "The relationship between residential yard management and neighborhood crime: An analysis from Baltimore City and County." *Landscape and Urban Planning* 147:78–87.

Troy, A. R., and J. M. Grove. 2008. "Property values, parks, and crime: A hedonic analysis in Baltimore, MD." *Landscape and Urban Planning* 87:233–45.

Troy, A. R., J. M. Grove, J. P. M. O'Neil-Dunne, M. L. Cadenasso, and S. T. A. Pickett. 2007. "Predicting opportunities for greening and patterns of vegetation on private urban lands." *Environmental Management* 40:394–412.

Turner, M. G., editor. 1987. *Landscape Heterogeneity and Disturbance*. New York: Springer-Verlag.

———. 1989. "Landscape ecology: The effect of pattern on process." *Annual Review of Ecology and Systematics* 20:171–97.

Tyndall, R. W. 1992. "Historical considerations of conifer expansion in Maryland Serpentine 'Barrens.'" *Castanea* 57:123–31.

Vitousek, P. M., and J. M. Melillo. 1979. "Nitrate losses from disturbed forests: Patterns and mechanisms." *Forest Science* 25:605–19.

Vogt, W. 1948. *Road to Survival*. New York: W. Sloane Associates.

Vygotsky, L. S. 1962. *Thought and Language*. Cambridge: MIT Press.

———. 1978. *Mind in Society: The Development of Higher Psychological Processes.* Cambridge: Harvard University Press.

Waters, E. R., J. L. Morse, N. D. Bettez, and P. M. Groffman. 2014. "Differential carbon and nitrogen controls of denitrification in riparian zones and streams along an urban to exurban gradient." *Journal of Environment Quality* 43:955.

Webb, C. O., D. D. Ackerly, M. A. McPeek, and M. J. Donoghue. 2002. "Phylogenies and community ecology." *Annual Review of Ecology and Systematics* 33:475–505.

Webster, J. R., J. D. Knoepp, W. T. Swank, and C. F. Miniat. 2016. "Evidence for a regime shift in nitrogen export from a forested watershed." *Ecosystems* 19:881–95.

Wells, J. E., et al. 2008. "Separate but equal? Desegregating Baltimore's golf courses." *Geographical Review* 98:151–70.

Westman, W. E. 1977. "How much are nature's services worth?" *Science* 197:960–64.

Whitman, T. S. 1997. *The Price of Freedom: Slavery and Manumission in Baltimore and Early National Maryland*. Lexington: University Press of Kentucky.

Whitney, G. G. 1994. *From Coastal Wilderness to Fruited Plain: An Ecological History of Northeastern United States*. Cambridge: Cambridge University Press.

Whittaker, R. H. 1956. "Vegetation of the Great Smoky Mountains." *Ecological Monographs* 26:1–80.

———. 1960. "Vegetation of the Siskiyou Mountains, Oregon and California." *Ecological Monographs* 30:279–338.

Wiens, J. A. 2000. "Ecological heterogeneity: An ontogeny of concepts and approaches." Pages 9–31 in M. J. Hutchings, E. A. John, and A. J. A. Stewart, editors, *The Ecological Consequences of Environmental Heterogeneity*. Malden, Mass.: Blackwell.

Wildermuth, T. A. 2008. *Yesterday's City of Tomorrow: The Minnesota Experimental City and Green Urbanism*. Urbana-Champaign: University of Illinois Press.

Wofford, J. 2009. "K-16 Computationally rich science education: A ten-year review of the Journal of Science Education and Technology (1998–2008)." *Journal of Science Education and Technology* 18:29–36.

Wolman, A. 1965. "The metabolism of cities." *Scientific American* 213:179–90.

Wong, S. 2013. "Architects and planners in the middle of a road war: The urban design concept team in Baltimore, 1966–71." *Journal of Planning History* 12:179.

Wrenn, D., and E. G. Irwin. 2015. "Time is money: An empirical examination of the dynamic effects of regulatory delay on residential subdivision development." *Regional Science and Urban Economics* 51:25–36.

Wright, H. E. 1977. "Quaternary vegetation history—Some comparisons between Europe and America." *Annual Review of Earth and Planetary Sciences* 5:123–58.

Wu, J. G., and O. L. Loucks. 1995. "From balance of nature to hierarchical patch dynamics: A paradigm shift in ecology." *Quarterly Review of Biology* 70:439–66.

Wu, J., and T. Wu. 2013. "Ecological resilience as a foundation for urban design and sustainability." Pages 211–29 in S. T. A. Pickett, M. L. Cadenasso, and B. McGrath, editors, *Resilience in Ecology and Urban Design: Linking Theory and Practice for Sustainable Cities.* New York: Springer.

Yoon, S. A., E. Anderson, J. Koehler-Yom, C. Evans, M. Park, J. Sheldon, I. Schoenfeld, D. Wendel, H. Scheintaub, and E. Klopfer. 2016. "Teaching about complex systems is no simple matter: Building effective professional development for computer-supported complex systems instruction." *Instructional Science* 45:99–121.

Zemlyanitskiy, L. T. 1963. "Characteristics of the soils in the cities." *Soviet Soil Science* 5:468–75.

Zhang, W., D. Wrenn, and E. G. Irwin. 2017. "Spatial heterogeneity, accessibility, and zoning: An empirical investigation of leapfrog development." *Journal of Economic Geography* 17:547–57.

Zhou, W., and M. L. Cadenasso. 2012. "Effects of patch characteristics and within patch heterogeneity on the accuracy of urban land cover estimates from visual interpretation." *Landscape Ecology* 27:1291–1305.

Zhou, W., M. L. Cadenasso, K. Schwarz, and S. T. A. Pickett. 2014. "Quantifying spatial heterogeneity in urban landscapes: Integrating visual interpretation and object-based classification." *Remote Sensing* 6:3369–86.

Zhou, W., G. Huang, S. T. A. Pickett, and M. L. Cadenasso. 2011a. "90 years of forest cover change in the urbanizing Gwynns Falls watershed, Baltimore, Maryland: Spatial and temporal dynamics." *Landscape Ecology* 26:645–59.

Zhou, W., S. T. A. Pickett, and M. L. Cadenasso. 2017. "Shifting concepts of urban spatial heterogeneity and their implications for sustainability." *Landscape Ecology* 32:15–30.

Zhou, W. Q., G. L. Huang, and M. L. Cadenasso. 2011b. "Does spatial configuration matter? Understanding the effects of land cover pattern on land surface temperature in urban landscapes." *Landscape and Urban Planning* 102:54–63.

Zhou, W., K. Schwarz, and M. L. Cadenasso. 2010. "Mapping urban landscape heterogeneity: Agreement between visual interpretation and digital classification approaches." *Landscape Ecology* 25:53–67.

Zhou, W., A. R. Troy, and J. M. Grove. 2008a. "Modeling residential lawn fertilization practices: Integrating high resolution remote sensing with socioeconomic data." *Environmental Management* 41:742–52.

———. 2008b. "Object-based land cover classification and change analysis in the Baltimore metropolitan area using multitemporal high resolution remote sensing data." *Sensors* 8:1613–36.

CONTRIBUTORS

DANIEL J. BAIN, Department of Geology and Environmental Science, University of Pittsburgh, Pittsburgh, PA 15260

LAWRENCE E. BAND, Departments of Environmental Science and Civil and Environmental Engineering, University of Virginia, Charlottesville, VA 22904

KENNETH T. BELT, USDA Forest Service, Northern Research Station, Suite 350, 5523 Research Park Drive, Baltimore, MD 21228

ALAN R. BERKOWITZ, Cary Institute of Ecosystem Studies, Box AB, Millbrook NY 12545

NEIL D. BETTEZ, Cary Institute of Ecosystem Studies, Box AB, Millbrook NY 12545

ADITI BHASKAR, Department of Civil and Environmental Engineering, Colorado State University, Fort Collins, CO 80523

SUSAN BLUNCK, Associate Clinical Professor, Department of Education, University of Maryland, Baltimore County

CHRISTOPHER G. BOONE, School of Human Evolution and Social Change, Global Institute of Sustainability, Arizona State University, PO Box 872402, Tempe, AZ 85287–2402

ANTHONY BRAZEL, School of Geographical Sciences and Urban Planning, Arizona State University, Tempe, AZ 85287

GRACE S. BRUSH, Department of Environmental and Health Engineering, Johns Hopkins University, 303 Ames Hall, 34th and Charles Sts., Baltimore, MD 21218

435

GEOFFREY L. BUCKLEY, Department of Geography, Ohio University, Athens, OH 45701–2979

MARY L. CADENASSO, Department of Plant Sciences, University of California, Davis, CA 95616

BESS CAPLAN, Cary Institute of Ecosystem Studies, Box AB, Millbrook NY 12545

JACQUELINE M. CARRERA, Carrera Consulting, 6520 Sandy Point Court, Rancho Palos Verdes, CA 90275

JANET COFFEY, Science Program, Gordon and Betty Moore Foundation, Palo Alto, CA

SCOTT L. COLLINS, Department of Biology, University of New Mexico, Albuquerque, NM 87131

SHAWN E. DALTON, Thrive Consulting, 25 Colonial Heights, Fredericton, NB E3B 5M2, Canada

EDWARD DOHENY, US Geological Survey, MD-DE-DC Water Science Center, 5522 Research Park Drive, Baltimore, MD 21228

JONATHAN M. DUNCAN, Pennsylvania State University, Department of Ecosystem Science and Management, University Park, PA 16802

MATTHEW GNAGEY, Goddard School of Business and Economics, Weber University, Ogden, UT 84408

JANICE L. GORDON, Johns Hopkins Bloomberg School of Public Health, Baltimore, MD

PETER M. GROFFMAN, City University of New York, Advanced Science Research Center at the Graduate Center, New York, NY 10031, and Cary Institute of Ecosystem Studies, Millbrook, NY 12545

J. MORGAN GROVE, USDA Forest Service, Northern Research Station, Suite 350, 5523 Research Park Drive, Baltimore, MD 21228

GORDON M. HEISLER, USDA Forest Service, Northern Research Station, 5 Moon Library, SUNY College of Environmental Science and Forestry, Syracuse, NY 13210.

ELENA G. IRWIN, Department of Agricultural, Environmental and Development Economics, Ohio State University, 316 Agricultural Administration Building, 2120 Fyffe Road, Columbus, OH 43210

NICHOLAS IRWIN, Department of Economics, University of Nevada, Las Vegas, Las Vegas, NV 89154

ANNA L. JOHNSON, Department of Biological Sciences, University of Pittsburgh, 4249 Fifth Avenue, Pittsburgh, PA 15260

SUJAY S. KAUSHAL, University of Maryland, College Park, 5825 University Research Court, Room #4048, College Park, MD 20740

SHARON E. KINGSLAND, History of Science and Technology Department, Johns Hopkins University, 3400 N. Charles St., Baltimore MD 21218

H. ALLEN KLAIBER, Department of Agricultural, Environmental and Development Economics, Ohio State University, 2120 Fyffe Road, Columbus, OH 43210

DEXTER LOCKE, National Socio-Environmental Synthesis Center (SESYNC), 1 Park Place, Suite 300, Annapolis, MD 21401

CHARLES LORD, Urgent VC, 831 Beacon Street, Suite 232, Newton Center, MA 02459

VICTORIA MARSHALL, Yale-National University of Singapore, 16 College Avenue West, #01–220, Singapore 138527

BRIAN MCGRATH, Parsons School of Design, 66 Fifth Avenue, New York, NY 10011

DAVID NEWBURN, Department of Agricultural and Resource Economics, University of Maryland, College Park, MD 20742

CHARLES H. NILON, University of Missouri—Columbia, School of Natural Resources, Fisheries and Wildlife, Columbia, MO 65211–7240

DAVID J. NOWAK, USDA Forest Service, Northeastern Research Station, 5 Moon Library, SUNY-CESF, Syracuse, NY 13210

JARLATH O'NEIL-DUNNE, Spatial Analysis Laboratory, Rubinstein School of Environment and Natural Resources, University of Vermont, Burlington, VT 05405

STEWARD T. A. PICKETT, Cary Institute of Ecosystem Studies, Box AB, Millbrook NY 12545

ERIN PIERCE, Cornell Lab of Ornithology, 159 Sapsucker Woods Road, Ithaca, NY 14850

RICHARD V. POUYAT, USDA Forest Service, 14th Street and Independence Ave., NW, Washington, DC 20250

MICHELE ROMOLINI, Center for Urban Resilience, Loyola Marymount University, 1 LMU Drive, Research Annex 119, Los Angeles, CA 90045

EMMA J. ROSI, Cary Institute of Ecosystem Studies, Box AB, Millbrook NY 12545

CHRISTOPHER M. SWAN, Department of Geography and Environmental Systems, University of Maryland, Baltimore County, 1000 Hilltop Circle, Baltimore, MD 21250

KATALIN SZLAVECZ, Johns Hopkins University, Department of Earth and Planetary Sciences, Olin Hall, 3400 N. Charles Street, Baltimore, MD 21218–2687

CHARLES TOWE, Department of Agricultural and Resource Economics, University of Connecticut, Storrs, Connecticut 06269

AUSTIN TROY, Department of Planning and Design, University of Colorado, Denver, Denver, CO 80208

CLAIRE WELTY, Center for Urban Environmental Research and Education, University of Maryland, Baltimore County, 1000 Hilltop Circle, TRC 102, Baltimore, MD 21250

DOUGLAS WRENN, Agricultural Economics, Sociology and Education, Penn State University, College of Agricultural Sciences, University Park, PA 16802

IAN D. YESILONIS, USDA Forest Service, Northern Research Station, Suite 350, 5523 Research Park Drive, Baltimore, MD 21228

WENDONG ZHANG, Department of Economics, Iowa State University, Ames, Iowa 50011

WEIQI ZHOU, Research Center for Eco-Environmental Sciences, Chinese Academy of Sciences, Beijing, 100085, People's Republic of China

INDEX